PLANT DEFENSE

T0188660

This book is dedicated to Beverley, for 30 years of love

PLANT DEFENSE

Warding off Attack by Pathogens, Herbivores, and Parasitic Plants

Dale R. Walters

Crop & Soil Systems Research Group
Scottish Agricultural College
Edinburgh, UK

A John Wiley & Sons, Ltd., Publication

This edition first published 2011
© 2011 by Blackwell Publishing Ltd.

Blackwell Publishing was acquired by John Wiley & Sons in February 2007. Blackwell's publishing programme
has been merged with Wiley's global Scientific, Technical, and Medical business to form Wiley-Blackwell.

Registered office
John Wiley & Sons Ltd, The Atrium, Southern Gate, Chichester, West Sussex, PO19 8SQ, United Kingdom

Editorial offices
9600 Garsington Road, Oxford, OX4 2DQ, United Kingdom
2121 State Avenue, Ames, Iowa 50014-8300, USA

For details of our global editorial offices, for customer services and for information about how to apply for
permission to reuse the copyright material in this book please see our website at
www.wiley.com/wiley-blackwell.

The right of the author to be identified as the author of this work has been asserted in accordance with the UK
Copyright, Designs and Patents Act 1988.

All rights reserved. No part of this publication may be reproduced, stored in a retrieval system, or transmitted, in
any form or by any means, electronic, mechanical, photocopying, recording or otherwise, except as permitted by
the UK Copyright, Designs and Patents Act 1988, without the prior permission of the publisher.

Wiley also publishes its books in a variety of electronic formats. Some content that appears in print may not be
available in electronic books.

Designations used by companies to distinguish their products are often claimed as trademarks. All brand names
and product names used in this book are trade names, service marks, trademarks or registered trademarks of their
respective owners. The publisher is not associated with any product or vendor mentioned in this book. This
publication is designed to provide accurate and authoritative information in regard to the subject matter covered.
It is sold on the understanding that the publisher is not engaged in rendering professional services. If professional
advice or other expert assistance is required, the services of a competent professional should be sought.

Library of Congress Cataloging-in-Publication Data

Walters, Dale.
   Plant defense: warding off attack by pathogens, herbivores, and parasitic plants /
Dale R. Walters. – 1st ed.
      p. cm.
   Includes bibliographical references and index.
   ISBN 978-1-4051-7589-0 (pbk. : alk. paper)   1. Plants–Disease and pest resistance.   I. Title.
   SB750.W35 2011
   632–dc22

                                                                              2010011214

A catalogue record for this book is available from the British Library.

Set in 10/12 pt Times by Aptara® Inc., New Delhi, India
Printed and bound in Singapore by Fabulous Printers Pte Ltd

1   2011

Contents

Preface

Plants are virtually stationary packages of food. They are sources of nourishment for thousands of fungi, bacteria, invertebrates, vertebrates, and even other plants. With so much of the biotic environment of our planet dependent on plants, it is surprising that plants exist at all. The fact that plants not only survive but thrive in nearly all environments on earth is testimony to their remarkable ability to deal with, what can at times be, a hostile environment. Indeed, plants possess a truly remarkable diversity of mechanisms to fend off attackers, and recent research shows just how complex and sophisticated these defence mechanisms can be. And to top it all, there are the internal signaling networks coordinating defence responses within the plant and the ability to warn neighboring plants.

This ability of plants to defend themselves is important not just for plants in their natural environment but also for plants under cultivation. Indeed, humans have made use of an increasing knowledge of plant defenses over the years to breed crop plants able to resist pest or pathogen attack. Induced resistance, where the plant can be primed for an intense defense response on pest or pathogen attack, offers the prospect of a more durable approach to disease control in crop plants. However, transferring this to practice will require an understanding of the effects of crop ecology in determining the expression of induced resistance. Of course, plants do not just need to fend off pests and pathogens, they also need to deal with attacks by parasitic plants and a range of vertebrate herbivores. Indeed, most plants will need to cope with attack by multiple enemies at the same time, and our understanding of how plants coordinate their responses to such attacks is increasing rapidly.

Although plants have to coordinate and mobilize their defenses to deal with simultaneous attacks by different organisms, these topics tend to be dealt with in separate, discipline-based textbooks, for example, plant pathology or entomology texts. This book is an attempt to remedy this deficiency. It deals with the range of different organisms plants need to fend off, how plants coordinate their defenses to deal with multiple attacks, the evolution of defense in plants, and the exploitation plant defenses in crop protection. I have written this book with senior undergraduates and postgraduates in mind and have included boxed readings dealing with particular topics in more detail, as well as a list of recommended reading, mostly books and review articles, at the end of each chapter.

I am very grateful to a number of colleagues for their encouragement during the preparation of this book, in particular Bill Spoor, Ian Bingham, and Adrian Newton. I am particularly grateful to Martin Heil, Adrian Newton, and Tony Reglinski, who kindly read, and provided comments on, some of the chapters. I am also grateful to Sergio Rasmann for providing some of the figures for Chapter 5. Thanks are also due to my wife Beverley, for love, encouragement, and incredible understanding—without this, I would never have completed the book.

I have been privileged to study plants for 31 years, and during that time, they have been a continual source of fascination and wonder. If I have managed to convey just a fraction of this in the following pages, I will be a happy man.

Dale Walters
Scottish Agricultural College
Edinburgh, UK

Chapter 1
Why Do Plants Need Defenses?

1.1 Plants as sources of food

All organisms need food to survive. To be more precise, they require a variety of chemical elements—the most important of which are carbon, nitrogen, and oxygen—to provide the building blocks for growth and development. This in turn requires a supply of energy, the only external supply of which comes from the sun. Plants are able to capture the energy from sunlight and convert it into chemical energy, thereby providing the means of financing the formation of carbohydrate from atmospheric CO_2 and water. This autotrophic ability of plants comes at a price. Because most organisms are not autotrophic, they must obtain their energy and building blocks for growth and development from consuming other organisms, including plants. In fact, plants are a direct source of food for an array of organisms that include invertebrates, vertebrates, fungi, bacteria, and even other plants.

The popularity of plants as food sources for so many organisms begs the question "what do plants offer other organisms by way of nutrition?" Clearly, plant tissues will provide a source of carbon and nitrogen, much of which will be in the form of carbohydrates, lipids, and proteins. They will also contain macroelements such as phosphorus, sulfur, calcium, and potassium, as well as various microelements such as iron, manganese, and zinc. However, the relative proportions of these components will vary depending on species. Moreover, different plant parts can have very different compositions (Figure 1.1). For example, fruits and phloem sap can be rich sources of carbohydrates, while seeds are usually good sources of fat. Some parts of the plant, such as bark, offer little in the way of nutrients, since they are composed largely of dead cells, with lignified walls. Nitrogen and protein content also varies between different parts of the plant, but in general, plants contain less nitrogen and protein than most of the organisms that use them as a food source. Typically, the total nitrogen content of plants is between 2 and 4% of their dry weight, while the nitrogen content of animals amounts to 8–14% of their dry bodyweight (Figure 1.2). The amount of nitrogen in insects is even greater than this and can be in the order of 30–40% of their dry weight (Southwood, 1973).

As indicated above, plants are used as food sources by a variety of organisms. Before we proceed further, it is worth considering the mechanisms used by these organisms to obtain the nutrients locked up in plant tissues.

Plant Defense, First Edition, by Dale Walters © 2011 by Blackwell Publishing Ltd.

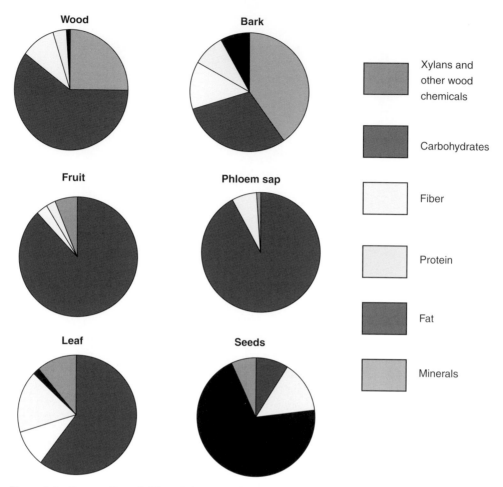

Figure 1.1   Composition of different plant parts that can serve as food for herbivores. (Adapted from Townsend *et al.* (2003), with permission of Blackwell Publishing Ltd.)

1.2   Organisms that use plants as food

1.2.1   Microorganisms

Plants are infected by a wide range of microorganisms. Some of these establish symbiotic associations with plant roots, such as *Rhizobium* bacteria found in nodules on roots of legumes and mycorrhizal fungi, which form intimate associations with the roots of most plant species. Other microorganisms are parasitic on plants and use plants as food sources, causing damage and sometimes plant death, in the process. Some of these microbes, including viruses, protozoa, and some fungi, are biotrophs. These grow and reproduce in nature only on living hosts. Powdery mildew and rust fungi (Figure 1.3a), for example, produce feeding structures called haustoria that invaginate the host plasma membrane, forming an intimate association with the plant cell. Other microbes, mostly fungi and

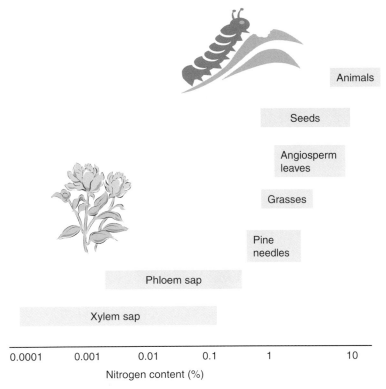

Figure 1.2   Variations in nitrogen concentration (dry weight percentage) of different plant parts compared with that in animals. Xylem and phloem sap concentrations are expressed as nitrogen weight/volume. (Adapted from Mattson (1980), with permission.)

bacteria, are necrotrophic. They secrete enzymes to cause disintegration of plant cells and, either alone or in combination with toxins, often lead to cell and tissue death (Figure 1.3b). The nutrients released in the process are then available for uptake by the pathogen.

1.2.2   Parasitic angiosperms

Plants are also parasitized by other plants. Indeed, parasitism among plants seems to have evolved many times during angiosperm evolution. It has been estimated that about 1% of angiosperms, some 3000 species in total, are parasitic on other plants (Parker & Riches, 1993). Parasitic angiosperms are distributed among 17 families, including the Viscaceae and the Cuscutaceae (Table 1.1), and include parasitic plants such as species of *Rhinanthus* and *Orobanche* (Figure 1.3c and 1.3d). There is considerable diversity in the extent to which parasitic angiosperms rely on the host for growth. Some, such as species of *Rhinanthus*, have functional roots and can therefore take up inorganic nutrients from the soil, while others, such as the mistletoes, have nothing that resembles a root nor functions as one (Hibberd & Jeschke, 2001). There is also considerable variation in the extent to which parasitic plants rely on the host for photoassimilates. Thus, parasitic plants such as *Rhinanthus minor* are able to photosynthesize and can grow with a carbon supply from the host, while others,

Figure 1.3 (a) Yellow rust (*Puccinia striiformis*) on wheat, (b) *Alternaria brassicae* on pods and stems of oilseed rape (*Brassica napus*), (c) the parasitic angiosperm *Rhinanthus minor*, (d) the parasitic angiosperm *Orobanche crenata*, (e) the plant parasitic nematode *Helicotylenchus*, (f) caterpillar of the large cabbage white butterfly, *Pieris brassicae*, (g) an aphid on a leaf, (h) moose, (i) cows grazing. (Image (c) is the copyright of Glyn Baker and is licensed for reuse under the Creative Commons Licence. Images (d), (e), (f) and (h) are reproduced courtesy of Lytton Musselman, the American Phytopathological Society, Rosemary Collier of the University of Warwick HRI, and the United States Geological Survey, respectively.)

Table 1.1 Main families of parasitic plants

Family	Approximate number of species	Type	Root/shoot
Santalaceae	400	Hemiparasitic	Largely root
Loranthaceae	700	Hemiparasitic	Mostly shoot
Viscaceae	400	Hemiparasitic	Shoot
Rafflesiaceae	500	Holoparasitic	Mostly root
Cuscutaceae	145	Holoparasitic	Shoot
Scrophulariaceae	1500	Hemi/holoparasitic	Root

such as *Cuscuta reflexa*, possess a very low photosynthetic capacity and are unable to grow without a carbohydrate supply from the host (Hibberd & Jeschke, 2001).

1.2.3   Nematodes

Nematodes are wormlike in appearance (Figure 1.3e) but are quite distinct from the true worms. Several hundred species are known to feed on living plants, obtaining their food with spears or stylets. Nematode feeding causes only slight mechanical damage to plants. The majority of the damage caused by nematodes appears to be caused by saliva injected into plants during feeding. Nematodes will puncture a cell wall, inject saliva into the cell, and withdraw part of the cell contents. Some nematodes feed rapidly and move on within a few seconds, while others feed more slowly and remain at the puncture for hours or even days. As long as the nematodes are feeding, they will inject saliva intermittently into the cell.

1.2.4   Insects

There are more species of insects than any other class of organisms on earth and nearly half of these, some 400,000 species, feed on plants (Schoonhoven *et al.*, 2005). These herbivorous insects harvest their food in a variety of different ways. Chewing insects possess "toothed" mandibles that cut, crush, and macerate plant tissues. Many feed externally on the plant, while others, such as leaf miners, harvest tissue layers between the upper and lower epidermis of the leaf. Chewing insects include species belonging to the orders Lepidoptera (moths and butterflies) (Figure 1.3f) and Orthoptera (e.g., grasshoppers). Some insects do not remove chunks of plant tissue, but rather suck fluids from the plant using specialized tubular mouthparts. Thus, insects in the order Hemiptera include aphids (Figure 1.3g), which feed on phloem sap. Other members of the Hemiptera, for example, whiteflies, feed on the contents of leaf mesophyll cells. Yet other insects make galls on their plant host. These insects manipulate the host tissues, providing themselves with both shelter and nutrients. A striking feature of relationships between insects and plants is the extent of food specialization among insect herbivores. Some insects, including many lepidopterous larvae, hemipterans, and coleopterans, occur on only one or a few closely related plant species and are termed monophagous. Others, such as the Colorado potato beetle, *Leptinotarsa decemlineata*, are oligophagous; these feed on a number of plant species, all belonging to

Table 1.2   Major categories of dietary specialization in herbivorous mammals, according to Eisenberg (1981)

Category	Diet
Nectarivores	Nectar and pollen
Gumivores	Exudates from trees
Frugivore/omnivore	Pericarp or fleshy outer covering of plant reproductive parts, invertebrates and small vertebrates
Frugivore/granivore	Reproductive parts of plants, including seeds
Frugivore/herbivore	Fleshy fruiting bodies and seeds of plants, storage roots, and some green leafy material
Herbivore/browser	Stems, twigs, buds, and leaves
Herbivore/grazer	Grasses

the same family. Yet other insect herbivores, for example, the aphid *Myzus persicae*, accept many plants belonging to different families. Such insects are polyphagous.

1.2.5   *Vertebrates*

Herbivory is not confined to insects. In fact, it is a common trait among mammals (Figure 1.3h and 1.3i), with roughly half of the 1000 or so genera of mammals including plants in their diet (Danell & Bergström, 2002). While the majority of herbivorous insects are mono- or oligophagous, feeding on a few plant species, vertebrate herbivores tend to be polyphagous and feed on a wider range of plant species. Vertebrate herbivores are larger than their invertebrate counterparts and are thus able to remove a greater amount of plant tissue with each mouthful (Danell & Bergström, 2002). The classification of animals into functional groups is usually achieved using diet composition. Sixteen major categories of dietary specialization in mammals have been proposed (Eisenberg, 1981), and of these, seven refer to herbivores (Table 1.2). Categories range from nectarivores that feed on nectar and pollen to gumivores that feed on exudates from trees. The most dominant group is numerically the frugivores/omnivores, which represent approximately 33% of vertebrate herbivore genera (Table 1.2).

1.3   Impact of infection and herbivory in natural and agricultural ecosystems

1.3.1   *Microorganisms*

Pathogenic microorganisms can exert a profound effect on the structure and dynamics of individual plant species and plant communities. The extent and type of damage to individual plants is related to the lifestyle of the pathogen, that is, whether it is a biotroph or a necrotroph. Necrotrophs destroy plant tissue and it seems obvious therefore, that loss of leaf tissue, for example, will decrease rates of photosynthesis, thereby reducing plant growth. In contrast, biotrophs do not kill plant tissue, although effects on photosynthesis can be just as profound. Thus, effects on chloroplast structure and function can lead to dramatic reductions in rates of photosynthesis in plants infected with rust or powdery mildew fungi (Walters & McRoberts, 2006). Photosynthesis can also be affected by other means. In plants infected with vascular wilt pathogens, such as *Verticillium albo-atrum*, blockage of xylem vessels can lead to water stress and partial closure of stomata, thus reducing rates of photosynthesis. Of course, the effects of pathogens on the host plant are not restricted to photosynthesis, and some pathogens, for example, can alter water and nutrient uptake, while others produce toxins, which affect host metabolism. Whatever the mechanism, pathogen infection can lead to greatly reduced plant growth and reproductive output. Thus, *Albugo candida* and *Peronospora parasitica*, both biotrophs, reduce reproductive output in *Capsella bursa-pastoris* (Alexander & Burdon, 1984), while the tobacco leaf curl virus reduces growth and seed production in its host, *Eupatorium chinense*, and is an important cause of plant mortality (Yahara & Oyama, 1993).

Pathogen infection can also lead to plant death. Thus, fungal damping off and root diseases can cause mass mortality of seedlings, especially under humid conditions. Damping off was responsible for 64–95% of seedling deaths of the tropical tree *Platypodium elegans* in the first 3 months after emergence (Augspurger, 1983). Pathogens can also cause death of older plants. In Australia, an epidemic of the root rot pathogen, *Phytophthora cinnamomi*, devasted a dry sclerophyll forest (Weste & Ashton, 1994), while the pathogen *Phacidium infestans* was a significant cause of mortality in 5- to 10-year-old *Pinus sylvestris* (Burdon *et al.*, 1994).

In crop production systems, losses due to pathogens can be substantial. In the period 1996–1998, global crop losses due to pathogens (fungi, bacteria, and viruses) were 12.6%, in spite of crop protection measures (Oerke & Dehne, 2004). Some crops seem to suffer more than others, and between 1996 and 1998, pathogens accounted for losses of 22% in global potato production. Even more devastating can be the spread of a pathogen into a new geographical area. For example, the soybean rust, *Phakopsora pachyrhizi*, was first reported in South America in 2001 (in Paraguay) and by 2003 was detected in most soybean-growing regions of Brazil, with losses estimated at 5% of total soybean production (Yorinori *et al.*, 2005).

1.3.2 *Parasitic angiosperms*

The diversion of host resources to parasitic plants can have large effects on host growth and reproductive output. Infection of *Poa alpina* with the annual hemiparasite, *R. minor*, reduced host biomass by more than 50% (Seel & Press, 1996), while the phloem-tapping mistletoe, *Tristerix aphyllus*, greatly reduced the production of buds, flowers, and fruits by its cactus host, *Echinopsis chilensis* (Silva & Martínez del Rio, 1996). Death of the host plant can occur, particularly in extreme cases, such as heavy infestation with mistletoe (Aukema, 2003).

Parasitic plants can exert a considerable impact on plant communities (Press & Phoenix, 2005). Thus, *Rhinanthus* species have been shown to reduce total productivity in European grasslands by between 8 and 73% (Davies *et al.*, 1997), while dwarf mistletoes can reduce volume growth of Douglas fir by up to 65% (Mathiasen *et al.*, 1990).

Striga is a genus of root hemiparasite with some 35 species, most of which are of no agricultural importance. However, those species that parasitize crop plants can be devastating. Yield losses in cereals infected by *Striga* can reach 100%, and fields can be so heavily infested that they are abandoned by farmers (Berner *et al.*, 1995). Some 40 million hectares of cereals are thought to be severely infested with *Striga* spp. in West Africa, and the Food and Agriculture Organization (FAO) estimates that annual yield losses in the savannah regions alone account for US$7 billion (Berner *et al.*, 1995).

1.3.3   Nematodes

Nematode infestation can lead to substantial reductions in plant growth. In clover, the stem nematode, *Ditylenchus dipsaci*, reduced establishment from seeds and led to a 30% reduction in shoot growth (Cook *et al.*, 1992), while the potato cyst nematodes, *Globodera pallida* and *Globodera rostochiensis*, reduced growth of potato roots within 1 day following inoculation onto root tips (Arnitzen *et al.*, 1994). Nematodes have also been linked to plant deaths. Thus, pathogenic nematodes have been identified as the probable cause of die-out of the dune grass, *Ammophila breviligulata*, on the mid-Atlantic coast of the USA (Seliskar, 1995), while the pine wood nematode, *Bursaphelenchus xylophilus*, was responsible for the deaths of some quarter of a million mature pine trees in a single location in Japan (Numata, 1989).

1.3.4   Insects

Given the existence of more than 300,000 species of herbivorous insects (Schoonhoven *et al.*, 2005), it is surprising that there is not more evidence of plant devastation. In fact, complete defoliation of vegetation occurs only sporadically. Some plants can compensate or overcompensate for sizeable amounts of damage, but even so, insect herbivory will reduce plant fitness (reproductive capacity) (Bigger & Marvier, 1998). We deal with the reasons for plant survival against such odds in later chapters. In the meantime, it is worth noting that some 10% of all annually produced biomass is consumed by insect herbivores (Barbosa & Schultz, 1987; Coupe & Cahill, 2003).

The extent of plant loss, however, is dependent on a number of factors, including vegetation type, timing of herbivory, and locality. Thus, herbivore pressure is likely to be greater in tropical dry forests than in temperate forests (Coley & Barone, 1996), with the result that rates of herbivory are significantly greater in forests in tropical regions than those in temperate zones (Coley & Aide, 1991; Figure 1.4). Variation in the damage caused by herbivory also exists between different species of plants. Thus, up to 50% of foliage production by Australian *Eucalyptus* trees can be lost as a result of insect herbivory, while other plant species, for example, *Juniperus* and *Rhododendron*, exhibit little damage from insects (Schoonhoven *et al.*, 2005). Considerable variation in damage resulting from insect herbivory also exists within the same genus. In a study of herbivory among different *Piper* species, some suffered little damage from insects, while other species lost up to 25% of their leaf area (Marquis, 1991; Figure 1.5).

Insect herbivory affects many plant parts, including leaves, roots, and seeds. Perhaps the most obvious signs of insect herbivory are seen on leaves. It is estimated that rates of defoliation caused by insects lie within the range 5–15% of leaf area per year (Landsburg & Ohmart, 1989), although this is thought to be an underestimate (Crawley, 1997). Less obvious to the observer is perhaps root herbivory, although this is more likely to have an impact on plant dynamics than leaf herbivory (Crawley, 1997). Whatever plant part

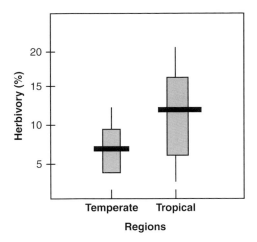

Figure 1.4   Rates of herbivory in temperate and tropical forests. Plots indicate mean ±SD and range. (From Schoonhoven *et al.* (2005), with permission of Oxford University Press.)

is used by insects as food, herbivory can have a sizeable impact on plant growth. For example, to study the effect of insect herbivory on tree growth in a eucalyptus forest, trees were sprayed with insecticide for several years (Morrow & LaMarche, 1978). Tree growth was substantially greater in sprayed trees, which had reduced insect loads, compared to unsprayed trees, which harbored greater insect numbers. Interestingly, while defoliating insects exerted little impact on acorn production by oaks (*Quercus robur*), the exclusion of sucking insects by spraying with insecticide increased acorn production consistently (Crawley, 1985).

Feeding by different types of insect herbivore can affect plant fitness more or less independently. For example, in *Lupinus arboreus*, the bush lupin, there was no statistical interaction between above ground and below ground herbivory, and both types had significant

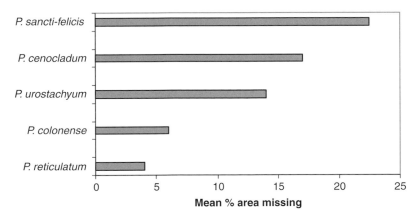

Figure 1.5   Rates of herbivory for different species of *Piper*. The data represent the mean percentage area missing for individual *Piper* species for a minimum of 50 freshly abscissed leaves per species. (From Schoonhoven *et al.* (2005), with permission of Oxford University Press.)

cumulative effects on plant fitness. When bush lupin plants were protected from chronic above ground herbivory, seed output over a 3-year period increased by 78%, while suppression of below ground herbivory increased mean seed production by 31%. Interestingly, root herbivory was associated with a greater risk of plant mortality (Maron, 1998).

Plants in agricultural ecosystems often suffer more damage as a result of insect herbivory than their natural counterparts. As a result, considerable quantities of insecticide are used to control insect pests. Nevertheless, it is estimated that some 15% of global crop production is lost annually to insect damage, despite the use of insecticides. It is estimated that some 9000 species of insects attack agricultural crops worldwide, although only about 450 of these are considered as serious pests (Pimentel, 1991). Most insect pests are specialist feeders, with 75–80% of lepidopterous pests being monophagous or oligophagous (Barbosa, 1993).

1.3.5   Vertebrates

As mentioned above, because of their size, vertebrate herbivores are likely to remove more plant tissue per mouthful than their invertebrate counterparts. However, invertebrate herbivores are probably ten times more abundant than vertebrate herbivores (Peters, 1983). Herbivores are estimated to remove about 10% of net primary production in terrestrial environments (Crawley, 1983), and indeed, if we look at the Nylsvley savanna in southern Africa, for example, vertebrate grazers and browsers are estimated to remove 6% and invertebrate herbivores (grasshoppers and caterpillars) are estimated to remove 5% of the above ground primary production (Figure 1.6). The amount of vegetation removed by vertebrate herbivores depends greatly on habitat and the herbivore. For example, although in arctic areas, vertebrate herbivores remove between 5 and 10% of net primary production (Mulder, 1999), lemmings and geese can remove as much as 90% (Cargill & Jefferies, 1984), while muskoxen remove only 1–2% of vegetation (Bliss, 1986). Their greater body

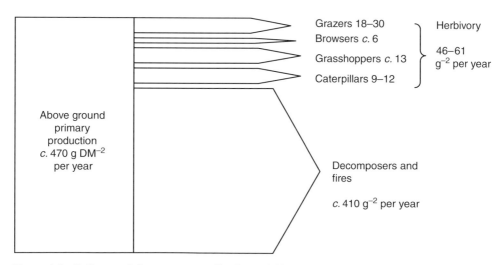

Figure 1.6   Pathways of disappearance of leaf material in the broad-leaved savanna at Nylsvley, South Africa, and their approximate magnitude. (Reproduced from Danell & Bergström (2002), with permission of Blackwell Publishing Ltd.)

size—polyphagy, individual bite size, mobility, and tolerance of starvation—suggests that vertebrate herbivores should exert a more immediate, and in the long term a more profound impact on plant populations than invertebrate herbivores (Danell & Bergström, 2002).

Some plants can tolerate vertebrate herbivory, and there is still debate about whether this tolerance is an evolutionary response to herbivory, and if so, whether grazing or browsing can benefit the plant in terms of increased fitness. The increase in net primary productivity of savanna grasses (McNaughton, 1979) and increased flower production in *Ipomopsis aggregata* (Paige, 1992) following herbivory are examples of the possible increase in plant fitness by vertebrate herbivory. However, vertebrate herbivory can deleteriously affect plant reproductive tissue and fecundity. Thus, simulating rodent damage to bilberry by branch cutting led to negative effects on flower production and berry development (Tolvanen *et al.*, 1993), while sheep grazing on a shrub in the Karoo rangeland in South Africa reduced both flower production and seed size (Milton, 1995). Equally, there are examples where vertebrate herbivory exerts minimal impact on plant fitness, as in the case of ungulate grazing of *Balsamorhiza sagittata* (Asteraceae), a dominant native perennial forb in western Montana, USA. Here, in comparison to insect herbivory, vertebrate grazing had little impact on plant fecundity (Amsberry & Maron, 2006).

Plant size and shape can be altered by herbivory (e.g., removal of the leading shoot or apical bud in woody species), with consequences for interplant competition and effects on other organisms. Even removal of a small amount of plant biomass can have a profound effect on plant shape. Thus, removal of the leading shoot of a tree by vertebrate grazing can alter the architecture of the whole tree for many years.

Although vertebrate herbivory does not often result in plant death, mortality can occur, especially when vertebrates feed on small plants or seedlings (Hulme, 1994), or when older trees are debarked, by, for example, elephants, deer, squirrels, voles, and hares (Danell & Bergström, 2002).

1.4   Conclusions

Plants are attacked by a wide range of organisms, from viruses and bacteria to large vertebrates. These interactions can have a considerable impact on natural plant populations and indeed are thought to represent a major selective force on the evolution of plant structure and function. In crop situations, attack by pathogens and herbivores can result in serious losses in yield and quality. However, as we shall see in the following chapters, plants are not defenseless against such attacks, no matter how large the attacker.

Recommended reading

Crawley MJ, 1997. Plant–herbivore dynamics. In: Crawley MJ, ed. *Plant Ecology*. Oxford: Blackwell Publishing Ltd., pp. 401–474.

Danell K, Bergström R, 2002. Mammalian herbivory in terrestrial environments. In: Herrera CM, Pellmyr O, eds. *Plant–Animal Interactions: An Evolutionary Approach*. Oxford: Blackwell Publishing Ltd., pp. 107–131.

Parker C, Riches CR, 1993. *Parasitic Weeds of the World: Biology and Control.* Wallingford: CAB International.

Schoonhoven LM, van Loon JJA, Dicke M, 2005. *Insect–Plant Biology.* Oxford: Oxford University Press.

References

Alexander HM, Burdon JJ, 1984. The effect of disease induced by *Albugo candida* (white rust) and *Peronospora parasitica* (downy mildew) on the survival and reproduction of *Capsella bursa-pastoris* (Shepherd's Purse). *Oecologia* **64**, 314–318.

Amsberry LK, Maron JL, 2006. Effects of herbivore identity on plant fecundity. *Plant Ecology* **187**, 39–48.

Arnitzen FK, Visser JHM, Hoogendoorn J, 1994. The effect of the potato cyst nematode *Globodera pallida* on *in vitro* root growth of potato genotypes differing in tolerance. *Annals of Applied Biology* **124**, 59–64.

Augspurger CK, 1983. Seed dispersal of the tropical tree *Platypodium elegans*, and the escape of its seedlings from fungal pathogens. *Journal of Ecology* **71**, 759–772.

Aukema JE, 2003. Vectors, viscin, and Viscaceae: mistletoes as parasites, mutualists, and resources. *Frontiers in Ecology and Environment* **1**, 212–219.

Barbosa P, 1993. Lepidopteran foraging on plants in agroecosystems: constraints and consequences. In: Stamp E, Casey TM, eds. *Caterpillars. Ecological and Evolutionary Constraints on Foraging.* New York: Chapman & Hall, pp. 523–566.

Barbosa P, Schultz JC, 1987. *Insect Outbreaks.* San Diego: Academic Press.

Berner DK, Kling JG, Singh BB, 1995. *Striga* research and control. A perspective from Africa. *Plant Disease* **79**, 652–660.

Bigger DS, Marvier MA, 1998. How different would a world without herbivory be? A search for generality in ecology. *Integrative Biology* **1**, 60–67.

Bliss LC, 1986. Arctic ecosystems: their structure, function and herbivore carrying capacity. In: Gudmundsson O, ed. *Grazing Research at Northern Latitudes.* New York: Plenum Press, pp. 5–26.

Burdon JJ, Wennstrom A, Muller WJ, Ericson L, 1994. Spatial patterning in young stands of *Pinus sylvestris* in relation to mortality caused by the snow blight pathogen *Phacidium infestans*. *Oikos* **71**, 130–136.

Cargill SM, Jefferies RL, 1984. The effects of grazing by lesser snow geese on the vegetation of a sub-arctic salt marsh. *Journal of Applied Ecology* **21**, 686–699.

Coley PD, Aide TM, 1991. Comparison of herbivory and plant defences in temperate and tropical broad-leaved forests. In: Price PW, Lewinsohn TM, Fernandes GW, Benson WW, eds. *Plant–Animal Interactions. Evolutionary Ecology in Tropical and Temperate Regions.* New York: John Wiley & Sons, Ltd., pp. 25–49.

Coley PD, Barone JA, 1996. Herbivory and plant defences in tropical forests. *Annual Review of Ecology and Systematics* **27**, 305–335.

Cook R, Evans DR, Williams TA, Mizen KA, 1992. The effect of stem nematode on establishment and early yields of white clover. *Annals of Applied Biology* **120**, 83–94.

Coupe MD, Cahill JF, 2003. Effects of insects on primary production in temperate herbaceous communities: a meta-analysis. *Ecological Entomology* **28**, 511–521.

Crawley MJ, 1983. *Herbivory: The Dynamics of Animal–Plant Interactions.* Oxford: Blackwell Publishing Ltd.

Crawley MJ, 1985. Reduction of oak fecundity by low density herbivore populations. *Nature* **314**, 163–164.

Crawley MJ, 1997. Plant–herbivore dynamics. In: Crawley MJ, ed. *Plant Ecology*. Oxford: Blackwell Publishing Ltd., pp. 401–474.

Danell K, Bergström R, 2002. Mammalian herbivory in terrestrial environments. In: Herrera CM, Pellmyr O, eds. *Plant–Animal Interactions: An Evolutionary Approach*. Oxford: Blackwell Publishing Ltd., pp. 107–131.

Davies DM, Graves JD, Elias CO, Williams PJ, 1997. The impact of *Rhinanthus* spp. on sward productivity and composition: implications for the restoration of species-rich grasslands. *Biological Conservation* **82**, 93–98.

Eisenberg JF, 1981. *The Mammalian Radiations: An Analysis of Trends in Evolution, Adaptation, and Behaviour*. London: Athlone Press.

Hibberd JM, Jeschke WD, 2001. Solute flux into parasitic plants. *Journal of Experimental Botany* **52**, 2043–2049.

Hulme PE, 1994. Seedling herbivory in grasslands—relative importance of vertebrate and invertebrate herbivores. *Journal of Ecology* **82**, 873–880.

Landsburg J, Ohmart C, 1989. Levels of insect defoliation in forests—patterns and concepts. *Trends in Ecology and Evolution* **4**, 96–100.

Maron JL, 1998. Insect herbivory above and below ground: individual and joint effects on plant fitness. *Ecology* **79**, 1281–1293.

Marquis RJ, 1991. Herbivore fauna of *Piper* (Piperaceae) in a Costa Rican wet forest, diversity, specificity and impact. In: Price PW, Lewinsohn TM, Fernandes GW, Benson WW, eds. *Plant–Animal Interactions. Evolutionary Ecology in Tropical and Temperate Regions*. New York: John Wiley & Sons, Ltd., pp. 179–199.

Mathiasen RL, Hawksworth FG, Edminster CB, 1990. Effects of dwarf mistletoe on growth and mortality of Douglas fir in the southwest. *Great Basin Newsletter* **50**, 173–179.

Mattson WJ, 1980. Herbivory in relation to plant nitrogen content. *Annual Review of Ecology and Systematics* **11**, 119–161.

McNaughton SJ, 1979. Grazing as an optimisation process: grass–ungulate relationships in the Serengeti. *American Naturalist* **113**, 691–703.

Milton SJ, 1995. Effects of rain, sheep and tephritid flies on seed production of two arid Karoo shrubs in South Africa. *Journal of Applied Ecology* **32**, 137–144.

Morrow PA, LaMarche JVC, 1978. Tree ring evidence for chronic insect suppression of productivity in subalpine *Eucalyptus*. *Science* **201**, 1244–1246.

Mulder CPH, 1999. Vertebrate herbivores and plants in the Arctic and subarctic: effects on individuals, populations and ecosystems. *Perspectives in Plant Ecology, Evolution and Systematics* **2**, 29–55.

Numata M, 1989. The ecological characteristics of aliens in natural and semi-natural stands. *Memoirs of Shukutoku University* **23**, 23–35.

Oerke E-C, Dehne H-W, 2004. Safeguarding production—losses in major crops and the role of crop protection. *Crop Protection* **23**, 275–285.

Paige KN, 1992. Overcompensation in response to mammalian herbivory: from mutualistic to antagonistic interactions. *Ecology* **73**, 2076–2085.

Parker C, Riches CR, 1993. *Parasitic Weeds of the World: Biology and Control*. Wallingford, UK: CAB International.

Peters RH, 1983. *The Ecological Implications of Body Size*. Cambridge: Cambridge University Press.

Pimentel D, 1991. Diversification of biological control strategies in agriculture. *Crop Protection* **10**, 243–253.

Press MC, Phoenix GK, 2005. Impacts of parasitic plants on natural communities. *New Phytologist* **166**, 737–751.

Schoonhoven LM, Van Loon JJA, Dicke M, 2005. *Insect–Plant Biology*. Oxford: Oxford University Press.

Seel WE, Press MC, 1996. Effects of repeated parasitism by *Rhinanthus minor* on the growth and photosynthesis of a perennial grass, *Poa alpina*. *New Phytologist* **134**, 495–502.

Seliskar DM, 1995. Coastal dune restoration—a strategy for alleviating dieout of *Ammophila breviligulata*. *Restoration Ecology* **3**, 54–60.

Silva A, Martínez del Rio C, 1996. Effects of the mistletoe *Tristerix aphyllus* (Loranthaceae) on the reproduction of its cactus host *Echinopsis chilensis*. *Oikos* **75**, 437–442.

Southwood TRE, 1973. The insect/plant relationship—an evolutionary perspective. In: Van Emden HF, ed. *Insect–Plant Relationships*. London: John Wiley & Sons, Ltd., pp. 3–30.

Tolvanen A, Laine K, Pakonen T, Saari E, Havas P, 1993. Above ground growth responses of the bilberry (*Vaccinium myrtillus* L.) to simulated herbivory. *Flora (Jena)* **188**, 197–202.

Townsend CR, Begon M, Harper JL, 2003. *Essentials of Ecology*. Oxford: Blackwell Publishing Ltd.

Walters DR, McRoberts N, 2006. Plants and biotrophs: a pivotal role for cytokinins? *Trends in Plant Science* **11**, 581–586.

Weste G, Ashton DH, 1994. Regeneration and survival of indigenous dry sclerophyll species in the Brisbane ranges, Victoria, after *Phytophthora cinnamomi* epidemic. *Australian Journal of Botany* **42**, 239–253.

Yahara T, Oyama K, 1993. Effects of virus infection on demographic traits of an agamospermous population of *Eupatorium chinense* (Asteraceae). *Oecologia* **96**, 310–315.

Yorinori JT, Frederick RD, Costamilan LM, Bertagnolli PF, Hartman GE, Godoy CV, Nunes J, 2005. Epidemics of soybean rust (*Phakopsora pachyrhizi*) in Brazil and Paraguay from 2001 to 2003. *Plant Disease* **89**, 675–677.

Chapter 2
What Defenses Do Plants Use?

2.1 Introduction

Although plants are attacked by a multitude of microorganisms, other plants, invertebrate, and vertebrate herbivores, most survive. The ability of plants to survive such assaults is due to the arsenal of defenses at their disposal, together with their ability to compensate for loss of tissue. This chapter deals with the defenses plants use to ward off attack. These defenses include structural mechanisms and biochemical responses that can slow down pathogen progress, ward off herbivores, and in some cases kill the invader. The variety of defenses used is staggering and ranges from the relatively simple to the very complex. The focus in this chapter is on defenses used against microbial pathogens, parasitic plants, nematodes, herbivorous insects, and vertebrate herbivores. There is also a brief foray into the realms of allelopathy—the defenses used by plants against neighboring plants.

Before we embark on our journey through plant defense, it is important to highlight definitions of two terms used throughout this book: *defense* and *resistance*. The latter term is used to describe the capacity of the plant to avoid or reduce damage caused by attackers. It is commonly used in the applied literature. Defense is used by some workers to mean that a particular trait has evolved or is maintained in the plant population as a result of selection exerted by attackers (Karban & Baldwin, 1997; Schoonhoven *et al.*, 2005). Although use of the term "resistance" is preferred in those cases where a defensive function for a particular trait is still unproved, as pointed out by Karban & Baldwin (1997), it is extremely difficult, if not impossible, to determine the specific selective factors that shape a trait.

2.2 Defenses used against pathogens

2.2.1 Background

In nature, most plants are resistant to most pathogenic agents to which they come into contact. In this case, plants are able to completely prevent penetration by the pathogen, and such plants are considered to be immune to that pathogen. Thus, wheat plants are not affected by pathogens of tomato plants, and vice versa. This is known as nonhost resistance. Of course, wheat can be infected by its own pathogens (pathogens with which it has coevolved) and so, for example, it can be infected by the black stem rust fungus, *Puccinia graminis*

Plant Defense, First Edition, by Dale Walters   © 2011 by Blackwell Publishing Ltd.

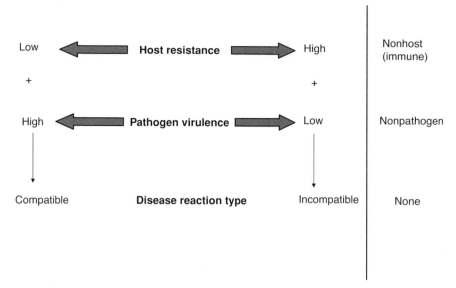

Figure 2.1   Relationship between host plants, pathogens, and disease reaction. (Reproduced from Lucas (1998), with permission of Blackwell Publishing Ltd.)

f.sp. *tritici*, while tomato plants can be infected by their pathogens, such as the leaf mold fungus, *Cladosporium fulvum*. Every plant will be host to a range of different pathogens, but the extent to which it can defend itself effectively against each of these pathogens will also be different. The plant might be highly resistant to one pathogen, allowing very little, if any pathogen growth in the tissue, but might be poorly resistant to another pathogen, allowing considerable growth of the pathogen in the tissue. This can also be looked at in terms of susceptibility, where high resistance would be equivalent to low susceptibility and vice versa (Figure 2.1). Resistance and susceptibility refer to the plant host, but, of course, the pathogen can also vary in its ability to infect the plant and cause disease. Any microbe incapable of causing disease in a plant is nonpathogenic on that plant. If the microbe can penetrate the plant, but its subsequent effects are insignificant, it is termed "avirulent." If, however, its effects on the host are significant, it is said to be "virulent" (Figure 2.1). An interaction between a plant and a pathogen (e.g., a susceptible host and a virulent pathogen) where symptoms are expressed clearly is known as a compatible interaction, while an interaction where no symptoms develop (e.g., a resistant host and an avirulent pathogen) is described as incompatible (Figure 2.1). As one might expect however, not everything is quite so clear-cut.

When attacked by a pathogen, the plant can use a variety of preexisting or induced mechanisms to defend itself. These defenses can be structural or chemical, and depending on the attacking pathogen, the defense might be partial or nearly complete. All plants possess this type of resistance, and its level will vary depending on the attacking pathogen. This resistance is controlled by many genes and is known as polygenic resistance. It is also known as quantitative, partial, horizontal, multigenic, field, durable, and minor gene resistance (Agrios, 2005; see also Chapter 6). Plants also possess a resistance that, rather than being controlled by many genes, is controlled by one or just a few genes. Here, each host plant carries one gene or genes (known as resistance (R) genes) for every pathogen capable of attacking it. In turn, the pathogen carries a matching gene or genes (known as avirulence (A) genes). If a plant possessing a specific R gene is attacked by a pathogen carrying a matching A gene, defenses are triggered in the plant, which can kill the pathogen and halt the infection process. Often, the host cell being attacked dies, together with a few surrounding cells, in a process known as a hypersensitive response. This resistance, controlled by just one or a few genes, is called race-specific, R gene, major gene, or vertical resistance (Agrios, 2005; Chapter 6). Regarding the pathogen, some have a wide host range and are unspecialized, while others are highly specialized. The latter may be divided into a number of formae speciales (singular, forma specialis, f.sp.), which are morphologically identical, or nearly so, but which are capable of attacking different host genera, for example, cereal powdery mildews, *Blumeria graminis*, formae speciales of which attack barley (f.sp. *hordei*) and wheat (f.sp. *tritici*). Within each forma specialis, there are frequently large numbers of physiologic races, each capable of attacking a different spectrum of host species, or cultivars, according to the genes for virulence they possess.

We will look at how these types of resistance are triggered later, but in the meantime, let us turn our attention to the mechanisms plants use to defend themselves.

2.2.2 Passive or preexisting defenses

2.2.2.1 Preexisting structural defenses

The cuticle
In order to enter a plant and get at the nutrients within, a pathogen must attach itself to the surface, for example, of the leaf or root, prior to penetration. It stands to reason therefore that these plant surfaces represent the first line of defense against microbial invaders. The aerial surfaces of plants such as leaves and stems are covered with a structure known as the cuticle (Figure 2.2). The cuticle is formed from cutin, a hydrophobic material comprising fatty acids and fatty esters (Brett & Waldron, 1990), and it acts as an important barrier to pathogens. Cuticles vary in thickness and composition depending on the plant species, and a thick cuticle might increase resistance to pathogens that penetrate leaves or stems directly. However, cuticle thickness and pathogen resistance are not always correlated, since many plants with thick cuticles are readily invaded by pathogens capable of direct penetration. Below the cuticle lie the epidermal cells, the walls of which can vary in thickness and

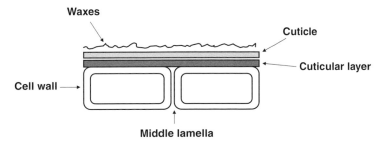

Figure 2.2 Cross-section of a leaf showing the barriers on the leaf surface: the cuticular layer, cuticle, and leaf surface waxes.

toughness depending on the plant. A thick, tough epidermal cell wall provides an almost impenetrable barrier to many fungal pathogens. Plant cells have a primary cell wall, which provides structural support, and many cells also form a secondary cell wall that develops inside the primary cell wall once the cell stops growing. The primary cell wall is composed mainly of cellulose, bundled into fibers known as microfibrils, which give strength and flexibility to the wall. The cell wall is also likely to contain two groups of branched polysaccharides, cross-linking glycans and pectins. The former include hemicellulose fibers that provide strength via cross-linkages with cellulose, while pectins form hydrated gels that help to cement neighboring cells together. Lignin, a heterogeneous polymer composed of phenolic compounds, is a component of many cell walls and provides rigidity to the cell. Lignified cell walls are highly impermeable to pathogens.

In contrast to leaves and stems, roots contain suberin, which is deposited along with associated waxes in the cell walls of epidermal, exodermal, and endodermal cells. The root epidermis is in direct contact with the soil environment and is often the site of penetration by soilborne pathogens. As such, it offers the first line of defense against root pathogens.

Stomata

A leaf surface can have up to 300 stomata/mm^2 occupying up to 2% of the total surface area of the leaf (Melotto *et al.*, 2008). However, although abundant, in reality, stomata account for a very small proportion of the leaf surface. Nevertheless, stomata offer potential entry points for pathogens. Indeed, many fungal and bacterial pathogens only enter the host plant via stomata. Many pathogens force their way through closed stomata, although some, for example, *P. graminis* f.sp. *tritici*, the cause of wheat stem rust, can only enter wheat plants via open stomata. Indeed, stomata have been considered as passive entry points, with bacteria, for example, freely able to invade the plant via open stomata. However, work on *Arabidopsis* has shown that stomata close in response to live bacteria. When leaves were inoculated with a suspension of *Pseudomonas syringae* pv. *tomato* (*Pst*) DC3000, 70% of stomata closed within 1 hour, although this closure was transient, with most stomata reverting to the open state after 3 hours (Melotto *et al.*, 2006). This work shows that stomatal guard cells can perceive live bacteria leading to significant stomatal closure and suggests that plants have evolved mechanisms to reduce entry of bacteria via stomata as an integral part of their defenses. Interestingly, *Pst* DC3000 produces a diffusible toxin (coronatine), which can reopen closed stomata (Melotto *et al.*, 2006).

2.2.2.2 Preexisting chemical defenses

Although passive structural defenses keep many pathogens out of plants, they are not always effective and many pathogens can breach such barriers. Plants therefore possess a second line of preexisting defenses, based on chemicals, which can stop a pathogen in its tracks. Plants are capable of producing a diverse array of antimicrobial compounds (Wink, 1999). These compounds are the products of secondary plant metabolism, in contrast to metabolites arising from primary metabolism such as respiration (Figure 2.3). Those chemicals produced constitutively are called phytoanticipins, while those produced in response to pathogen attack are called phytoalexins (see Section 2.2.3.2) (Morrissey & Osbourn, 1999; Wittstock & Gerschenzon, 2002). Phytoanticipins are synthesized by unchallenged plants during normal growth and development, and accumulate in specific

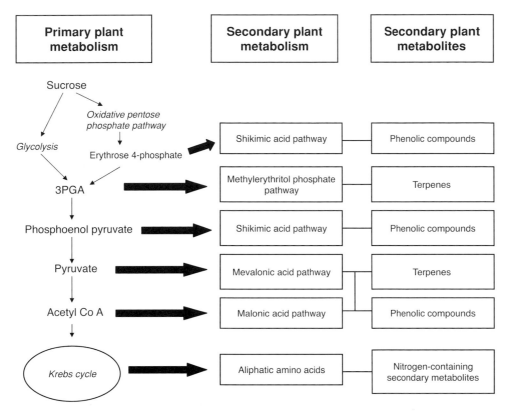

Figure 2.3 Plant secondary metabolism. This figure summarizes the major pathways involved in plant secondary metabolism and the main families of compounds formed. (Redrawn from Scott (2008), with permission of Blackwell Publishing Ltd.)

tissues, organs, or specialized structures, where they are usually sequestered in the vacuole or other subcellular compartments (Morrissey & Osbourn, 1999).

An aside—a quick excursion into plant chemistry

Three groups of compounds are produced by plants to act in defense: phenolics, terpenoids, and nitrogen-containing organic compounds (Figure 2.3). Defensive compounds can be virtually ubiquitous (e.g., terpenoids) or unique to one or two plant families (e.g., coumarins) (Table 2.1).

Terpenes are synthesized via the mevalonate pathway or the methylerythritol phosphate pathway (Figures 2.3 and 2.4). The former is located in the cytosol and leads to the formation of sesquiterpenes, triterpenes, sterols, and polyterpenes, while the latter pathway is located in the plastids and results in the formation of isoprene, monoterpenes, diterpenes, and carotenoids (Figure 2.5) (Lichtenthaler, 1999). Most terpenes are made from the basic unit isoprene, which is converted into isopentyl pyrophosphate.

Phenolic compounds are the second biggest group of plant secondary metabolites involved in plant defense. They are synthesized from the shikimic acid or malonic acid pathways (Figure 2.3) and possess an aromatic ring with one or more hydroxyl groups, and a number of other constituents. They range from simple phenolics with 6 or 7 carbon

Table 2.1 Examples of secondary compounds used in chemical defense in plants

Compound class	Plant distribution	Approximate number of chemicals	Example
Nonprotein amino acids	2500 species, 130 families (especially in seeds of legumes)	600	Canavanine
Alkaloids	20% of angiosperms	10,000	Nicotine
Glucosinolates	300 species, 15 families (Cruciferae and a number of other families)	100	Methyl glucosinolate
Cyanogenic compounds	2500 species, 130 families	60	Linamarin
Terpenoids (includes monoterpenes, diterpenoids, sesquiterpene lactones, saponins, and cardenolides)	Widely distributed	8,000	Pulegone Momilactone A Rishitin
Phenolics (simple phenols, flavonoids, quinines)	Widely distributed	5,000	Benzoic acid Scopoletin Coumarin
Iridoid glycosides	57 families	600	Aucubin

Source: Modified from Harborne (1993) and Strauss & Zangerl (2002), with permission of Blackwell Publishing Ltd.

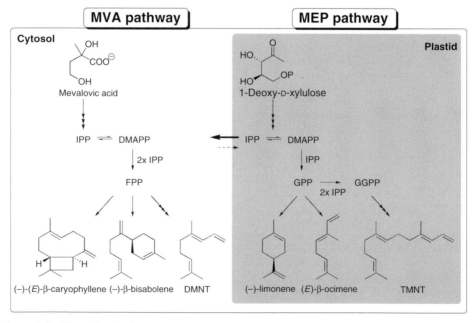

Figure 2.4 Biosynthesis of terpenes. MEP, methyl erythritol phosphate; IPP, isopentenyl diphosphate; FPP, farnesyl diphosphate; GPP, geranyl diphosphate; DMAPP, dimethylallyl pyrophosphate; DMNT, (*E*)-4,8-dimethyl-1,3,7-nonatriene; TMNT, (*E,E*)-4,8,12-trimethyl-1,3, 7,11-tridecatetraene. (Reproduced by courtesy of Dr. Stefan Garms.)

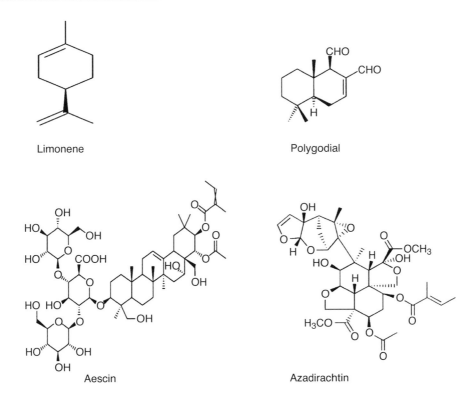

Limonene

Polygodial

Aescin

Azadirachtin

Figure 2.5 Examples of terpenoids: limonene (a monoterpenoid), polygodial (a sesquiterpenoid), azadirachtin (a triterpenoid), and aescin (a saponin).

atoms, such as catechol and salicylic acid, through coumarins with 9 carbon atoms, and on to more complex compounds with 15 carbon atoms, for example, flavonoids (Table 2.2 and Figure 2.6). The flavonoid nucleus is usually attached to a sugar molecule to yield a glycoside, which tends to be stored in the vacuole.

Nitrogen-containing organic compounds are synthesized from common amino acids and include cyanogenic glycosides, glucosinolates, and alkaloids. Cyanogenic glycosides are made through conversion of amino acids to oximes, which are then glycosylated. After synthesis, they are kept in separate cellular compartments from the enzymes that break them down and only come into contact when cells are ruptured following attack. The result is the liberation of hydrogen cyanide. Glucosinolates occur mainly in the Brassicaceae and contain sulfur as well as nitrogen. They are broken down by the enzyme myrosinase, leading to the formation of isothiocyanates (mustard oils), nitriles, and other compounds. Because of this, glucosinolates and myrosinase are kept in different cellular compartments to prevent autotoxicity. Alkaloids are cyclic compounds (Figure 2.7), produced from either amino acids (e.g., lysine, tyrosine, and ornithine) or purines and pyrimidines. Plants of the genus *Nicotiana* produce the alkaloid nicotine, which is made from ornithine. Other alkaloids include the tropane alkaloids (e.g., atropine) found in members of the Solanaceae, and quinolizidine alkaloids, which are derived from lysine and are commonly found in lupins.

Concentrations of secondary compounds produced can vary enormously between plants and even within the same plant. For example, it would seem prudent for the plant to allocate

Table 2.2   Major classes of phenolic compounds in plants

Basic skeleton	Number of carbon atoms	Class	Examples
C_6	6	Simple phenols	Catechol, hydroquinone
C_6-C_1	7	Phenolic acids	p-Hydroxybenzoic acid salicylic acid
C_6-C_2	8	Phenylacetic acids	p-Hydroxyphenylacetic acid
C_6-C_3	9	Hydroxycinnamic acids	Caffeic acid, ferulic acid
		Phenylpropenes	Myristicin, eugenol
		Coumarins	Umbelliferone
		Isocoumarins	Bergenin
C_6-C_4	10	Naphthoquinones	Juglone
C_6-C_1-C_6	13	Xanthones	Mangiferin
C_6-C_2-C_6	14	Stilbenes	Lunularic acid
		Anthraquinones	Emodin
C_6-C_3-C_6	15	Flavonoids	Quercitin
		Isoflavonoids	Genistein
$(C_6$-$C_3)_2$	18	Lignans	Podophyllotoxin
$(C_6$-$C_3)_n$	9_n	Lignins	

Source: Modified from Schoonhoven *et al.* (2005), with permission of Oxford University Press.

most of its defensive chemicals to vulnerable tissues or plant parts that are important for successful reproduction and hence plant fitness. There is also increasing evidence from a number of plant species that chemicals are located where they are most readily perceived by attackers such as insect herbivores—the plant surface, including leaf waxes, trichomes, and so on (Harborne, 1993). Even within a tissue, for example, a leaf, concentrations of defensive chemicals can vary, with important consequences for the attacker. Concentrations of defensive chemicals can also be affected greatly by age of the tissue, and environmental factors such as light, water, and other soil factors, for example, mineral nutrition. Examples of variations in concentrations of chemicals and differences in location within the plant are given in the various sections below.

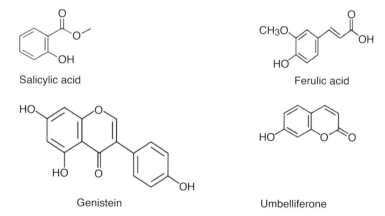

Salicylic acid   Ferulic acid

Genistein   Umbelliferone

Figure 2.6   Some plant phenolics: salicylic acid (a phenolic acid), ferulic acid (a hydroxycinnamic acid), umbelliferone (a coumarin), and genistein (an isoflavonoid).

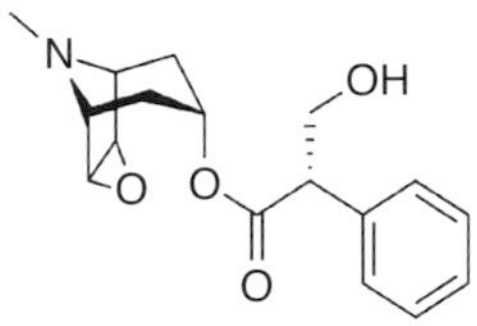

Scopolamine 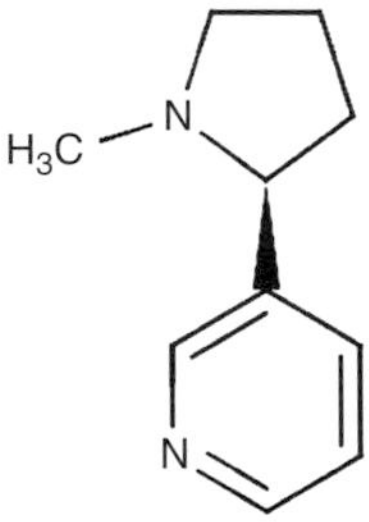   Nicotine

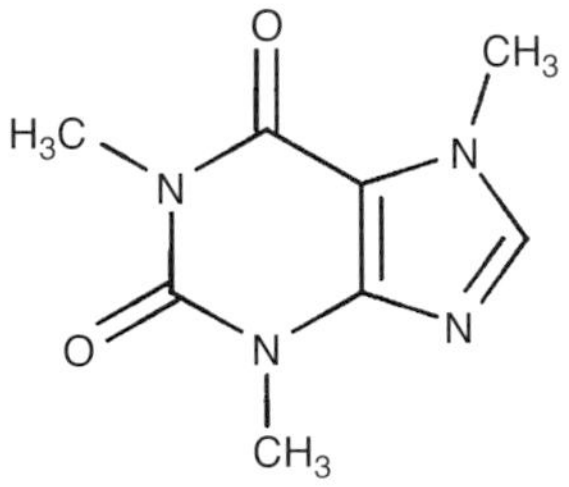

Caffeine

Figure 2.7 Examples of alkaloids: scopolamine (a tropane alkaloid), nicotine (formed from ornithine and nicotinic acid), and caffeine (a purine alkaloid).

From this quick excursion into plant chemistry, it will be obvious that plants possess a formidable array of chemicals that can be deployed in defense against different attackers. Specific examples of the involvement of chemicals in plant defense against a range of attackers are given in the various sections below.

Chemicals on plant surfaces
Although it might be expected that preexisting chemical defenses would only be encountered once the pathogen has entered the plant, there are examples where the pathogen encounters such defenses while still on the plant surface. The classic example involves onion smudge, caused by the fungal pathogen *Colletotrichum circinans*. Onion cultivars with yellow/brown or red scales contain the phenolic compounds protocatechuic acid and catechol. *C. circinans*, like most fungal pathogens, requires liquid water for spore germination. However, these two fungitoxic compounds diffuse into water droplets on the surface of the onions, killing the pathogen, thereby protecting the plant from infection. Onion cultivars with colored scales tend therefore to be resistant to this pathogen. Interestingly, onion cultivars with white scales do not contain protocatechuic acid and catechol and are susceptible.

Chemicals within plant cells
Once inside the plant, the pathogen can be faced with a formidable chemical arsenal. Many plants accumulate phenolic compounds, tannins, or fatty acid-like compounds such as dienes in cells of young fruits, leaves, or seeds. These accumulated chemicals are thought

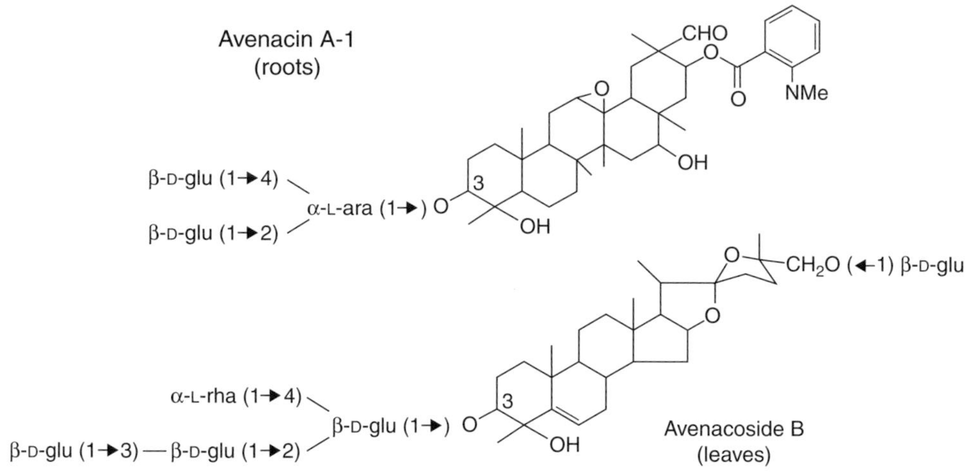

Figure 2.8 Representatives of the two different families of saponins synthesized by oat: the triterpenoid saponin avenacin A-1, which is the major saponin present in the roots (top); the steroidal saponin avenacoside B, which is found in the leaves (bottom). (Reproduced from Osbourn *et al.* (2003), with permission of Blackwell Publishing Ltd.)

to provide resistance in young tissues to various pathogens. Many dicotyledonous plant species produce glycosylated triterpenes and steroids known as saponins. Interestingly, among the monocots, although oats (*Avena* spp.) can produce saponins, it seems that most cereals and grasses cannot (Hostettmann & Marston, 1995). Oats produce two different families of saponins, steroidal avenacosides found in leaves and triterpenoid avenacins found in roots (Figure 2.8; Osbourn *et al.*, 2003). Avenacin A-1 accumulates in epidermal cells of the root tip and is compartmentalized in the vacuoles. It is a powerful antifungal chemical, conferring resistance to a range of soil-inhabiting fungi (Field *et al.*, 2006).

Some constitutively produced chemicals are stored as inactive precursors and are converted into the biologically active form following pathogen attack. A good example involves glucosinolates produced by members of the Brassicaceae. In *Arabidopsis*, glucosinolates can be stored in vacuoles of specialized cells called S-cells, while the enzymes responsible for converting them to the active form (myrosinases) are located in myrosin cells. Following pathogen attack, damage to the tissues leads to breakdown of this compartmentalization and the myrosinases hydrolyze the glucosinolates to unstable aglucones. These, in turn, are converted to toxic chemicals such as isothiocyanates and nitriles (Wittstock & Halkier, 2002; Field *et al.*, 2006).

In addition to the chemicals discussed above, several preformed plant proteins have been reported to play a role in plant defense against pathogens. One example is the lectins, proteins that bind with free sugar or with sugar residues such as polysaccharides, that are either free or attached to cell membranes. Indeed, many plant lectins have a high affinity for oligosaccharides, which are uncommon in plants, for example, those which bind chitin, a component of the cell walls of fungi and the exoskeleton of insects, but which is not found in plants (Debenham, 2005). This suggests that lectins play a role in plant defense, possibly via recognition of the attacking fungal pathogen or insect pest (the latter aspect is dealt with later in this chapter). In addition, various plants contain enzymes called ribosome-inactivating proteins (RIPs), which inhibit protein synthesis and can therefore block virus replication. RIPs have been shown to have broad-spectrum antiviral activity, affecting RNA, DNA, plants, and animal viruses (Wang & Turner, 2000). It is important to appreciate that the distinction between passive and active defense is not always clear-cut, and some of these plant proteins, for example, certain lectins, can also be produced following pathogen attack.

2.2.3 Active or inducible defenses

If a plant is to successfully ward off a pathogen, rapid mobilization of its defenses is necessary. This requires early recognition of the pathogen by the host plant. Amazingly, the plant starts to receive signals from the pathogen as soon as contact is established. Once it has breached the outer layers of the plant (e.g., cuticle and cell wall), it is subject to molecular recognition by individual plant cells. Plants have evolved two types of immune receptors to detect nonself molecules. First, there are pattern-recognition receptors (PRRs) that are located on the cell surface. PRRs sense microbes by perception of pathogen-associated molecular patterns (PAMPs; also referred to as MAMPs for microbe-associated molecular patterns, since they are not restricted to pathogenic microbes). MAMPs are conserved microbial molecules that include components of fungal cell walls such as chitin, lipopolysaccharides from Gram-negative bacteria, and short peptides derived from bacterial flagellin (Zipfel, 2008). This first level of immunity represents a basal resistance and is known as PAMP-triggered immunity (PTI) or innate immunity (Chisholm *et al.*, 2006). PTI can protect plants against entire groups of pathogens. A virulent pathogen can successfully infect the host plant by evading recognition or suppressing MAMP-triggered signaling. The pathogen probably manages to achieve the latter by secreting virulence effectors (De Wit, 2007). In turn, some plants have evolved resistance (R) proteins capable of recognizing these effector proteins, resulting in a second line of defense known as effector-triggered immunity (ETI) (Pieterse & Dicke, 2007). In an ongoing evolutionary arms race, pathogens have evolved effectors capable of suppressing ETI (Zipfel, 2008).

If the pathogen-derived elicitors are recognized by the plant, alarm signals are generated leading eventually to the mobilization of defenses. Some of these alarm signals are intra-cellular, although alarm signals can be transmitted to adjacent cells, systemically within the plant to other tissues, and even outside the plant to neighboring plants. Signaling for the mobilization of defenses is dealt with in Chapter 3. The defenses induced following host recognition of the pathogen can be structural and chemical. Whether a plant turns out to be resistant or susceptible to infection seems to be determined by the speed and magnitude with which these defense mechanisms are activated and expressed, and by their effectiveness against individual pathogens with different modes of attack.

2.2.3.1 Inducible structural defenses

Cell wall appositions

Although, as indicated above, the plant cell wall provides an almost impenetrable barrier to many pathogens, others have developed the ability to breach this structural barrier. However, all is not lost, because plants can reinforce the cell wall in response to pathogen attack through the formation of cell wall appositions (CWAs). For example, when a

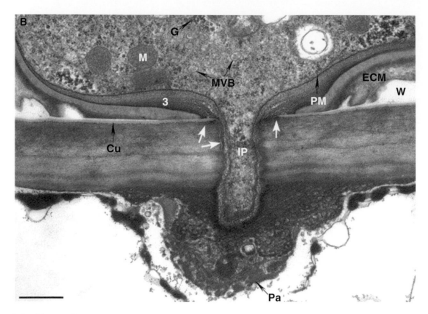

Figure 2.9 Transmission electron micrograph showing penetration of sorghum epidermal cells by appressoria of *C. sublineolum* 42 hours after inoculation. Note that the infection peg (IP) has been encased by a host papilla (Pa) (bar = 1 μm). ECM, extracellular matrix; W, leaf epicuticular wax layer; M, mitochondria; G, glycogen granules; MVB, polyribosomes and multivesicular bodies. (From Wharton *et al.* (2001), with permission of APS.)

nonpathogenic or avirulent fungal pathogen attempts to infect a leaf, a plug of material, known as a papilla, is deposited beneath the site of attempted penetration (Figure 2.9). In the interaction between barley and the powdery mildew fungus, *B. graminis* f.sp. *hordei*, these papillae are composed of phenolics, callose, peroxidases, and cell wall material. Also, reactive oxygen species (ROS) such as hydrogen peroxide (H_2O_2) and the superoxide anion (O_2^-) can also accumulate in the papillae, coincident with attempted penetration by the fungus (Thordal-Christensen *et al.*, 1997; Hückelhoven & Kogel, 1998). This suggests reinforcement of the cell wall via oxidative cross-linking. In barley genotypes carrying the *mlo* gene for resistance, papilla formation appears to occur earlier in response to powdery mildew infection than in genotypes not carrying this gene. This suggests that papilla formation is an important component of *mlo*-dependent resistance in barley, although whether the processes involved in papilla formation are direct determinants of resistance is not known (Schulze-Lefert & Vogel, 2000).

Thickening and modification of the host cell wall also occur in other plant–pathogen interactions. Thus, in roots of many plants, cortical cells respond to attempted fungal penetration by formation of a lignituber, where fungal hyphae are enveloped by callose, which is then impregnated with phenolic compounds.

As well as impeding fungal penetration of plant tissues, it has been suggested that CWAs might also increase the resistance of plant cell walls to attack by fungal enzymes. Further, the phenolic compounds that impregnate CWAs can exert direct toxic effects against invading pathogens. This all suggests that CWAs are important in plant defense against fungal pathogens. However, what is not known is whether any of the proposed functions of CWAs actually determine the outcome of a host–pathogen interaction (Lucas, 1998).

Lignification

Lignin is a complex biopolymer made up of several different monomers (Figure 2.10). It is very difficult to break down and metabolize and represents a formidable barrier to pathogens. When it is deposited in response to pathogen infection, it is referred to as defense lignin (Nicholson & Hammerschmidt, 1992) and has a different composition to that deposited during plant development (Garcion *et al.*, 2007). For example, defense lignin deposited in squash is rich in *p*-coumaraldehyde units, while that deposited during development is rich in guaiacyl/syringyl units (Stange *et al.*, 2001). In *Arabidopsis*, two genes involved in lignin biosynthesis are differentially regulated during development and in response to pathogen challenge. Thus, the cinnamoyl-CoA reductase gene *AtCCR2* is induced during the incompatible interaction with the bacterial pathogen *Xanthomonas campestris* pv. *campestris*, but is weakly expressed during development, while the related gene *AtCCR1* is expressed strongly in tissue undergoing lignification (Lauvergeat *et al.*, 2001).

Cork layers

The structural defenses induced by plants following pathogen attack are not limited to cell wall modifications. A common inducible structural defense is the formation of cork layers. Following infection, especially by fungi and bacteria, many plants form several layers of cork cells beyond the point of infection. These cork layers have several functions: (1) they inhibit progress of the pathogen through the plant tissue, (2) they limit the diffusion of any toxic compounds produced by the pathogen, and (3) they prevent the flow of water and nutrients to the site of pathogen infection, thereby starving the pathogen. Equally effective in ridding the plant of an invading pathogen is the formation of an abscission layer in young leaves, usually on stone fruit trees. Following infection, the middle lamella between two layers of cells is dissolved throughout the thickness of the leaf. This completely cuts off the infection area from the rest of the leaf. Deprived of water and nutrients, this area of leaf dies, shrivels, and sloughs off, taking the pathogen with it. This leads to the appearance of "shot holes" on leaves (Figure 2.11).

Tyloses

Another inducible structural defense is the formation of tyloses. These are overgrowths of the protoplast of parenchyma cells adjacent to xylem vessels, which protrude into the lumen of the vessels through half-bordered pits (Figure 2.12). Tyloses form in many plants during invasion by vascular wilt pathogens such as *Verticillium*. If tyloses form quickly ahead of the pathogen, they can block its spread in the xylem. Moreover, since tyloses have cellulosic walls, they cannot be penetrated by vascular wilt pathogens. There are reports of resistance to vascular wilt pathogens associated with the rapid formation of tyloses. For example, rapid occlusion of xylem vessels by tyloses has been implicated in the resistance of cotton to *Verticillium dahliae* (Mace, 1978; Harrison & Beckman, 1982). Interestingly, however, tyloses could not be detected in cotton plants inoculated with the vascular wilt pathogen *Fusarium oxysporum* f.sp. *vasinfectum* (Shi *et al.*, 1992).

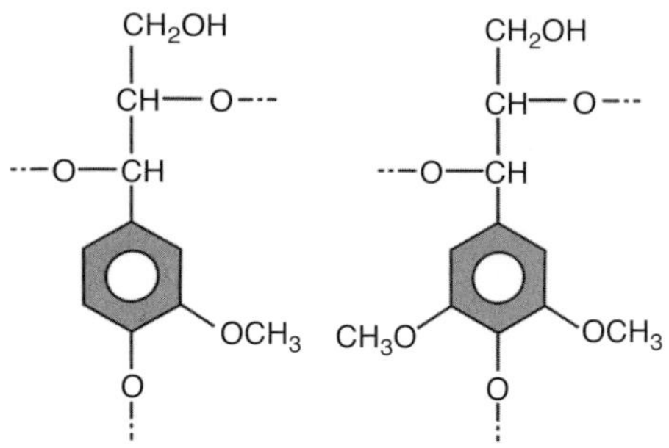

Coniferyl alcohol   Sinapyl alcohol

(a)

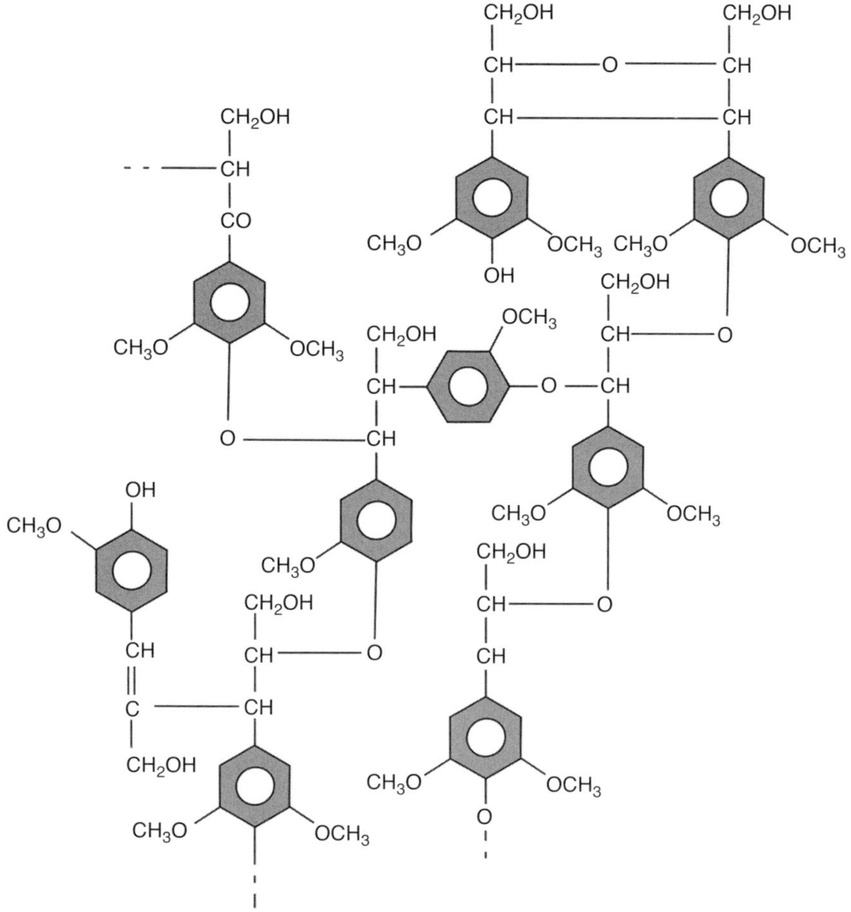

(b)

Figure 2.10   The structure of lignin. Lignin is a biopolymer made up of a number of different monomers. Two of these monomers are coniferyl alcohol and sinapyl alcohol (a). These monomers are polymerized to form the complex structure of lignin, an example of which is shown in (b). (Reproduced from Scott (2008) with permission of Blackwell Publishing Ltd.)

Figure 2.11 Shot holes in a taro leaf resisting attack by the pathogen *Phytophthora colocasiae.*
(Courtesy of Janice Uchida, University of Hawaii.)

2.2.3.2 Inducible chemical defenses

Phytoalexins

Phytoalexins are low-molecular-weight compounds with antimicrobial properties, which
are synthesized by and accumulate in plant cells following microbial infection (Paxton,
1981). In contrast to preformed inhibitors, phytoalexins are not present in healthy tissues.
Rather, they are only formed after infection or injury. More than 300 molecules have been
identified as phytoalexins, from some 900 species representing 40 different plant families
(Harborne, 1999). Although they are structurally diverse, phytoalexins can be grouped into
structural families and related by their biosynthetic pathways. Indeed, because the number
of major biosynthetic pathways is small relative to the wide chemical diversity of phytoalexins, this provides a useful way to organize and classify them (Figure 2.13). There is a
close association between some structures and plant taxa. For example, sesquiterpenes are
mainly produced by members of the Solanaceae, sulfur-containing indoles by the Brassicaceae and isoflavonoids by the Papilionoideae subfamily of the Leguminosae (Harborne,
1999). In contrast, some phytoalexins are shared by widely divergent plant species, for example, stilbenes, which occur in pine, grapevine, and peanut. Incredibly, several related and
unrelated phytoalexins can be produced in a single species. Thus, 16 different phytoalexins
have been isolated from rice, although whether all of them are involved in defense against
pathogens is unknown.

Proving that phytoalexins play a role in defense against pathogens has been a difficult
task. One problem is that phytoalexin induction is nonspecific because, in addition to

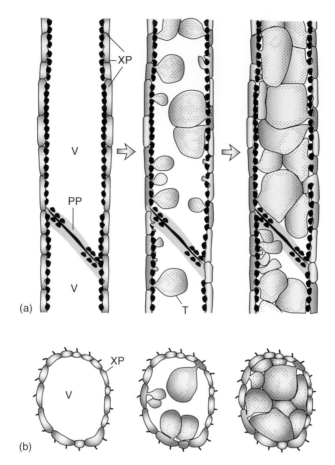

Figure 2.12   Schematic drawing of tyloses in a xylem vessel: (a) longitudinal section through the xylem vessel; (b) cross-section through the xylem vessel. V, vessel; PP, perforation plate; XP, xylem parenchyma. (Reproduced from Agrios. Copyright 2005, with permission of Elsevier.)

microbes, a range of chemical and physical agents can trigger phytoalexin biosynthesis. In this regard, phytoalexins could be considered as stress metabolites. For a phytoalexin to play a role in defense, it must fulfill several criteria: (a) it must accumulate in response to infection; (b) it must be inhibitory to the invading pathogen; (c) it must accumulate to inhibitory concentrations in the vicinity of the pathogen at the time the latter ceases growth; (d) variation in the rate of accumulation of the compound should lead to a corresponding variation in plant resistance; (e) variation in sensitivity of the pathogen should result in a corresponding variation in its virulence.

Phytoalexins are considered as relatively weak antifungal and antibacterial agents (Kuć, 1995), which raises the issue of their actual concentration in the close vicinity of the pathogen. However, timing of phytoalexin production appears to be more important than the final concentration that accumulates. In some classic work, the phytoalexin phaseollin was shown to accumulate earlier in French bean reacting hypersensitively to the fungal pathogen *Colletotrichum lindemuthianum*, than in plants infected by a compatible isolate of the pathogen (Figure 2.14). The more rapid production of phaseollin in the incompatible

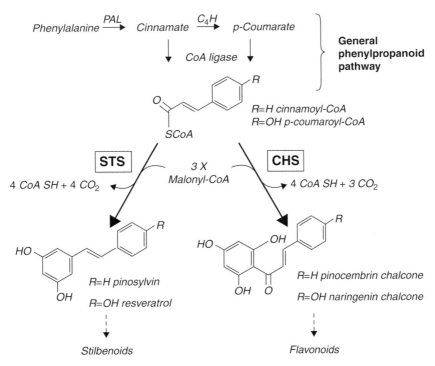

Figure 2.13 Pathways for stilbenoid and flavonoid biosyntheses. Stilbene synthase (STS) and chalcone synthase (CHS), respectively, lead to stilbenoid and flavonoid biosynthesis from a cinnamoyl-CoA/p-coumaroyl-CoA with three malonyl-CoAs. PAL, phenylalanine ammonia-lyase; C_4H, cinnamate 4-hydroxylase. (From Kodan *et al.* (2002), with permission of the National Academy of Sciences, USA.)

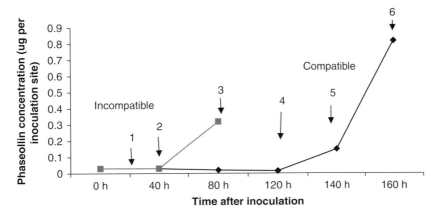

Figure 2.14 Accumulation of phaseollin in beans inoculated with compatible and incompatible races of the pathogen *C. lindemuthianum*. 1, appressorium formation; 2, hypersensitive response visible; 3, hypersensitive response complete; 4, 1% lesions in compatible interaction; 5, 80% lesions in compatible interaction; 6, 100% lesions in compatible interaction. (Redrawn from Bailey & Deverall (1971), with permission of Elsevier.)

interaction meant that considerably less accumulated than in the compatible interaction. In the latter, by the time phaseollin levels began to increase, the pathogen was established in the host tissue (Bailey & Deverall, 1971). A similar situation was observed in the incompatible interaction between sorghum and the fungal pathogen *Colletotrichum sublineolum* (Figure 2.15a). Accumulation of 3-deoxyanthocyanidin phytoalexins was more rapid in the incompatible compared to the compatible interaction. The phytoalexin response in this interaction was quite complex, with differences in both the timing of production of individual phytoalexins and the concentrations produced (Figure 2.15b–2.15e). Infection vesicles and primary hyphae of the fungus were severely distorted in the incompatible interaction, while the fungus was able to grow and colonize host tissue substantially in the compatible interaction (Lo *et al.*, 1999).

Although there is much evidence of a correlation between phytoalexin accumulation and resistance to pathogen infection, providing proof that phytoalexin accumulation is actually responsible for resistance is another matter. Stilbene synthase (STS) is the enzyme that catalyzes the formation of the phytoalexin resveratrol. When the stilbene synthase gene was introduced into tobacco, it led to increased resistance to *Botrytis cinerea* (Hain *et al.*, 1993), and when introduced into a range of other crop plants, resistance to a number of pathogens was increased (Zhu *et al.*, 2004). Conversely, the *pad3* mutant of *Arabidopsis* is defective in biosynthesis of the phytoalexin camalexin, and such plants are considerably more susceptible to *Alternaria brassicicola* than the parental line (Thomma *et al.*, 1999).

Virulent pathogens tend to be more tolerant of phytoalexins of their host plants than avirulent or nonpathogenic organisms. An excellent example is *Nectria haematococca*, a pathogen of pea. Virulent strains of the fungus produce the enzyme pisatin demethylase, which inactivates the major pea phytoalexin, pisatin. When strains of *N. haematococca* producing low and high amounts of the enzyme were crossed, only progeny strains with high amounts of pisatin demethylase could cause significant lesion development on peas. These data suggest that pathogenicity was correlated with ability to degrade the phytoalexin (Kistler & Van Etten, 1984). However, the current view is that, while the ability to degrade phytoalexins can limit lesion development, additional factors are required for full pathogenicity.

Interestingly, a major mechanism allowing various pathogens to tolerate phytoalexins involves multidrug efflux pumps. When different drug extrusion systems were subjected to targeted mutations, virulence of a number of pathogens on their hosts was decreased, including *Erwinia amylovora* on apple trees (Burse *et al.*, 2004) and *Magnaporthe grisea* on rice and barley (Urban *et al.*, 1999). Conversely, pathogen sensitivity to phytoalexins can be greatly increased by inhibiting multidrug efflux systems using chemicals (Tegos *et al.*, 2002).

Antifungal proteins

Nearly 40 years ago, work on tobacco reacting hypersensitively to tobacco mosaic virus (TMV) showed the appearance of novel proteins accumulating in response to the infection. These proteins were called pathogenesis-related proteins (PRs) and were shown to occur in plant species from at least 13 families following infection by oomycetes, fungi, bacteria, and viruses, as well as nematode and insect attack (Datta & Muthukrishnan, 1999). Several PRs were shown subsequently to have biochemical functions that suggested they might possess antimicrobial activity. Currently, PRs have been assigned to 17 families of induced proteins (Table 2.3) and identified biochemical functions include β-1,3-glucanase

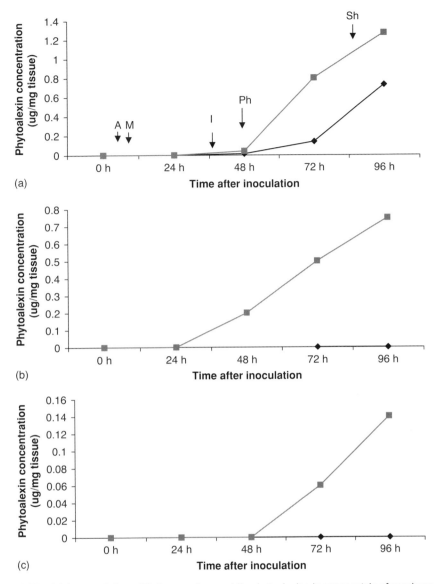

Figure 2.15 (a) Accumulation of 3-deoxyanthocyanidin phytoalexins in mesocotyls of sorghum cultivars SC748-5 (resistant, ▪) and BTx623 (susceptible, ◆) after inoculation with *C. sublineolum*. The time course of phytoalexin accumulation was compared to the development of the fungus. A, formation of appressoria; M, melanization of appressoria; I, formation of infection vesicles; Ph, emergence of primary hyphae from infection vesicles; Sh, branching of secondary hyphae from primary hyphae (in susceptible tissue only). (b–e) Accumulation of 3-deoxyanthocyanidin phytoalexins in mesocotyls of sorghum cultivars SC748-5 (resistant, ▪) and BTx623 (susceptible, ◆) after inoculation with *C. sublineolum:* (b) luteolinidin; (c) 5-methoxyluteolinidin; (d) apigeninidin; (e) caffeic acid ester of arabinosyl-5-*O*-apigeninidin. (Reproduced from Lo *et al.* (1999), with permission of Elsevier.)

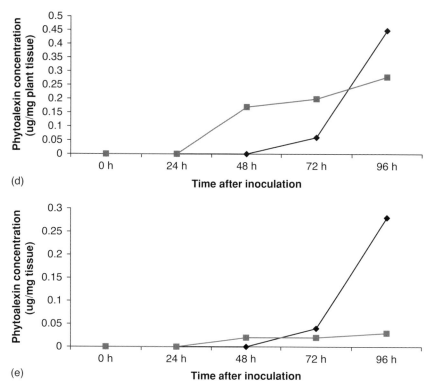

(d)

(e)

Figure 2.15   (*Continued*).

(PR-2), endochitinase (PR-3, -4, -8, -11), proteinase inhibitors (PR-6), defensins (PR-12), and thionins (PR-13). PRs were classified originally on the basis that they were induced following pathogen infection, while those proteins occurring in the absence of pathogen infection were to be called PR-like proteins. However, this distinction became blurred over the years, with the term "pathogenesis-related proteins" being used as a collective term for all microbe-induced proteins and their homologues, including many (such as peroxidase) that are generally present constitutively. Because of this, some workers use the term "inducible defense-related proteins" to indicate those proteins which are largely not detectable in uninfected plant tissues, but are induced following pathogen infection (Van Loon *et al.*, 2006).

The importance of PRs in plant resistance to pathogen infection has been tested by overexpression in various plants. For example, a threefold increase in resistance of flax to *Fusarium culmorum* and *F. oxysporum* was obtained by overexpression of a potato β-1,3-glucanase. The protection obtained was demonstrated to result from a direct effect of the enzyme against growth of the fungus (Wróbel-Kwiatkowska *et al.*, 2004). Many of these experiments are conducted under controlled conditions and results can often be different in the field. Thus, genes encoding a chitinase and a β-1,3-glucanase were isolated from a wheat variety resistant to *Fusarium graminearum* and transferred to a susceptible wheat variety. One line expressing this gene combination exhibited a delay in the spread of infection in the glasshouse, but not in the field (Anand *et al.*, 2003).

Table 2.3 Recognized families of pathogenesis-related proteins

Family	Type member	Properties	Gene symbols
PR-1	Tobacco PR-1a	Unknown	*Ypr1*
PR-2	Tobacco PR-2	β-1,3-Glucanase	*Ypr2, [Gns2 ("Glb")]*
PR-3	Tobacco P, Q	Chitinase type I, II, IV, V, VI, VII	*Ypr3, Chia*
PR-4	Tobacco "R"	Chitinase type I, II	*Ypr4, Chid*
PR-5	Tobacco S	Thaumatin-like	*Ypr5*
PR-6	Tomato inhibitor I	Proteinase inhibitor	*Ypr6, Pis ("Pin")*
PR-7	Tomato P_{69}	Endoproteinase	*Ypr7*
PR-8	Cucumber chitinase	Chitinase type III	*Ypr8, Chib*
PR-9	Tobacco "lignin-forming peroxidase"	Peroxidase	*Ypr9, Prx*
PR-10	Parsley "PR1"	Ribonuclease-like	*Ypr10*
PR-11	Tobacco "class V" chitinase	Chitinase type I	*Ypr11, Chic*
PR-12	Radish Rs-AFP3	Defensin	*Ypr12*
PR-13	Arabidopsis THI2.1	Thionin	*Ypr13, Thi*
PR-14	Barley LTP4	Lipid transfer protein	*Ypr14, ltp*
PR-15	Barley OxOa (germin)	Oxalate oxidase	*Ypr15*
PR-16	Barley OxOLP	Oxalate oxidase-like	*Ypr16*
PR-17	Tobacco PRp27	Unknown	*Ypr17*

Source: Reproduced from Van Loon *et al.* (2006), with permission.

Other proteins associated with plant defense

Various phenolic compounds have been shown to accumulate at a faster rate in plants resistant to pathogens compared to their susceptible counterparts (Agrios, 2005), and, as already discussed, lignin formation is a commonly observed active plant defense. An important precursor for the biosynthesis of phenolics and lignin is the amino acid phenylalanine, with its C_6C_3 skeleton. After pathogen attack, phenylalanine is converted to *trans*-cinnamic acid and then enters the phenylpropanoid pathway (Figure 2.13). This pathway is responsible not just for the formation of phenolics and lignin but also for the formation of many phytoalexins.

The starting point for this pathway involves the deamination of phenylalanine to *trans*-cinnamic acid, in a reaction catalyzed by the enzyme phenylalanine ammonia lyase (PAL). Although a number of phytoalexins can be readily formed from *trans*-cinnamic acid, extension of the C_6C_3 skeleton by enzymes known as polyketide synthases, such as chalcone synthase (CHS) and STS, leads to the flavonoid and stilbenoid families of phytoalexins, respectively (Figure 2.13). Activities of these enzymes have been shown to increase in incompatible interactions between plants and pathogens. For example, in the incompatible interaction between coffee and the rust fungus *Hemileia vastatrix*, two peaks of PAL activity were detected (Figure 2.16). The first peak occurred 2 days after inoculation and was associated with early accumulation of phenolic compounds and the beginning of hypersensitive cell death, while the second peak at 5 days after inoculation appeared to be related to later accumulation of phenols and lignification of plant cell walls (Silva *et al.*, 2002). In the interaction between sorghum and *C. sublineolum*, discussed in the section on phytoalexins above, synthesis of the 3-deoxyanthocyanidin phytoalexins involves the action of CHS. A clear difference was observed in the pattern of CHS mRNA accumulation in the

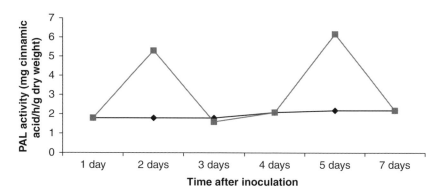

Figure 2.16 PAL activity in coffee plants resistant to the rust fungus, *H. vastatrix* (race II): control plants (◆); inoculated plants (■). The peaks in PAL activity in inoculated plants 2 and 5 days after inoculation are significantly different from the control plants. (Reproduced from Silva *et al.* (2002), with permission of Elsevier.)

incompatible and compatible interactions. In the former, CHS mRNA started to accumulate before fungal penetration into the plant had been completed and could be detected from 24 hours after inoculation. In the compatible interaction, accumulation of CHS mRNA did not start until 36 hours after inoculation, at which time the pathogen had already invaded host cells (Lo *et al.*, 1999). Although the increase in CHS mRNA in the incompatible interaction, occurring after just 24 hours, is rapid, even more rapid induction of defenses can occur following application of a pathogen-derived elicitor to plant cell cultures. Thus, when a fungal elicitor is applied to cell cultures of French bean, *Phaseolus vulgaris*, mRNA for a number of defense-related enzymes, including PAL, CHS, and chitinase, starts accumulating within 5 minutes. For chitinase, there is a 10-fold increase in mRNA after just 10 minutes, rising to a 30-fold accumulation after 30 minutes, making this one of the earliest defense-related events recorded, with kinetics comparable to the most rapid gene activation systems in animal cells (Hedrick *et al.*, 1988).

Not all proteins associated with plant defense are enzymes. For example, structural proteins rich in the amino acid hydroxyproline, and thus termed "hydroxyproline-rich glycoproteins" (HRGPs), accumulate in plant cell walls following pathogen attack. Although HRGP mRNA accumulation occurs more rapidly in incompatible than compatible interactions, research on cell cultures suggests that the response is slower than that observed for defense-related enzymes such as PAL and CHS. Further, HRGP mRNA accumulation occurs at a distance from the site of pathogen infection, unlike other defense responses, for example, phytoalexin accumulation, which occur at the infection site (Showalter *et al.*, 1985).

The hypersensitive response
The hypersensitive response (HR) is a common reaction in plants attacked by avirulent pathogens, where plant cells in and around the site of attempted penetration collapse and die. It is easy to see that the HR, with its localized necrosis, would be an effective defense against obligately biotrophic pathogens, since they require living host cells for growth and reproduction. However, the role of the HR in defense against necrotrophic pathogens is questionable. Necrotrophs can live off dead plant tissue, and indeed, it has been suggested

that they can induce host cell death as part of their invasion strategy (e.g., Van Baarlen *et al.*, 2004). However, superficially similar host reactions occur in response to attack by necrotrophic fungi and bacteria (e.g., Brown & Mansfield, 1988). Since the HR appears therefore to operate against both biotrophs and necrotrophs, host cell death alone cannot determine the fate of the pathogen. Rather, death of host cells is accompanied by a series of events including changes in oxidative metabolism, phytoalexin accumulation, lignification of cell walls, and production of hydrolytic enzymes such as chitinases and glucanases. This generates a hostile environment for the pathogen, restricting its growth through a combination of cutting off its food supply, direct toxicity as a result of the accumulation of chemicals, and creation of a physical barrier.

Considerable controversy still surrounds the question whether the HR is a primary determinant of resistance. Some research has shown that host cell death occurs prior to inhibition of pathogen growth. Thus, in the interaction between potato containing a specific *R* gene and an incompatible race of *Phytophthora infestans*, death of plant cells occurs several hours before pathogen growth is affected (Tomiyama, 1971). However, there is also evidence that pathogen growth is inhibited before host cell death occurs. For example, haustoria of the rust fungus *Puccinia coronata* f.sp. *avenae* were found to die before death of host cells (Prusky *et al.*, 1980). The latter evidence suggests that the HR is probably a symptom of incompatibility rather than a primary determinant of resistance. As suggested by Lucas (1998), this controversy surrounding the relationship between death of host cells and resistance obscures an important point—what actually triggers the HR? There is much evidence from incompatible host–pathogen interactions for a period of pathogen growth prior to any visible host reaction. Such data suggest that induction of active host defense might depend on host recognition of signal molecules emanating from the growing pathogen. In this context it is interesting to note that virulent, but not avirulent, pathogens can escape the *avr*-based plant recognition mechanism and can suppress defense responses and inhibit the HR (Nomura *et al.*, 2005). We will return to this in Chapter 3, when the signaling associated with plant defense, including the HR, is discussed.

2.2.4   Defenses used against pathogens—the next step

In this section, we have looked at the various mechanisms, passive and active, which plants can use to defend themselves against pathogen attack. So far, the discussion has focused on the tissue being attacked. But what about other tissues and organs away from the attack site? Are these parts of the plant affected in any way? Do they get advance warning of potential danger? Also, and most importantly, how does the plant coordinate its defense in the face of pathogen attack? These aspects are discussed in Chapter 3.

2.3   Defenses used against parasitic plants

2.3.1   Background

Some parasitic plants infect the root of the host plant, while others infect the shoot. Irrespective of the site of infection however, parasitic plants will attempt to establish a connection with the vascular system of the host, and the degree of parasitism that results depends on the extent of the vascular connection established. Facultative parasitic plants, such as *Rhinanthus minor*, can grow perfectly well in the absence of a host, but tend to grow better in a parasitic relationship. Obligate parasitic plants can only survive in a parasitic relationship and include those that can photosynthesize (hemiparasites), such as *Striga* spp. and dodder (*Cuscuta reflexa*), and those devoid of chlorophyll and nonphotosynthetic, such as the broomrapes (*Orobanche* spp.).

Because food reserves in seeds of parasitic plants are limited, they need to be very close to the host plant for germination. The seeds detect the presence of specific compounds in the soil, released by the host, and which act as germination cues. On contact with the host root, the parasite responds to further signals from the host and produces a haustorium. The parasite then uses the haustorium to penetrate the host tissue via a combination of enzymic degradation of cell walls and mechanical pressure, eventually linking up with the host vascular tissue. Although in some cases the parasitic plant can form a direct, continuous link between its own xylem and that of the host, in most cases the parasite vascular tissue simply abuts that of the host and nutrients and water are either transferred across this interface or move into the parasite via transfer cells.

2.3.2   Preattachment defense mechanisms

Since seeds of parasitic plants require the presence of germination stimulants exuded from the host root to germinate and then locate the host root, the ability to produce low concentrations of such chemicals might represent an effective defense mechanism. Indeed, this trait has been found in accessions of various legumes and in sunflower (Pérez-de-Luque *et al.*, 2008) and has been used to breed sorghum for enhanced resistance to *Striga* (Haussmann *et al.*, 2000). Equally effective at this stage in development of the parasitic plant would be the secretion by the host of germination inhibitors. Indeed, sunflower (*Helianthus annuus*) produces simple 7-hydroxylated coumarins, which appear to play a defensive role against *Orobanche cernua* by preventing successful germination. Sunflower varieties resistant to *O. cernua* were found to accumulate higher levels of the coumarins and to excrete greater quantities into the soil than susceptible varieties (Serghini *et al.*, 2001). Once seeds of the parasitic plant have germinated and located the host root, formation of the haustorium requires perception of a haustorium induction factor secreted by the host. Some wild sorghum accessions are known to possess a low ability to induce haustorium formation in *Striga asiatica* (Rich *et al.*, 2004), while there is evidence that a wild relative of maize, *Tripsacum dactyloides*, produces a signal that inhibits haustorial development in *Striga hermonthica* (Gurney *et al.*, 2003).

2.3.3   Prehaustorial defense mechanisms

Once the parasitic plant has made contact with the host root and haustorium formation has been initiated, penetration of the host can occur. The penetration progress of *Orobanche* spp. can be halted at various stages: in the cortex of the host root, at the endodermis, and inside the central cylinder (Figure 2.17) (Pérez-de-Luque *et al.*, 2008). Stopping the progress of the parasite in the cortex is associated with the reinforcement of the cell walls by protein cross-linking, deposition of callose, and suberization, as well as accumulation of phenolic compounds in the apoplast at the point of infection (Pérez-de-Luque *et al.*,

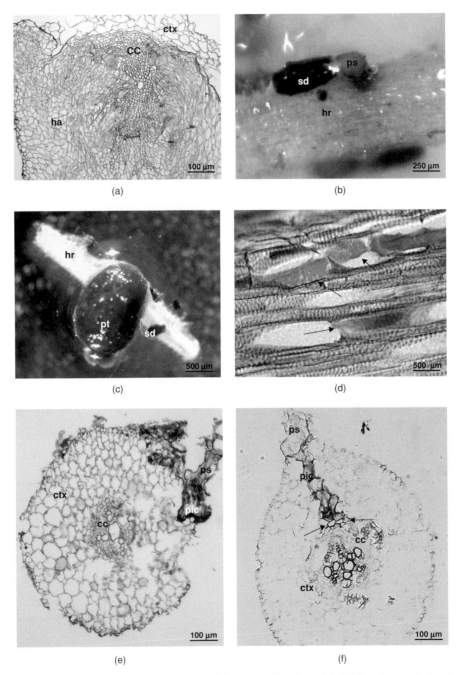

Figure 2.17 Resistance to *Orobanche* spp. (a) Cross-section of a vetch (*Vicia sativa*) root showing the haustorium (endophyte) of *O. crenata* in a compatible interaction. The black line separates parasite from host tissues. (b) A failed penetration attempt of *Orobanche foetida* on chickpea (*Cicer arietinum*) showing darkening around the attachment point (arrow). (c) A dead *O. crenata* tubercle on vetch at a preliminary growing stage. (d) Mucilage (indicated by arrows) filling xylem vessels of vetch during an incompatible interaction with *O. crenata*. (e) *O. crenata* penetration attempt halted in the cortex of faba bean (*V. faba*). (f) *O. crenata* penetration attempt halted at the endodermis of vetch. Arrows indicate lignified endodermal and pericycle cell walls.

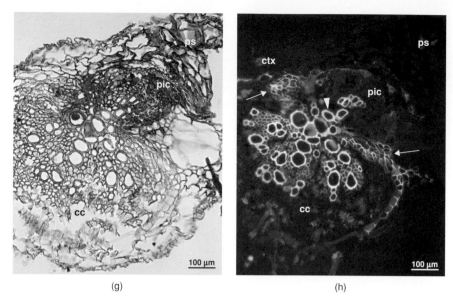

(g) (h)

Figure 2.17 (*Continued*) (g) *O. crenata* penetration attempt halted inside the central cylinder of *Medicago truncatula*. (h) *O. crenata* penetration attempt halted inside the central cylinder of *M. truncatula* but observed by epifluorescence (excitation at 450–490 nm) showing fluorescence of host cell walls surrounding the parasite tissues (arrows) and changes in the fluorescence of cell walls from host xylem vessels near the parasite (arrowhead). cc, host central cylinder; ctx, host cortex; ha, haustorium; hr, host root; pic, parasite intrusive cells; ps, parasite seedling; pt, parasite tubercle; sd, parasite seed. (Reproduced from Pérez-de-Luque *et al.* (2008), with permission of Blackwell Publishing Ltd.)

2008). Halting parasite progress at the endodermis involves lignification of endodermal and pericycle cell walls, while stopping the invader within the central cylinder is associated with accumulation of phenolic compounds, since strong fluorescence was observed in surrounding tissues and xylem vessels (Figure 2.17) (Pérez-de-Luque *et al.*, 2005). In *Vicia faba* expressing resistance to *Orobanche crenata*, penetration of host tissue is stopped by cell wall reinforcement. Here, reinforcement of cortical cell walls is via callose deposition, and should the parasite breach this barrier, endodermal cell walls become lignified, thereby preventing further penetration into the central cylinder (Pérez-de-Luque *et al.*, 2007). Interestingly, in a resistant sunflower accession, *O. crenata* does not progress beyond the cortex, due to suberization of cell walls and further cell wall reinforcement as a result of protein cross-linking (Echevarría-Zomeño *et al.*, 2006). To complete the coordinated defense response, phenolic compounds are secreted into the apoplast, thereby creating a toxic environment for the invading parasite.

2.3.4 Posthaustorial defense mechanisms

Once the parasite has established vascular connections with the host, it might seem unlikely that host defenses could be anything more than a token gesture. However, host defenses can still be effective at this stage. In *Orobanche*, once the host tissue is penetrated, a tubular link to the host xylem is formed. This is known as a tubercle and is the structure through which the parasite obtains water and nutrients. In some plants, necrosis and death of the tubercle can occur and in some cases it has been associated with the presence of gels or gums within host xylem vessels (Figure 2.17) (e.g., Pérez-de-Luque *et al.*, 2005). This is similar to the mechanism used by plants against various vascular wilt pathogens (Beckman, 2000). There is evidence that the host might also deliver toxic phenolic compounds into the parasite via the xylem (Lozano-Baena *et al.*, 2007). Thus, not only would the functioning of the tubercle be compromised by the presence of gels and gums, but parasite cells could also be killed by uptake and accumulation of host-produced phenolics.

2.4 Defenses used against nematodes

2.4.1 Background

The lifestyles pursued by plant parasitic nematodes can be crudely divided into ectoparasitic and endoparasitic (Jasmer *et al.*, 2003). Ectoparasitic nematodes do not enter roots and feed by inserting a stylet into epidermal cells or cells deeper within the root (Wyss, 1997). Many such nematodes, for example, *Trichodorus* and *Tylenchus* spp., inject saliva into the host cell, liquefying the cytoplasm around the stylet, thereby facilitating rapid ingestion. This type of relationship could be considered to be necrotrophic, since the feeding behavior usually kills the host cell (Trudgill, 1991). Endoparasites, on the other hand, feed and reproduce within the plant and two types can be distinguished: migratory and sedentary. Migratory endoparasites, for example, *Pratylenchus* and *Radopholus* spp., invade the plant, moving and feeding inter- or intracellularly. Feeding may be necrotrophic, although in some nematode species it is partly biotrophic, since it involves favorable changes in host cells next to the feeding site (Trudgill, 1991). Sedentary endoparasites migrate through root tissue and eventually transform selected host cells into a specialized feeding structure (Decraemer & Hunt, 2006). Some, such as root-knot and cyst nematodes, can be considered to be obligate biotrophs, and must invade roots of a susceptible plant to complete their lifecycle. Within the root, they transform a number of cells into metabolically active transfer cells, which sustain the nematode throughout its life. This enables the nematodes to lose their locomotory musculature and to develop into saccate, reproductive females (Fuller *et al.*, 2008). Root-knot and cyst nematodes exhibit different migratory behaviors within the root. Root-knot nematodes migrate through the root intercellularly, reaching the zone of cell division, where they transform a few vascular parenchyma cells into "giant" cells. In contrast, cyst nematodes migrate through the root intracellularly, and on reaching the zone of elongation, they transform a cell at the edge of the vascular system into a feeding site known as a syncytium (Fuller *et al.*, 2008). Thereafter, as many as 200 surrounding cells might be recruited to the expanding syncytium via partial dissolution of their cell walls (Figure 2.18). In spite of the differences in structure and formation, both types of feeding site act as nutrient sinks and transfer cells, providing nutrition to the nematode. The nematode extrudes a feeding tube from its stylet tip, enabling the cytosol and its contents to be removed without damaging the host cell (Trudgill, 1991).

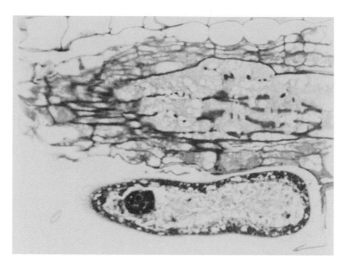

Figure 2.18 In a compatible interaction between a cyst nematode and its host plant, the nematode induces metabolically active multinucleate feeding cells within the vascular cylinder of the host root. This feeding site is called a syncytium and is characterized by cell wall dissolution between an initially infected cell and its neighbors, which become incorporated during syncytium expansion. The image shows a bright-field micrograph of a toluidine blue-stained longitudinal section through a syncytium 5 days after an *Arabidopsis* root had been inoculated with the cyst nematode *Heterodera schachtii*. (Reproduced from De Almeida Engler *et al.* (1999), with permission.)

As with resistance to microbial pathogens, resistance to nematodes can be determined by a number of genes or a single dominant resistance (*R*) gene (see Section 2.2.1) and several nematode-resistance genes have been isolated from plants, all conferring resistance against sedentary endoparasites (Williamson & Kumar, 2006). Defense against nematodes can involve both physical and chemical barriers, which are either constitutively present (passive or preexisting) or induced by infection (active).

2.4.2   Passive or preexisting defenses

Passive defenses, both physical and chemical, are likely to hinder the invasion of plant roots by a nematode or to perturb its development (Giebel, 1982). Thus, the plant might produce chemicals capable of exerting a direct effect on the nematode, as with the roots of *Tagetes patula* and *Tagetes erecta*, which contain α-terthienyl and derivatives of bithienyl, both of which limit populations of *Meloidogyne* and *Pratylenchus* (Giebel, 1982). Roots of banana plants resistant to the burrowing nematode *Radopholus similis* were found to contain high levels of lignin, flavonoids, dopamine, caffeic esters, and ferulic acids (Valette *et al.*, 1998). The constitutive phenolics were associated with limiting penetration of host roots by the nematode, while the high level of vascular lignification and suberization of endodermal cells were thought to restrict invasion by, and prevent multiplication of, the parasite in the vascular tissues. These results were confirmed and extended by Wuyts *et al.* (2007), who found that, compared to susceptible banana varieties, resistant varieties had constitutively higher levels of lignin in the vascular bundle and cell wall-bound ferulic acid esters in the cortex. Ferulic acids bound to cell walls protect them from enzymic attack, and since plant parasitic nematodes are known to secrete cell wall-degrading enzymes during the infection process, the presence of ferulic acids bound to the cell walls might well hinder the activities of these enzymes (Wuyts *et al.*, 2007). Interestingly, altering the phenylpropanoid profile of the plant has been shown to affect nematode development. Thus, in transgenic *Arabidopsis* containing elevated levels of syringyl lignin, fecundity and development of *Meloidogyne incognita* were significantly reduced (Wuyts *et al.*, 2006). These authors speculated that the increased levels of syringyl lignin in the vascular bundles might impede the flow of nutrients to the nematode's giant cells or hamper the nematode in feeding from its giant cells, thereby affecting reproduction and development. Indeed, because of lignification, most endoparasitic nematodes, except cyst nematodes, do not cross the endodermis in plant roots (Wuyts *et al.*, 2006).

According to Giebel (1982), passive resistance might also occur in plants that do not contain sufficient amounts of substances required for nematode reproduction and development. The result would be a failure of females to reach maturity. This is especially important for sedentary endoparasites, where the lack of appropriate food disrupts normal nematode development, leading to an increase in the ratio of males to females (Dropkin & Nelson, 1960).

2.4.3   Active or inducible defenses

2.4.3.1   Phenylpropanoid metabolism

As with plant interactions with microbial pathogens, nematode attack can trigger the expression of active defenses in the host. In common with other plant–pathogen interactions, a positive correlation has been established between plant resistance and products of the phenylpropanoid pathway. For example, the phytoalexin glyceollin I was found to increase in roots of a resistant soybean variety following inoculation with the cyst nematode *Heterodera glycines* (Huang & Barker, 1991). The phytoalexin accumulated immediately adjacent to the head of the nematode, and increases could be detected from just 8 hours after nematode penetration. Later work on this plant–nematode interaction showed that activities of PAL and 4-coumaryl-CoA ligase, both enzymes of the phenylpropanoid pathway, were enhanced in a resistant, but not in a susceptible variety, following nematode inoculation (Edens *et al.*, 1995). Further, transcription of enzymes later in the pathway, leading to glyceollin synthesis, was also increased in soybeans inoculated with *H. glycines*, with a much greater increase observed in the resistant, compared to the susceptible variety. In roots of a resistant banana variety, flavonoids were found to accumulate early in the cell walls of all tissues, including the vascular tissues (Valette *et al.*, 1998), while lignification of the endodermis and cells within the vascular bundle was induced in infected roots and appeared to represent part of a general defense response to protect the vascular system (Wuyts *et al.*, 2007). In oat interacting with the nematodes *Pratylenchus neglectus*, *Heterodera avenae*, or *Ditylenchus dipsaci*, a novel flavone glycoside, *O*-methyl-apigenin-*C*-deoxyhexoside-*O*-hexoside, was identified and shown to inhibit nematode invasion (Soriano *et al.*, 2004).

2.4.3.2   Hypersensitive response

Although the induction of a localized HR in resistant hosts following nematode infection appears to be similar to those induced by microbial pathogens, the timing and localization of the response vary with the particular host–nematode interaction (Williamson & Kumar, 2006). In plants carrying an *R* gene, avirulent cyst and root-knot nematodes elicit a range of responses. Thus, in the interaction between avirulent juveniles of *Globodera* and tomato plants carrying the *Hero A* resistance gene, the syncytium is induced, but is subsequently surrounded by a layer of necrotic cells and starts to degrade (Sobczak *et al.*, 2005). This abnormal development of the feeding sites results in mostly males developing to maturity, leading to a severe reduction in reproduction of the invading nematodes. A more rapid HR often characterizes resistance to root-knot nematodes. In the *Mi-1*-mediated resistance in tomato, the HR is induced rapidly following nematode invasion, with the result that a feeding site cannot be established and either the juvenile nematodes leave the root or they die (Williamson, 1998; Melillo *et al.*, 2006). At present, it is not known whether these differences in timing of the HR reflect when the nematode elicitor is recognized or the timing or pathway of the response.

2.5   Defenses used against herbivorous insects

2.5.1   Background

Insects are not the only invertebrates to feed on plants. So too do molluscs (slugs and snails), worms, and millipedes. Indeed, slugs and snails can cause serious damage to plants, and mollusc herbivory can act as an important selective force on the abundance and distribution of plant species (Bruelheide & Scheidel, 1999). However, since on most plants the largest category of herbivores tends to be insects (Root & Cappuccino, 1992), this section focuses on insect herbivory.

Insects obtain their nourishment from plants using one of two modes of feeding: either they imbibe their meal in liquid form or they bite off pieces of plant tissue and chew it (Schoonhoven *et al.*, 2005). Insects that bite and chew plant tissue, such as grasshoppers, are known as mandibulates, since they possess mandibles (jaws) capable of cutting and grinding their food. Those which can pierce plant tissues and imbibe liquid food, such as aphids, possess haustellate mouthparts. Their mode of obtaining food places a size restriction on sap-feeding insects, and as a result they tend to be smaller than biting and chewing insects and also tend to cause less mechanical damage to plants. Chewing insects can ingest large amounts of plant tissue, including toxic materials contained within that tissue. On the other hand, since phloem sap has a smaller ratio of toxic compounds to nutrients than many other plant tissues, sap-feeding insects can often avoid ingesting such toxic materials (Schoonhoven *et al.*, 2005).

As we saw in Chapter 1, although the nutritional quality of plants and the limited availability of various plant parts deter many insects from using plants as a food source, plants are on the menu for a great many herbivorous arthropods. However, insects can be prevented from feeding by plant defenses, ranging from physical barriers to toxins and antifeedants. As with defenses used against pathogens, some are always present (constitutive), while others are inducible, coming into operation following insect attack. However, it is important to remember that the distinction between constitutive and inducible defenses is not absolute, and many constitutive defenses are also inducible, as we see below. Further, although the defenses considered in the sections below are divided into physical and chemical, there can be interactions between the two, as in the case of resins and latex exuded from secretory canals.

2.5.2 Physical barriers

2.5.2.1 Waxes on the leaf surface

As with pathogens, one of the first lines of defense is the outer surface of the plant. The waxy cuticle of plants is important not just in defense against pathogens, but it can also protect plants against insect herbivory. Cuticles with slippery wax layers can present a serious problem to many herbivorous insects, which can experience difficulty in getting a good grip on the leaf surface. For example, young leaves of *Eucalyptus globulus* possess epicuticular wax, which makes the leaves slippery and reduces adherence of herbivorous psyllids. This in turn could make it difficult for the insects to feed (Brennan & Weinbaum, 2001). However, the presence of epicuticular wax can be a double-edged sword, since it can also impair attachment of predatory insects on leaves, leading to increased populations of herbivorous insects (Eigenbrode *et al.*, 1999).

More than half of all angiosperm species contain "extra" chemicals such as sterols and flavonoids, mixed in with the cuticular wax (Martin & Juniper, 1970). It seems likely that these substances have a protective role.

2.5.2.2 Trichomes

Leaves of many plant species are covered with densely packed fine spines or hairs, called trichomes. The trichome cover of a leaf surface, known as its pubescence, might deter insects by physically barring them from the underlying plant surface. Thus, heavy pubescence can prevent small sap-sucking insects from reaching the surface with their mouthparts. The leafhopper *Empoasca fabae* has a proboscis that is between 0.2 and 0.4 mm long and cannot reach the mesophyll cells and vascular bundles of hairy soybean leaves (Lee, 1983). This effect of trichomes was demonstrated clearly when removal of trichomes on pods of mustard (*Brassica hirta*) increased feeding damage by the flea beetle *Phyllotreta cruciferae* (Lamb, 1980).

Trichomes can also be induced following insect herbivory. For example, prior herbivory of wild radish led to increased numbers of trichomes on newly produced leaves (Agrawal, 1999). This effect was associated with impaired growth of generalist noctuid larvae on newly produced leaves, but had no effect on *Pieris rapae*, a specialist feeder. Work on *Brassica nigra* found that trichome density was increased in young leaves following damage of older leaves by three insect herbivores (Figure 2.19) (Traw & Dawson, 2002). Here, plants responded more quickly to damage by *P. rapae* than to damage by the cabbage looper *Trichoplusia ni*, despite similar levels of damage to the leaves and the fact that both insects exhibited similar feeding styles.

However, as with epicuticular waxes, the presence of trichomes is not always effective. For example, the parasitic wasp *Encarsia formosa* is much less efficient in finding its host, whitefly larvae, on hairy leaves than on glabrous or hairless leaves (Van Lenteren *et al.*, 1995).

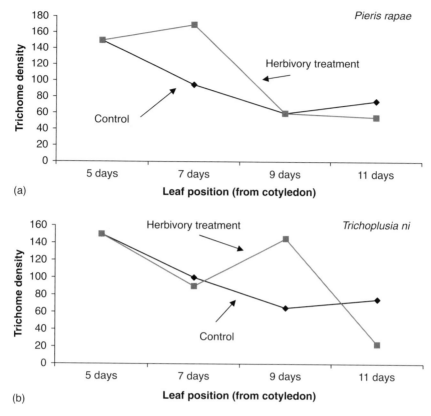

Figure 2.19   The effect of herbivory of older leaves by *P. rapae* and *T. ni* on trichome density (trichome number per square centimeter) on younger leaves of *B. nigra*. (Redrawn from Traw & Dawson (2002), with kind permission of Springer Science and Business Media.)

In some cases, trichomes are associated with secretion of noxious chemicals. This is dealt with in Section 2.5.3. An interesting point worth noting about the chemical constituents of trichomes is that they are almost invariably multifunctional, with effects not just on insect herbivores but also against mammalian herbivores and microbial pathogens (Harborne, 1993).

2.5.2.3   Secretory canals

Many plant species produce secretions such as resins and latex, which are stored under pressure in networks of canals in stems and leaves, where they follow the vascular bundles (Fahn, 1979). When these plants are damaged, there is a rapid release of liquid from the injured tissue. The liquid exuded, such as resins or latex, can gum up insect mouthparts, impede movement of the insect on the plant surface, and in the case of small insects, can cause death by asphyxiation (Becerra, 1994a). In addition, the physical effects of the secretions can be reinforced by the presence of toxic chemicals (see Section 2.5.3.1). Although these canals and their secretions are effective against many insect herbivores (Dussourd & Eisner, 1987), others are little affected. For example, phloem-sucking insects can direct their feeding organs to avoid secretory canals, while other insects can deactivate canals by cutting veins or trenches in the plant tissue (Johnson, 1992; Becerra, 1994b).

2.5.2.4 *Leaf toughness and leaf folding*

Leaf toughness can be gauged by measuring punch strength or by calculating fracture toughness. Punch strength can be obtained by measuring the force required to pass a rod of a given diameter through the leaf lamina, while fracture toughness can be calculated from the work required to shear a leaf using an apparatus based on scissors or a guillotine (Dominy *et al.*, 2008). Punch strength is dependent on both the amount and the distribution of materials in the leaf. In many plant species, fracture toughness is determined mainly by the extent of development of fibers around the vascular bundles (Lucas *et al.*, 2000). Whether punch strength or fracture toughness is more closely correlated with inhibition of herbivory is still unclear. Nevertheless, leaf toughness might reduce the suitability of leaves as a food source for insect herbivores in a number of ways. For example, polymers such as cellulose and lignin in secondary tissues are indigestible, and this might reduce the rate of leaf consumption, as well as being inappropriate for growth and development of the herbivore.

In a study of the defensive characteristics of tree species in a lowland tropical forest, Coley (1983) found that leaf toughness was most highly correlated with herbivory. However, there is great variation in leaf toughness among plant types. For example, leaves of grasses are considerably tougher than leaves of herbaceous dicots, while the fully expanded leaves of woody plants are still tougher (Bernays, 1991). Among grasses, leaf toughness in C_4 species is greater than in C_3 species. This seems to be due partly to the Kranz anatomy that is a distinctive feature of C_4 leaf structure, where leaf veins are surrounded by thick-walled bundle sheath cells. There is evidence that physical constraints, of which leaf toughness is an example, can deter insects from feeding and ovipositing on C_4 plants (Scheirs *et al.*, 2001). Interestingly, leaves of trees in tropical forests are substantially tougher than leaves of temperate trees, a difference attributed in part to the greater selective pressure of insect herbivory in tropical regions (Dyer & Coley, 2002). More recent work has also shown that in tropical lowland rain forests (TLRF), monocots have tougher leaves than dicots (Dominy *et al.*, 2008). Based on these results, it was predicted that monocots in TLRF would experience lower rates of herbivory than dicots. It should be pointed out that not all studies of dicots have shown correlations between leaf toughness and herbivory. For example, no relationship between leaf toughness and herbivory was found in four species of Dipterocarpaceae (Blundell & Peart, 1998), while no significant correlation between toughness and the rate of herbivory was found in a tropical dry forest (Filip *et al.*, 1995).

In TLRF, young leaves on many plants remain tightly folded or rolled until a late stage in elongation of the leaf. Indeed, an examination of species, where young leaves remained tightly folded or rolled until they had reached more than 50% of their mature length, showed that 56% of monocots but only 3.3% of dicots fell into this category (Grubb & Jackson, 2007). Subsequent work in TLRF at six geographical sites showed that substantially more leaf area was lost to invertebrate herbivory in dicots than in monocots (Grubb *et al.*, 2008), confirming the prediction made by Dominy *et al.* (2008). It was hypothesized that leaf toughness and the late folding and rolling of monocot leaves can protect monocot leaves against herbivorous insects, although the relative importance of each will vary depending on species (Box 2.1).

Box 2.1   Here come the cavalry—enlisting the help of others in defense

In a world fraught with danger from multifarious attackers, it makes sense to get some help to strengthen one's defenses. Plants are no exception here, and indeed, there are many examples of plants that have enlisted insects, such as ants, to aid their defense against herbivores. The coevolution of plants and ants involves a system of rewards and services and has led to a number of elaborate mutualistic interactions known as ant-guard systems. The ants provide a service in terms of protection from herbivory, while the plants provide rewards such as extrafloral nectar, specialized food bodies, and nest sites (Beattie & Hughes, 2002).

Some 93 angiosperm families possess extrafloral nectaries—secretory tissues located on leaves, twigs, or the external surfaces of flowers, which are not part of the plant's pollination system. The nectar produced contains mostly sugars, as well as amino acids and lipids, and attracts ants that either defend the food or are predatory on herbivorous insects. In fact, if ants are excluded from plants possessing extrafloral nectaries, herbivore damage increases, demonstrating clearly the benefits associated with the presence of the ants (Table A). Herbivory can actually induce the formation of extrafloral nectaries. Thus, in broad bean (*V. faba*), leaf damage led to an increased number of extrafloral nectaries, which would attract predatory insects (Mondor & Addicott, 2003), while in *Ricinus communis*, herbivory resulted in a severalfold increase in nectar production by extrafloral nectaries (Wäckers *et al.*, 2001).

Table A   Some effects of ants on plants bearing extrafloral nectaries: all effects are significantly different

Type of damage to plants	Plant species	Plants without ants	Plants with ants
Destruction of stigmas by grasshoppers	*Ipomoea leptophylla*	74%	48%
Seed destruction by bruchid beetles	*Ipomoea leptophylla*	34%	24%
Mean number of seeds per plant	*Ipomoea leptophylla*	45	403
Number of sightings of parasitic flies on inflorescences (wet season)	*Costus woodsonii*	872	128
Mean number of seeds per inflorescence (wet season)	*Costus woodsonii*	183	612
Mean number of insect predators per capitulum	*Helianthella quinquenervis*	7.6	2.9
Seed predation	*Helianthella quinquenervis*	43.5%	27.6%
Number of mature fruits per branch	*Catalpa speciosa*	0.85	1.11
Average number of psyllid nymphs per shoot	*Acacia pycnantha*	351	145
Shoot tips destroyed by psyllid nymphs	*Acacia pycnantha*	34%	7%

Reproduced from Beattie & Hughes (2002), with permission of Blackwell Publishing Ltd.

Domatia are genetically determined plant structures that appear to be specific adaptations for ant occupation. They are formed by hypertrophy of internal tissue, creating internal cavities that are attractive to ants (Beattie & Hughes, 2002). The protective function of ants associated with domatia has been demonstrated in various studies. For example, removal of adult herbivores by ant colonies inhabiting domatia on the plant *Maieta guianensis* led to increased fruit set (Figure A), while eggs and juveniles of insect herbivores were removed by ant colonies associated with domatia on *Macaranga* in southeast Asia (Fiala *et al.*, 1991). Some plants offer both homes and a nectar reward. Thus, as shoots form on *Acacia drepanolobium*, the thorns swell and nectarines are formed on the leaves. Ants (*Crematogaster* spp.) live in these thorns (Stanton *et al.*, 2002) and protect the tree from herbivorous insects and have even been shown to deter mammalian herbivores such as giraffes.

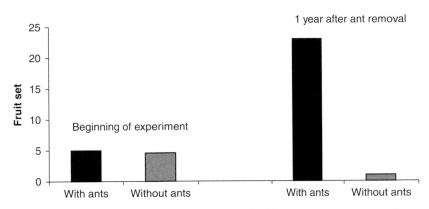

Figure A   Increased fruit set following removal of adult herbivores by ant colonies inhabiting leaf-pouch domatia in *M. guianensis*. Plants with ants had approximately 45 times the number of fruits than plants from which ants were chemically excluded. (Reproduced from Beattie & Hughes (2002), with permission of Blackwell Publishing Ltd.)

Plants can also attract bodyguards in other ways. Various plants are known to release volatile chemicals following damage, thereby attracting predators and parasitoids of herbivorous insects. We will deal with this topic more fully when we deal with communication in Chapter 3.

2.5.3   *Chemical defenses*

Despite the physical barriers on the plant surface, many insect herbivores still manage to make a meal of plant tissue. However, once they bite into or pierce the plant, they enter a world where chemical defenses abound. Plants produce an enormous variety of chemicals with roles in defense against herbivory. As mentioned earlier, these chemicals are referred to as "secondary plant metabolites," reflecting that they have no known metabolic function and as such are not strictly essential for the daily functioning of plants (Fraenkel, 1959). It is estimated that plants have the capacity to make more than 100,000 different organic compounds, providing them with a huge armory of defensive compounds to use against attackers. Many of these secondary compounds are responsible for the flavorings and spicy nature of different human foods, for example, chillies, tea, and coffee, and many are used in medicine.

Two criteria must be met if a plant trait (e.g., possession of a particular chemical) can be considered to be defensive (Karban & Baldwin, 1997). First, the chemical or other attribute must affect the extent to which a plant is attacked, and second, plants possessing the attribute must have greater fitness than plants without it or with smaller amounts of it. Defensive compounds are known as "allelochemicals" and many are toxic, disrupting important biochemical processes in herbivores. However, not all allelochemicals are toxic and some deter herbivores from eating the plant; such chemicals are known as "deterrents" (Bernays & Cornelius, 1992).

As discussed earlier in this chapter, plants produce a range of secondary compounds to act in plant defense, including phenolics, terpenoids, and nitrogen-containing organic compounds (Table 2.1). Examples of these compounds in relation to defense against insect herbivores are given in the following sections.

2.5.3.1   Terpenes

As described earlier, terpenes can be grouped into monoterpenes, diterpenes, sesquiterpenes, and steroids. Many monoterpenes are volatile and are widespread components of essential oils (Table 2.4). Examples include geraniol, limonene, and pinene. Plants in the family Asteraceae produce many monoterpenes with potent insecticidal activity. For example, flowers of *Chrysanthemum cinerariafolium* produce chrysanthemic acid and pyrethric acid, collectively known as pyrethrins. These act on the nervous system of insects, although they have little effect on mammals. The naturally occurring compounds have been used as the basis for synthesis of numerous pyrethrins that are used as commercial insecticides. Diterpenes include the clerodanes, which are potent feeding deterrents to many insects (Caballero *et al.*, 2001), while the triterpenes include the liminoids and cucurbitacins. Azadirachtin from the neem tree, *Azadirachta indica*, is a liminoid with potent activity as an insect feeding deterrent (Mordue & Blackwell, 1993). The cucurbitacins are bitter-tasting and deter feeding in many insects, although a number of specialized insect herbivores use these compounds as host-recognition cues (Tallamy *et al.*, 1997; Abe & Matsuda, 2000). The sesquiterpenes represent the largest group of terpenes and are common components of plant essential oils. Sesquiterpene lactones, which possess a five-membered lactone ring, are widely distributed in the Compositae and include compounds with potent toxicity to insects. Glaucolide A, a sesquiterpene lactone of *Veronia*, has been shown to be an effective feeding deterrent. Some plants, for example, mint, produce a range of monoterpenes and sesquiterpenes, which accumulate in trichomes and are released upon damage.

Diterpenoid glycosides, especially those with geranyl linalool (GL) carbon skeletons (GL-DTGs), are abundant in various *Nicotiana* species. Two DTGs (hydroxygeranyllinalool (HGL)-DTGs) were shown to function as antifeedants in *N. attenuata* against the tobacco hornworm (Snook *et al.*, 1997). In subsequent work, suppression of HGL-DTG formation in

Table 2.4   Major classes of plant terpenoids

Terpenoid category and general formula	Plant product	Principal types
Hemiterpenoids (C_5H_8)	Essential oils	Tuliposides
Monoterpenoids ($C_{10}H_{16}$)	Essential oils	Iridoids
Sesquiterpenoids ($C_{15}H_{24}$)	Essential oils, resins	Sesquiterpene lactones
Diterpenoids ($C_{20}H_{32}$)	Resins, bitter extracts	Clerodanes, tiglianes, gibberellins
Triterpenoids ($C_{30}H_{48}$)	Resins, latex, corks, cutins	Sterols, cardenolides, phytoecdysteroids, cucurbitacins, saponins
Tetraterpenoids ($C_{40}H_{64}$)	Pigments	Carotenes, xanthophylls
Polyterpenoids ($C_5H_8)_n$	Latex	Rubber, balata, gutta

Source: Reproduced from Schoonhoven *et al.* (2005), with permission of Oxford University Press.

N. attenuata led to a dramatic impairment in resistance to the tobacco hornworm, suggesting that HGL-DTGs function as direct defenses in this plant (Jassbi *et al.*, 2008).

As mentioned in Section 2.5.2.3 above, production of resin or latex is common in plants and the physical effects of these secretions in gumming up mouthparts are often reinforced by the occurrence of toxic chemicals in the exudates. For example, the latex produced by chicory (*Cichorium intybus*) is rich in bitter-tasting sesquiterpene lactones (lactupicrin and 8-deoxylactucin), and feeding experiments with *Schistocerca gregaria* demonstrated that these are antifeedants at a concentration of 0.2% dry weight (Rees & Harborne, 1985). Interestingly, roots, stems, and leaves contain high concentrations of these lactones, although no part of the plant contains less than the threshold required to deter insect herbivores. Many milkweeds (*Asclepias* spp.) have concentrations of cardenolides (toxic steroids; see paragraph below) that are considerably higher in their latex than in their leaves (Zalucki *et al.*, 2001), while the latex of many species, including milkweeds, contains cysteine proteases (Agrawal *et al.*, 2008), which are known to digest peritrophic membranes of insects. Leaves of a number of latex-producing plants, including papaya and wild fig, exhibited strong toxicity and growth inhibition against lepidopteran larvae (Figure 2.20) (Konno *et al.*, 2004). Latex from these plants contained cysteine proteases, and when the latex was washed off, leaves of these plants lost their toxicity. Indeed, papain, a cysteine protease in latex of the papaya tree was found to be a crucial factor in the defense against herbivorous insects (Konno *et al.*, 2004).

Plants of the genus *Bursera* produce an array of monoterpenes that are distributed in a network of canals in the stems and leaves (Mooney & Emboden, 1968). In common with other canal-bearing plant species, damage results in rapid release of fluids from the site of injury. In some *Bursera* species however, the resins are under considerable pressure and, when a leaf is damaged, liquid can be released in a spectacular squirt—a so-called squirt-gun defense. Amazingly, in some species of *Bursera*, this squirt can last for several seconds and travel up to 2 m (Becerra *et al.*, 2001). Apparently, an interaction exists between the mechanical and chemical components of this defense. For example, some *Bursera* species have powerful squirts but a simple chemical composition of the released liquid (just one or two monoterpenes). In contrast, nonsquirting species have more complex chemical mixtures in the fluid released from canals upon injury, comprising, for example, of sesquiterpenes and diterpenes (Becerra *et al.*, 2001).

The final group of terpenes with insecticidal activity is the steroids. Because insects cannot make the steroid nucleus in any quantity, they need to get cholesterol or sitosterol from their diet to synthesize steroid hormones, for example, ecdysone, the moulting hormone. Some plant species produce a group of compounds known as phytoecdysones or phytoecdysteroids, which can mimic the activity of moulting hormones in insects. These compounds disrupt moulting and can perturb normal insect development. Thus, silkworms fed phytoecdysones are unable to remove the old cuticle during moulting, often leading to death (Kubo *et al.*, 1983). Although phytoecdysones have been found in more than 100 plant families, representing ferns, gymnosperms, and angiosperms, whether they have a physiological role in plants is not known (Dinan, 2001). It is tempting therefore to speculate that their primary function is in defense against insect herbivores. However, although arguments can be made in favor of such a role (Harborne, 1993), experimental evidence is lacking (Dinan, 2001).

In a classic example of plant–animal coevolution, certain butterflies (the Monarch butterfly, *Danaus plexippus*, and some other danaid butterflies) use toxins from their host plant

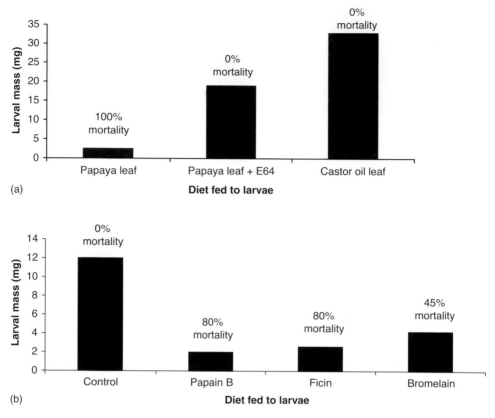

Figure 2.20 (a) Effects of feeding papaya leaves, papaya leaves painted with a cysteine protease inhibitor (E64), and leaves of the castor oil plant, on growth and mortality of first-instar larvae of the Eri silkworm, *Samia ricini*. The castor oil plant is the normal host of *S. ricini*. (b) Effects of feeding *S. ricini* first-instar larvae with papain, ficin, or bromelain, on larval growth and mortality. (Adapted from Konno *et al.* (2004), with permission of Blackwell Publishing Ltd.)

to defend themselves against predators. These butterflies feed on milkweeds, which produce several cardenolides (cardiac glycosides) as a passive defense against insect feeding. Monarch butterflies have become adapted to these toxins, which are sequestered during feeding and stored out of harm's way. However, because these cardenolides are bitter-tasting and toxic to higher animals, butterflies feeding on milkweeds tend to be avoided by birds (blue jays), although some bird species have apparently been able to overcome this defense (Fink & Brower, 1981). It might seem odd that plants should continue to produce toxins if they are then used by insects as part of their own defensive system. However, it is important to remember that such toxins will provide protection from many different insects and mammalian herbivores.

2.5.3.2 *Phenolics*

Flavonoids are the largest and most diverse group of phenolic compounds in plants, and as a result, most herbivorous insects will encounter them when feeding. Flavonoids share a basic C_6-C_3-C_6 structure (Table 2.2) and can be subdivided into several groups, including flavones, flavanones, flavonols, and chalcones (Harborne, 1979). While some flavonoids absorb light in the visible region of the spectrum and are responsible for yellow and cream colors in plants, other compounds are colorless and exert effects on insect herbivores as feeding deterrents. Thus, various flavone and flavonol glycosides accumulate in leaves of most angiosperms and some, for example, rutin and isoquercitrin, are toxic to a number of insect species. When isoquercitrin is added to the diet of the tobacco budworm, *Heliothis virescens*, larval growth is reduced by 90% (Hedin *et al.*, 1983), while in tomato, a major trichome component is rutin, which, when added to the diet of fruitworms, inhibits larval growth (Isman & Duffey, 1982). A particularly powerful feeding deterrent is phaseolin; just 0.03 ppm of this compound reduced feeding by larvae of the beetle *Costelytra zealandica* by 50% (Lane *et al.*, 1985). It should be pointed out however that some flavonoids are used by various insect species in host recognition and to stimulate feeding (Simmonds, 2001).

2.5.3.3   *Nitrogen-containing organic compounds*

Glucosinolates

During insect feeding on cruciferous plants, the usual compartmentation of glucosinolates and their breakdown enzyme, myrosinase, is disrupted. This releases isothiocyanates or mustard oils, and nitriles. Both glucosinolates and isothiocyanates act as allelochemicals, mediating interactions between the plant and herbivorous insects. Thus, glucosinolates are unpalatable and toxic to many generalist insects and to several specialists that live on noncruciferous crops (Renwick, 2002).

In *Arabidopsis*, larvae of the generalist lepidopteran *Helicoverpa armigera* (the cotton bollworm) are known to avoid feeding on the midvein and periphery of rosette leaves, feeding instead on the inner lamina (Figure 2.21). This feeding pattern was thought to be due to glucosinolates because larvae behaved differently on leaves of mutant plants unable

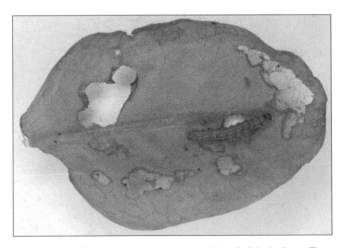

Figure 2.21   Feeding pattern of *H.* larvae on mature rosette leaf of *A. thaliana*. Ten second-instar larvae were allowed to feed for 5 hours. (From Shroff *et al.* (2008), with permission of the National Academy of Sciences, USA.)

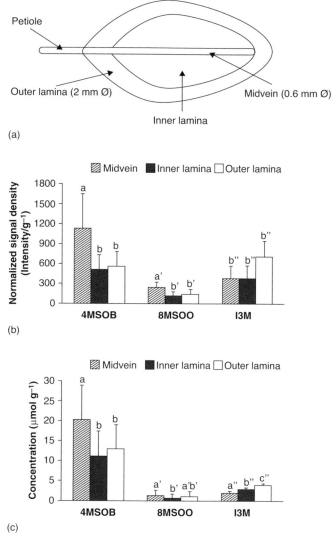

Figure 2.22 Comparison of the relative amounts of three glucosinolates in three different leaf regions of *A. thaliana* leaves defined as in (a). Normalized ion intensities were calculated for the midvein, the inner lamina, and the outer lamina areas from mass spectrometric imaging (b), and glucosinolate concentrations were measured by high-performance liquid chromatography (HPLC) analysis (c). Error bars represent scanning electron microscopy (SEM) from five mature leaves from different plants. Bars marked by different letters show significant difference at $P < 0.05$. (From Shroff *et al.* (2008), with permission of the National Academy of Sciences, USA.)

to activate glucosinolate defenses. Elegant work by Shroff *et al.* (2008) found that the major glucosinolates in *Arabidopsis* are more abundant in the tissues of the midvein and periphery of the leaves than the inner lamina (Figure 2.22). It would appear therefore that the distribution of glucosinolates within leaves can control the feeding preference of cotton bollworm larvae.

Despite the toxicity of isothiocyanates, several species of Lepidoptera feed successfully on glucosinolate-containing plants. Larvae of the specialist insect, *P. rapae*, are a good example. It is now known that they are biochemically adapted to the glucosinolate–myrosinase system of defense in their *Brassica* hosts. In larvae of *P. rapae*, a gut protein was discovered that, following ingestion of plant material, prevents the formation of isothiocyanates by redirecting glucosinolate hydrolysis toward nitrile formation. The nitriles so formed are then excreted with the feces (Wittstock *et al.*, 2004). Interestingly, in *Plutella xylostella*, another specialist lepidopteran feeding on glucosinolate-containing plants, a different mechanism is used to deal with this defense. Here, the insect uses a glucosinolate sulfatase, which essentially prevents the action of myrosinase on glucosinolates, thereby avoiding the production of toxic hydrolysis products (Ratzka *et al.*, 2002). Some insects have not only been able to deal with the glucosinolate–myrosinase defense system, but can also use glucosinolates as attractants. Indeed, glucosinolates are strong feeding and oviposition stimulants to numerous specialist insects that feed on cruciferous plants (Chew & Renwick, 1995).

Cyanogenic glycosides

Cyanogenic glycosides have been identified in more than 200 plant species and indeed all plants probably have the ability to make them, although in most plants they are metabolized and do not accumulate (Schoonhoven *et al.*, 2005). Cyanogenic glycosides are synthesized via the conversion of amino acids to oximes, and are subsequently glycosylated. They are then stored in separate cellular compartments from the enzymes that break them down. When cells are damaged, for example, by insect feeding, the contents of the different cellular compartments mix and the cyanogenic glycosides are broken down to yield hydrogen cyanide. Cyanide is a respiration inhibitor and is toxic at low doses.

The bitter taste of these glycosides appears to act as a deterrent to insects. However, since most insects are capable of detoxifying hydrogen cyanide, any protective effect of cyanide might be due to its further metabolism to β-cyanoalanine, which is toxic (Harborne, 1993).

2.5.3.4 Arthropod-inducible proteins

Plants can produce proteins that play a critical role in defense by targeting the digestive system of insects, thus impairing the ability of the insect to digest and absorb food (Zhu-Salzman *et al.*, 2008). These proteins include protease inhibitors (PIs; also known as proteinase inhibitors), which, by inhibiting proteolysis in the insect gut, decrease its access to essential amino acids. PIs are categorized according to the proteases they inhibit (e.g., cysteine proteases) and tend to be found in high concentrations in tissues such as seeds, the loss of which would exert a negative impact on plant fitness. Other arthropod-inducible proteins (AIPs) include enzymes such as arginase and threonine deaminase (TD). Work by Chen *et al.* (2005) showed that arginase and TD, induced in tomato by feeding of larvae of *Manduca sexta* and ingested by the larvae, disrupted digestion of insects by degrading existing amino acids required for insect growth. Work such as this supports the hypothesis that breakdown of amino acids in the digestive system of insects by plant enzymes plays an important role in plant defense against herbivorous insects.

A number of other enzymes can be considered as AIPs, including (1) polyphenol oxidases, which are widespread in plants and are inducible by wounding and herbivory, (2) lipoxygenases, some of which are involved in biosynthesis of jasmonic acid and subsequent

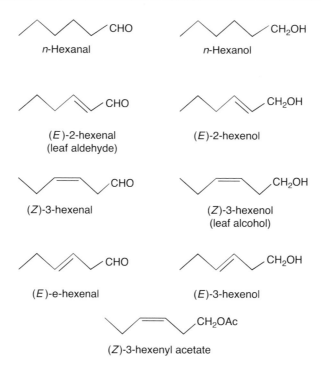

n-Hexanal

n-Hexanol

(*E*)-2-hexenal
(leaf aldehyde)

(*E*)-2-hexenol

(*Z*)-3-hexenal

(*Z*)-3-hexenol
(leaf alcohol)

(*E*)-e-hexenal

(*E*)-3-hexenol

(*Z*)-3-hexenyl acetate

Figure 2.23   Structures of GLVs commonly found in plants. (Reproduced from Matsui (2006), with permission of Elsevier.)

defense against insects (see Chapter 3), and (3) enzymes capable of disrupting the insect redox system, including NADH oxidase. Perturbation of the redox status of the insect gut could cause proliferation of oxyradicals that damage proteins, lipids, and DNA (see Zhu-Salzman *et al.*, 2008).

2.5.3.5   Volatile compounds

All plants emit a cocktail of volatile hydrocarbons. Although undamaged plants emit such compounds, the emission rate is much lower than that from damaged tissues. Volatile compounds emitted from plants can be of a general nature, such as green leaf volatiles (GLVs), which are commonly given off following damage, as well as chemicals that are specific to plant taxa. GLVs are six-carbon (C_6) aldehydes, alcohols, and their esters (Figure 2.23), so-called because of the distinctive scent produced when the leaves are damaged (Matsui, 2006). They are formed by oxidation of leaf lipids, specifically by the action of lipoxygenase on linolenic or linoleic acids, and the composition, and especially the relative proportions, of the individual components in GLVs is likely to be unique for a particular species. This can be used by some insects in host selection, allowing discrimination between host and nonhost plants. In fact, GLVs have a wide variety of effects on insects, ranging from stimulating feeding to exerting a synergistic effect on the action of sex hormones (Matsui, 2006).

The emission of plant volatiles can increase substantially following insect damage (Figure 2.24), and it is worth noting that it is not only GLVs that are released. Terpenoids are

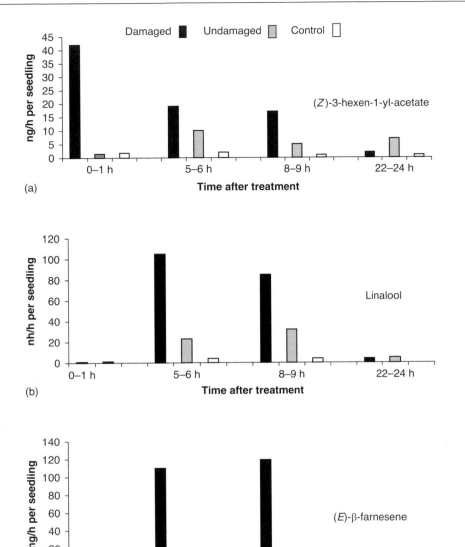

Figure 2.24   Amounts of three components of leaf volatiles—(Z)-3-hexen-1-yl-acetate, linalool, and (E)-β-farnesene—emitted by corn seedlings after damage, followed by treatment with caterpillar regurgitant (to mimic herbivory) at various times after treatment ("damaged"). Some of volatiles are also released systemically by "undamaged" leaves of injured plants. Volatiles released by undamaged plants were used as a "control." (Redrawn from Turlings & Tumlinson (1992) and Schoonhoven *et al.* (2005), with permission of Oxford University Press.)

often found in volatiles emitted after damage, and their relative proportions in the volatile blend can vary depending on the insect herbivore. For example, the volatile mixture released by apple infested with the spider mite *Panonychus ulmi* contains nearly 50% of the terpenoid 4,8-dimethyl-1,3-(*E*),7-nonatriene, while leaves infested with a different spider mite, *Tetranychus urticae*, produce a volatile blend, containing only 9% of this compound (Dicke, 1994).

The volatiles released by plants upon damage by insect herbivores can affect plant defense both directly and indirectly. These aspects are dealt with in Chapter 3.

2.6   Defenses used against vertebrate herbivores

2.6.1   Background

Despite the poor nutritive value of plants, the fact that they cannot escape attack by running away makes them a popular source of food. Indeed, as indicated in Chapter 1, herbivory is common among mammals, with more than half of all mammalian genera including plants in their diet. The 560 genera classified as herbivores include mammals specialized to feed on nectar and pollen, through to others feeding on plant exudates and yet others specialized to feed on stems, buds, and leaves. In a tropical rain forest community in Venezuela, almost 70% of the mammalian species present were herbivores, while in terms of weight, mammalian herbivores accounted for some 93% of the mammalian biomass (Eisenberg, 1980). Interestingly, in this community, 22% of the mammals were frugivores/granivores, 25% were frugivores/browsers, 27% were browsers/frugivores and browsers, while 21% were frugivores/carnivores and frugivores/omnivores, combining herbivory and carnivory. Feeding on nectar tends to be restricted to small mammals (<100 g in weight), while feeding on grasses and woody plants tends to be the province of larger mammals, some of which can exceed a ton in weight (Danell & Bergström, 2002). However, specializing on a particular type of food comes with certain constraints. For example, because of the difficulty of climbing and moving in trees, arboreal herbivores cannot be too large and have an upper limit of 13–15 kg (Cork & Foley, 1991). The majority feed on low-fiber, high-energy plant material such as fruits and flowers (Cork, 1994). In contrast, small, forest-inhabiting marsupials tend to feed on sap, nectar, and invertebrates, avoiding foliage (Cork & Foley, 1991). Mammalian herbivores are more polyphagous than insect herbivores, feeding on a wide range of plants. This, together with their mobility, greater body size, and individual bite size, means that mammalian herbivores can exert a profound effect on plants and plant populations. However, as we have already seen for other types of attackers, plants have a range of physical and chemical means of defending themselves against mammalian herbivores.

2.6.2   Physical defenses

As we have seen previously, the first line of defense is the outer surface of the plant. This can be used just as effectively to ward off vertebrate attackers as it can to fend off other types of attackers. Thorns, spines, and prickles—here collectively termed thorns—are a common defense against herbivory in thousands of plant species, especially in arid regions

(Grubb, 1992). Thorns can be found on stems, branches, and leaves, and in some plant species, for example, acacia trees, the thorns are very large, while in other plants, for example, thistles, the thorns can be smaller and softer. The role of thorns in defense against mammalian herbivores is well established (Grubb, 1992). Large thorns are a deterrent to many large, browsing mammals, since they can inflict painful wounds to mouthparts. In Africa, the large thorns on acacias target large browsers such as giraffes and antelope, although some browsers can still reach the foliage, despite the presence of vicious-looking thorns. Many thorny plants are aposematic; that is, the unpalatability of the plants is associated with conspicuous coloration. It has been suggested that such coloration deters large herbivores (e.g., Lev-Yadun, 2001). A sinister aspect of thorns as a defense against mammalian herbivores is the finding that they can carry bacteria (Halpern *et al.*, 2007). Therefore, in addition to inflicting a painful wound, thorns can also deliver potentially harmful bacteria into the mouthparts or other parts of the browsing mammal.

Thorns can also be induced by herbivory, as with browsing of the shrub *Hormathophylla spinosa* by ungulates, and browsing of bramble (*Rubus fructicosus*) by fallow deer (Bazely *et al.*, 1991; Gomez & Zamora, 2002). In an interesting study by Kato *et al.* (2008), the effects of heavy browsing of Japanese nettles by Sika deer were examined. This study was based in a park where a population of Sika deer had been maintained for 1200 years. Wild nettles in the park exhibited smaller leaves, and between 11 and 223 times more stinging hairs per leaf, and between 58 and 630 times greater stinging hair densities, than nettles from areas with no evidence of browsing from Sika deer. These data suggest that extremely high densities of stinging hairs evolved through natural selection due to heavy browsing by Sika deer.

2.6.3   *Chemical defenses*

As we have already seen in this chapter, plants possess a formidable arsenal of chemicals for use in defense, and in addition to providing protection against microbial pathogens and herbivorous insects, can also provide an effective defense against mammalian herbivores. When plant tissues are eaten, the chemicals deter feeding in some ways. Thus, they may be bitter-tasting, poisonous, or have antinutritional effects.

Although the chemicals used in defense against mammalian herbivores are secondary metabolites belonging to the phenolics, terpenoids, and nitrogen-containing compounds, products of primary metabolism can also be used. For example, although most of the 200 or more fatty acids produced by plants are nutritionally acceptable to mammals, some can be hazardous when eaten in large quantities. Thus, erucic acid, a major fatty acid of oilseed rape, can cause adverse effects following ingestion. As a result, only oilseed rape varieties low in erucic acid are used in animal feeds (Harborne, 1991).

2.6.3.1   *Phenolic compounds*

As outlined in Section 2.2.2.2, phenolic compounds are aromatic structures with one or more hydroxyl groups. They are formed biosynthetically from phenylalanine via the shikimic acid pathway (Figure 2.3) and range from simple compounds, such as phenol itself, through to complex anthocyanin pigments and the even more complex, polymeric condensed tannins

(Table 2.2). Because of their toxicity in plants, phenolics are conjugated with sugar, for example, to form glycosides, and are stored in the vacuole.

Mammalian herbivores are liable to encounter phenolics throughout the processes of selecting their food, and then eating and digesting it. Thus, anthocyanin pigments in the skin of fruits provide a visual signal that the fruit is ripe and ready to eat. In many fruits, for example, *Ribes*, *Rubus*, and *Vaccinium*, the bright colors when ripe are due to the presence of anthocyanins. The brightly colored fruits stand out against a uniform green background, attracting herbivores, which then disperse the seeds at a distance from the plant. However, bright colors can also represent a warning to herbivores not to eat the fruit, as with the purple black fruits of the deadly nightshade, *Atropa belladonna*, which contains alkaloids poisonous to mammals.

Some plants contain allergenic substances, many of which are phenolic and which appear to protect plants against herbivory. In many cases, the compounds are photosensitizers because their toxic effects are increased with exposure to light (Towers, 1980). Examples include urushiol, a catechol derivative, which is the main irritant in the oil of poison ivy, *Toxicodendron radicans*, and hypericin, a polyhydroxy-extended quinone from species of *Hypericum*, which causes facial eczema in sheep in Australia (Towers, 1980). Another group of phenols, the isoflavones, has long-term effects on grazing animals due to their estrogenic properties. A good example is 7-hydroxy-4′-methoxyisoflavone (formononetin), which occurs in various species of *Trifolium*. In Australia, it is the cause of serious infertility in ewes. Following ingestion, formononetin is metabolized to 7,4′-dihydroxyisoflavan (equol), which perturbs normal estrus, resulting in "clover disease" (Shutt, 1976). Another isoflavone with estrogenic properties is coumestrol, which was extracted from alfalfa, *Medicago sativa*, and ladino clover, *Trifolium repens*. The structures of these compounds mimic the steroidal nucleus of the natural female hormone estrone. In fact, all isoflavones are estrogenic when given by parenteral injection to animals. A survey of *Trifolium* showed that 18 species have a high isoflavone content (~1% dry weight), presenting a considerable hazard to farm animals.

Whether it is purely accidental that isoflavones possess estrogenic properties, or whether they have been deliberately produced by the plant to interfere with reproduction in mammalian herbivores, is not known. Interestingly, and perhaps significantly, the isoflavone skeleton also provides the basis of disease resistance in legumes, since the phytoalexins formed by leguminous plants are nearly all reduced forms of formononetin or genistein (Harborne, 1993) (Figure 2.25).

Any benefit to the plant from the possession of isoflavones with estrogenic properties, for example, reducing the fertility of grazing animals to limit damage during feeding, is relatively long-term. More immediate effects may be provided via the antifeedant properties of various phenolics. Studies with the snowshoe hare, *Lepus americanus*, and the mountain hare, *Lepus timidus*, have shown that simple phenolics are effective feeding deterrents to hares. In *Alnus crispa*, chemical protection is provided by pinosylvin and its methyl ester. Protection is allocated to those organs which are dormant during the winter, that is, the buds and staminate catkins, in preference to more expendable parts of the plant such as internodal tissues. As a result, concentrations of pinosylvin methyl ester in buds and catkins are severalfold higher than in internodal tissues (Bryant *et al.*, 1983). In Northern Europe, *L. timidus* also feeds selectively. Here, species of *Populus* and *Salix* are the main food in winter and young trees are protected by increased levels of phenolic glycosides, mainly salicin.

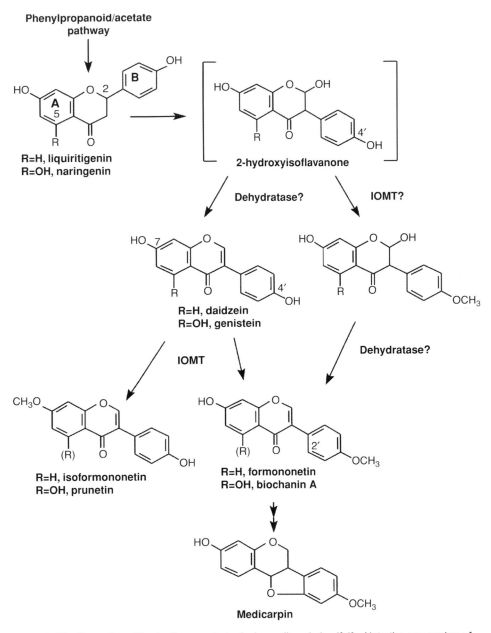

Figure 2.25 Formation of the isoflavone phytoalexin medicarpin in alfalfa. Note the conversion of genistein to formononetin and then to medicarpin. IOMT, isoflavone O-methyltransferase. (Reproduced from He & Dixon (2000), with permission.)

Concentrations of the glycosides fall in mature tall willow species, but remain high in low-growing willows, as an adaptation against browsing mammals (Tahvanainen *et al.*, 1985).

Tannins are defined as those polyphenols which have the ability to bind with proteins, that is, have tanning activity on animal skins. Indeed, the name "tannin" comes from the Celtic word for oak, referring to the ancient source of tannins for treating animal skins to make leather (Scott, 2008). Tannins are deposited on the surface waxes of the plant or are stored in the vacuole. The latter ensures that they do not interfere with the normal metabolic processes of the plant and are only released when cells are damaged. Tannins interfere with glycoproteins in the saliva, giving them a bitter, astringent taste. They also interact with digestive proteins, making food indigestible. It is no surprise therefore that mammalian herbivores can be severely affected by a high tannin content in food. Thus, in Africa, plants with a high tannin content are avoided by colobus monkeys (Oates *et al.*, 1977), while in the South African savanna, kudu, impala, and goats do not feed on plants with more than 5% dry weight of tannin in the leaves (Cooper & Owen-Smith, 1985).

2.6.3.2   Terpenoids

Terpenoids have their biosynthetic origin in the isopentenyl and dimethylallyl pyrophosphates and are mainly cyclic unsaturated hydrocarbons, with varying degrees of oxygenation in the substituent groups attached to the basic carbon skeleton (see Section 2.2.2.2 and Table 2.4). They are classified according to the number of five-carbon (C_5) units that are present, from the monoterpenoids (C_{10}) to the sesqui- (C_{15}), di- (C_{20}), and tri- (C_{30}) terpenoids.

Terpenoids have been implicated in many interactions between plants and animals and are recognized as pheromones, defense agents, or signal molecules. Some are highly toxic to animals, and others can interfere with hormonal control of animal growth and reproduction (Table 2.5). Among the most deadly terpenoids are the cardenolides, which occur in the milkweed family (see Section 2.5.3.1). Digitoxin is a glycosylated cardenolide, which

Table 2.5   Plant terpenoids and their effects on higher animals

Terpenoid class	Biological effects
Monoterpenes	Feeding deterrence (e.g., camphor) to red deer, snowshoe hares, and voles
Sesquiterpenes	Toxic to livestock (e.g., ngaione)
Sesquiterpene lactones	Vertebrate poisons (hymenoxin), allergenic (parthenolide), feeding deterrence to deer and rabbits (glaucolide A)
Diterpenoids	Poisonous (atractyloside, grayanotoxins), irritant, and cocarcinogenic (phorbol esters)
Cardenolides	Heart poisons
Triterpenes	Poisonous (lantadenes A and B0), antifeedant (papyriferic acid)
Saponins	Hemolytic, bloat production in ruminants
Phytosterols	Estrogenic (miroestrol)
Cucurbitacins	Poisonous to rabbits (cucurbitacin A)

Source: From Harborne (1991).

interferes with Na^+- and K^+-activated ATPases, including an ATPase in the heart. The result of a mammalian herbivore eating a cardenolide-containing plant is often heart failure. Indeed, just 23 g of dried milkweed leaf can kill a 45 kg sheep (Seiber *et al.*, 1984). Cardenolides are bitter-tasting, and as a result milkweed is normally unpalatable to livestock. However, accidental feeding with milkweed does occur, resulting in fatalities (Seiber *et al.*, 1984).

Other classes of triterpenoid, for example, the limonoids and cucurbitacins, are extremely bitter-tasting and might act as antifeedants. An association between toxicity and antifeedant ability is clear from work on the defensive role of papyriferic acid in protecting the paper birch, *Betula papyrifera*, from various herbivores, such as snowshoe hares, moose, and rodents (Reichardt *et al.*, 1984). Papyriferic acid is present in juvenile growth-phase internodes of paper birch at concentrations of up to 30% of dry weight—25 times the level found in mature internodes. This makes the juvenile tissue unpalatable to snowshoe hares. In fact, papyriferic acid was clearly identified as a feeding deterrent by feeding it to hares in oatmeal at 2% of the dry weight.

2.6.3.3   Nitrogen-containing compounds

The best-known nitrogen-containing secondary compounds of plants are the alkaloids. They are a huge group of related compounds produced from either amino acids or purines and pyrimidines. There are estimated to be more than 10,000 different alkaloids, found in about 20% of angiosperm species. They are generally absent from ferns, mosses, and gymnosperms (Harborne, 1991).

There is good evidence that alkaloids are efficiently used as defensive agents and may be moved around the plant to those parts requiring greater protection during growth and development. A good example is in *Senecio vulgaris*, which contains pyrrolizidine alkaloids (Hartmann *et al.*, 1989). Thus, in stems, alkaloids are concentrated near the surface, with severalfold more alkaloid present in epidermal cells than in mesophyll cells. Further, the alkaloid moves up the plant from roots and leaves during flowering, thereby protecting the inflorescence from grazing. In the coffee plant, *Coffea arabica*, the purine alkaloid caffeine is concentrated in tissues vulnerable to herbivory. Thus, young leaves can contain as much as 4% dry weight of caffeine, while older leaves, which become thicker and harder, contain considerably less. Similarly, soft, young coffee beans contain approximately 2% caffeine, while older, ripe beans contain nearly 10 times less alkaloid (Frischnecht *et al.*, 1986).

A number of plant species are protected from herbivory by cyanogenic glycosides (see Section 2.2.2.2). Within cells, cyanogenic glycosides are stored separately from the enzymes that cleave them, and following damage by herbivory, the glycoside and enzyme are brought together, with the release of hydrogen cyanide. The latter is a potent respiratory inhibitor, and although it is usually thought to be responsible for any protection from herbivory, other products from the breakdown of the glycoside (e.g., acetone and benzaldehyde) are equally effective in deterring herbivory (Jones, 1988).

Nitrogen-containing metabolites are all ultimately derived from one or other of the protein amino acids. This means that there is always competition for precursors required for basic plant processes such as protein synthesis. Thus, concentrations of alkaloids, for example, produced under herbivory, tend to be low. However, the low concentrations produced are offset by the potent toxicity of these compounds.

2.6.3.4 Other chemicals

Many plants hyperaccumulate metals, and various hypotheses for the functional significance of such hyperaccumulation have been proposed, including protection from both herbivores and pathogens (e.g., Boyd & Martens, 1992). Increasing evidence suggests that selenium (Se) can protect plants from a variety of herbivores and pathogens, including lepidopteran larvae, aphids, and fungal pathogens (e.g., Hanson *et al.*, 2003, 2004). Interestingly, it appears that the Se in hyperaccumulators is concentrated in organs and tissues that are most susceptible to both herbivory and pathogens (Freeman *et al.*, 2006). Results from a recent study support a protective function for hyperaccumulated Se against herbivory by prairie dogs in their native habitat (Quinn *et al.*, 2008). Field surveys indicated that Se hyperaccumulators were abundant on densely populated prairie dog colonies and suffered little herbivory compared to other nongrass herbaceous species on the site. Prairie dogs preferred to feed on plants without Se when given a choice between high or low Se plants of the same species, suggesting that accumulated Se deterred prairie dog herbivory.

2.6.3.5 A final word on chemical defenses against vertebrate herbivory

There is no doubt that most plants possess formidable chemical barriers to herbivory. However, what is important is not whether a plant contains a particular chemical, but rather the concentration of the chemical in a plant tissue and the ability of the herbivore to metabolize the chemical. In fact, mammalian herbivores are very well equipped to detoxify most chemicals used in protection. Crucially, detoxifying such chemicals is energetically expensive, placing a limit to the extent to which resources can be mobilized to carry out detoxification.

2.7   Defenses used against neighboring plants—allelopathy

2.7.1   Background

So far in this chapter, we have dealt with the defenses used by plants to ward off other organisms that used them as a source of food. However, plants grow alongside other plants, and this means that they compete with one another for the resources required for their growth, development, and reproduction, for example, light, water, and nutrients. Such competition has led to the development of various defenses for use against neighboring plants. Where such defense is chemical, it is known as allelopathy. The chemicals can be released into the air, water, or soil around the plant, where it exhibits an inhibitory effect on neighboring plants or seeds. Often, chemical defenses are invoked when one type of plant competes with another, different type of plant. Of course, competition can also occur between individuals of the same species, and this is referred to as autotoxicity. As pointed out by Scott (2008), establishing whether allelopathy exists requires three issues to be addressed: (1) the observations must be shown not to be simply the result of a direct effect of competition for resources; (2) the active agent must be identified; and (3) autotoxicity needs to be discussed.

  Most of the chemicals that have been positively identified in studies of allelopathy are of relatively simple structure and are either volatile terpenes or phenolic compounds

(Harborne, 1993). Because of this, it has been suggested that these chemicals first arose as a result of herbivore pressure and any allelopathic effects are a secondary function (Whittaker, 1972). Regardless of whether allelopathy is the primary or secondary function of such chemicals, there are examples where there is good evidence of allelopathy, and these are discussed below.

2.7.2 Allelopathy and the black walnut

It has been known for some time that plants growing close to a black walnut tree (*Juglans nigra*) turn chlorotic and eventually die. In fact, direct evidence of this phenomenon was obtained in 1925, when tomato and alfalfa planted in a region up to 27 m of the trunk of a walnut tree died (Massey, 1925). Massey found that plants growing outside the radius of root growth of the walnut were unaffected and assumed that plants within this radius were killed by a toxin diffusing from the roots of the walnut tree. It is now known that black walnut releases a compound called hydrojuglone from its roots and leaves. Once in the soil, hydrojuglone is oxidized to juglone, which is extremely toxic. Since it is not very mobile in the soil, the greatest juglone concentrations are found in very close proximity to the roots of the walnut tree, hence the observations made by Massey (1925). The production and release of juglone by black walnut inhibit the growth of most neighboring plants, although some plants, for example, *R. fructicosus* and *Poa pratensis*, are well able to grow under walnut trees (Harborne, 1993).

2.7.3 Allelopathy and the Californian chaparral

The Californian chaparral refers to an area of vegetation along the coast of Southern California. A number of shrubs dominate the flora of this region, including *Salvia leucophylla* (sagebrush) and *Artemisia californica*. Nothing grows within a couple of meters of these shrubs, and beyond this zone, any plants growing show limited growth and stunted development. Careful research showed that biological factors (e.g., shade, drought, nutrients, root competition, and insects) were not responsible for these effects, which instead appeared to be caused by volatile terpenes. These compounds were found to be present in the leaves of these shrubs; in the soil surrounding the plants, they could be transported into roots and seeds of various plants, and they inhibited germination of the seeds of grasses, which normally grow in adjacent areas (Muller, 1970). The two most potent terpenes identified in these shrubs were 1,8-cineole and camphor. Interestingly, the vegetation of the chaparral undergoes a change roughly every 25 years, when natural fires destroy the shrubs and the terpenes in the soil, allowing the annual plants to grow. Over a period of 6–7 years, the shrubs regrow and the terpenes produced exert their inhibitory effects, leading to the reappearance of the typical chaparral zonation of vegetation.

2.7.4 Allelopathy and spotted knapweed

It had long been suspected that spotted knapweed (*Centaurea maculosa*) could kill neighboring plants. Indeed, work using tissue culture showed that spotted knapweed produces catechin in two forms (+ and −), with (−)-catechin possessing herbicidal properties (Scott, 2008). This compound is released into the soil by the spotted knapweed and kills neighboring plants. In Europe, plants that have coevolved with spotted knapweed have developed a tolerance to (−)-catechin. However, spotted knapweed was introduced into North America where native plants possess no such tolerance. Hence, spotted knapweed is considered an exotic, invasive species in the USA.

Work by Prithiviraj *et al.* (2007) found that the effects of (±)-catechin on *Arabidopsis* were concentration-dependent. At high concentrations, (±)-catechin is phytotoxic, while at lower concentrations, it leads to increased plant biomass and induced pathogen defense responses. The authors suggested that the elicitation of ROS by low concentrations of (±)-catechin stimulates cell division and cell expansion, thereby leading to increased plant growth. Whatever the mechanism(s) underlying these plant responses to (±)-catechin, the work highlights the potential for plant compounds to act as toxins against other plants at higher concentrations, and as signals for facilitation of plant growth and disease resistance at lower concentrations.

2.8   Conclusions

In this chapter, we have dealt with defenses singly. In reality however, when plants are attacked, it is likely that, in terms of active or inducible defenses, the attacker will encounter more than one defense mechanism. In some cases, these defenses will be specific to the type of attacker, for example, a pathogenic microbe or an insect, but not always. For example, when *Arabidopsis thaliana* was attacked by the specialist *Brassica* pest, the aphid *Brevicoryne brassicae*, not only was gene expression related to glucosinolate biosynthesis increased, so too was the expression of genes responsible for biosynthesis of the phytoalexin camelexin (Kusnierczyk *et al.*, 2008). As a phytoalexin, camelexin is usually thought of as effective against pathogenic bacteria and fungi. However, camelexin was found to impair the fitness of *B. brassicae.* This might not be the complete story, and it might be that the aphid can metabolize camelexin, since the levels of camelexin detected in aphid-infested plants were lower than those measured following fungal or bacterial infection. Whatever the full story, it is clear that the mobilization of defenses by plants is amazingly complex.

Because plants are a source of food for many other organisms, including other plants, their survival depends on their ability to ward off attack and if that fails, to tolerate the infection or infestation sufficiently to allow growth and reproductive development. This latter ability of plants does not lie within the scope of this book, although further information on this topic can be obtained from a number of reviews (e.g., Bingham *et al.*, 2009). What is clear, however, is that plants possess a remarkable variety of defenses that are effective at warding off most attackers. In terms of chemical defenses, the spectrum produced by plants is incredibly wide, although it will not have escaped notice that the same classes of chemicals are deployed against different types of attackers, for example, microbial pathogens, insects, and mammalian herbivores. Indeed, the roles of these chemicals are truly multifaceted, since they provide protection, not just against other organisms, but also against harsh environmental conditions. However, although the classes of chemicals used in defense against different attackers might be the same, in many cases, specific chemicals or combinations of chemicals are used to deal with specific types of attackers. Most of our knowledge of plant defenses comes from research conducted on particular plant–attacker interactions. However, in the real world, plants can face different types of attackers at the same time. How plants deploy and coordinate their defenses under such conditions is therefore of great importance. This is dealt with in Chapter 4. Dealing effectively with multiple attackers requires good systems of communication within (and even outside) the plant. This is the topic of the next chapter.

Recommended reading

Agrios GN, 2005. *Plant Pathology*, third edition. London: Elsevier Academic Press.
Danell K, Bergström R, 2002. Mammalian herbivory in terrestrial environments. In: Herrera CM, Pellmyr O, eds. *Plant–Animal Interactions: An Evolutionary Approach*. Oxford: Blackwell Publishing Ltd., pp. 107–131.
Harborne JB, 1993. *Introduction to Ecological Biochemistry*, fourth edition. London: Elsevier Academic Press.
Lucas JA, 1998. *Plant Pathology and Plant Pathogens*. Oxford: Blackwell Publishing Ltd.
Palo RT, Robbins CT, 1991. *Plant Defenses against Mammalian Herbivory*. Boca Raton, FL: CRC Press.
Parker C, Riches CR, 1993. *Parasitic Weeds of the World: Biology and Control*. Wallingford: CAB International.
Schoonhoven LM, van Loon JJA, Dicke M, 2005. *Insect–Plant Biology*. Oxford: Oxford University Press.

References

Abe M, Matsuda K, 2000. Feeding responses of four phytophagous lady beetle species (Coleoptera: Coccinellidae) to cucurbitacins and alkaloids. *Applied Entomology and Zoology* **35**, 257–264.
Agrawal AA, 1999. Induced responses to herbivory in wild radish: effects on several herbivores and plant fitness. *Ecology* **80**, 1713–1723.
Agrawal AA, Lajeunesse MJ, Fishbein M, 2008. Evolution of latex and its constituent defensive chemistry in milkweeds (*Asclepias*): a phylogenetic test of plant defense escalation. *Entomologia Experimentalis et Applicata* **128**, 126–138.
Agrios GN, 2005. *Plant Pathology*, third edition. London: Elsevier Academic Press.
Anand A, Zhou T, Trick HN, Gill BS, Bockus WW, Muthukrishnan S, 2003. Greenhouse and field testing of transgenic wheat plants stably expressing genes for thaumatin-like protein, chitinase and glucanase against *Fusarium graminearum*. *Journal of Experimental Botany* **54**, 1101–1111.
Bailey JA, Deverall BJ, 1971. Formation and activity of phaseollin in the interaction between bean hypocotyls (*Phaseolus vulgaris*) and physiological races of *Colletotrichum lindemuthianum*. *Physiological Plant Pathology* **1**, 435–449.
Bazely DR, Myers JH, Dasilva KB, 1991. The response of numbers of bramble prickles to herbivory and depressed resource availability. *Oikos* **61**, 327–336.
Beattie AJ, Hughes L, 2002. Ant–plant interactions. In: Herrera CM, Pellmyr O, eds. *Plant–Animal Interactions: An Evolutionary Approach*. Oxford: Blackwell Publishing Ltd., pp. 211–235.
Becerra JX, 1994a. Squirt-gun defense in *Bursera* and the chrysomelid counterploy. *Ecology* **75**, 1991–1996.
Becerra JX, 1994b. Chrysomelid behavioural counterploys to secretive canals in plants. In: Jolivet P, Cox P, Petitpierre D, eds. *Novel Aspects of the Biology of Chrysomelidae (Coleoptera)*. Dordrecht: Kluwer Academic Press, pp. 327–330.

Becerra JX, Venable DL, Evans PH, Bowers WS, 2001. Interactions between chemical and mechanical defenses in the plant genus *Bursera* and their implications for herbivores. *American Zoologist* **41**, 865–876.

Beckman CH, 2000. Phenolic storing cells: keys to programmed cell death and periderm formation in wilt disease and in general defense responses in plants? *Physiological and Molecular Plant Pathology* **57**, 101–110.

Bernays EA, 1991. Evolution of insect morphology in relation to plants. *Philosophical Transactions of the Royal Society, London* **B333**, 257–264.

Bernays EA, Cornelius M, 1992. Relationship between deterrence and toxicity of plant secondary compounds for the alfalfa weevil *Hypera brunneipennis. Entomologia Experimentalis et Applicata* **64**, 289–292.

Bingham IJ, Walters DR, Foulkes MJ, Paveley ND, 2009. Crop traits and the tolerance of wheat and barley to foliar disease. *Annals of Applied Biology* **154**, 159–173.

Blundell AG, Peart DR, 1998. Distance-dependence in herbivory and foliar condition for juvenile *Shorea* trees in Bornean dipterocarp rain forest. *Oecologia* **117**, 151–160.

Boyd RS, Martens SN, 1992. The raison d'être for metal for metal hyperaccumulation by plants. In: Baker AJM, Proctor J, Reeves RD, eds. *The Vegetation of Ultramafic (Serpentine) Soils.* Andover: Intercept, pp. 279–289.

Brennan EB, Weinbaum SA, 2001. Stylet penetration and survival of three psyllid species on adult leaves and "waxy" and "de-waxed" juvenile leaves of *Eucalyptus globules. Entomologia Experimentalis et Applicata* **100**, 355–366.

Brett C, Waldron K, 1990. *Physiology and Biochemistry of Plant Cell Walls. Topics in Plant Physiology 2.* London: Unwin Hyman Ltd.

Brown IR, Mansfield JW, 1988. An ultrastructural study, including cytochemistry and quantitative analyses, of the interactions between pseudomonads and leaves of *Phaseolus vulgaris* L. *Physiological and Molecular Plant Pathology* **33**, 351–376.

Bruelheide H, Scheidel U, 1999. Slug herbivory as a limiting factor for the geographical range of *Arnica montana. Journal of Ecology* **87**, 839–848.

Bryant JP, Wieland GD, Reichardt PB, Lewis VE, McCarthy MC, 1983. Pinosylvin methyl ether deters snowshoe hare feeding on Green Alder. *Science* **222**, 1023–1025.

Burse A, Weingart H, Ullrich MS, 2004. The phytoalexin inducible multidrug efflux pump AcrAB contributes to virulence in the fire blight pathogen, *Erwinia amylovora. Molecular Plant-Microbe Interactions* **17**, 43–54.

Caballero C, Castañera P, Ortego F, Fontana G, Pierro P, Savona G, Rodriguez B, 2001. Effects of ajugarins and related neoclerodane diterpenoids on feeding behaviour of *Leptinotarsa decemlineata* and *Spodoptera exigua* larvae. *Phytochemistry* **58**, 249–256.

Chen H, Wilkerson CG, Kuchar JA, Phinney BS, Howe GA, 2005. Jasmonate-inducible plant enzymes degrade essential amino acids in the herbivore midgut. *Proceedings of the National Academy of Sciences, USA* **102**, 19237–19242.

Chew FS, Renwick JJA, 1995. Host-plant choice in *Pieris butterflies.* In: Cardé RT, Bell WJ, eds. *Chemical Ecology of Insects.* New York: Chapman & Hall, pp. 214–238.

Chisholm ST, Coaker G, Day B, Staskawicz BJ, 2006. Host–microbe interactions: shaping the evolution of the plant immune response. *Cell* **124**, 803–814.

Coley PD, 1983. Herbivory and defensive characteristics of tree species in a lowland rain forest. *Ecological Monographs* **53**, 209–233.

Cooper SM, Owen-Smith N, 1985. Condensed tannins deter feeding by browsing ruminants in a South African savanna. *Oecologia* **67**, 142–146.

Cork SJ, 1994. Digestive constraints on dietary scope in small and moderately sized mammals: how much do we really understand? In: Chivers DJ, Langer P, eds. *The Digestive System in Mammals: Food, Form and Function.* Cambridge: Cambridge University Press, pp. 337–369.

Cork SJ, Foley WJ, 1991. Digestive and metabolic strategies of arboreal mammalian folivores in relation to chemical defences in temperate and tropical forests. In: Palo RT, Robbins CT, eds. *Plant Defences against Mammalian Herbivory*. Boca Raton, FL: CRC Press, pp. 133–166.

Danell K, Bergström R, 2002. Mammalian herbivory in terrestrial environments. In: Herrera CM, Pellmyr O, eds. *Plant–Animal Interactions: An Evolutionary Approach*. Malden, MA: Blackwell Publishing Ltd., pp. 107–131.

Datta SK, Muthukrishnan S, 1999. *Pathogenesis-Related Proteins in Plants*. Boca Raton, FL: CRC Press.

De Almeida Engler J, De Vleesschauwer V, Burssens S, Celenza JL, Inze D, Van Montagu M, Engler G, Gheysen G, 1999. Molecular markers and cell cycle inhibitors show the importance of cell cycle progression in nematode-induced galls and syncytia. *The Plant Cell* **11**, 793–808.

De Wit PJGM, 2007. How plants recognise pathogens and defend themselves. *Cellular and Molecular Life Sciences* **64**, 2726–2732.

Debenham M, 2005. Waiting to fight? Lectins' passive defence role in plants. *Biologist* **52**, 74–79.

Decraemer W, Hunt DJ, 2006. Structure and classification. In: Perry RN, Moens M, eds. *Plant Nematology*. King's Lynn, UK: CABI, pp. 3–32.

Dicke M, 1994. Local and systemic production of volatile herbivore-induced terpenoids: their role in plant-carnivore mutualism. *Journal of Plant Physiology* **143**, 465–472.

Dinan L, 2001. Phytoecdysteroids: biological aspects. *Phytochemistry* **57**, 325–329.

Dominy NJ, Grubb PJ, Jackson RV, Lucas PW, Metcalfe DJ, Svenning J-C, Turner IM, 2008. In tropical lowland rainforests monocots have tougher leaves than dicots, and include a new kind of tough leaf. *Annals of Botany* **101**, 1363–1377.

Dropkin VH, Nelson PE, 1960. The histopathology of root-knot nematode infections in soybeans. *Phytopathology* **50**, 442–447.

Dussourd DE, Eisner T, 1987. Vein-cutting behaviour: insect counterploy to the latex defense of plants. *Science* **237**, 898–901.

Dyer LA, Coley PD, 2002. Tritrophic interactions in tropical versus temperate communities. In: Tscharntke T, Hawkins BA, eds. *Multitrophic Level Interactions*. Cambridge: Cambridge University Press, pp. 67–88.

Echevarría-Zomeño S, Pérez-de-Luque A, Jorrín J, Maldonado AM, 2006. Pre-haustorial resistance to broomrape (*Orobanche cumana*) in sunflower (*Helianthus annuus*): cytochemical studies. *Journal of Experimental Botany* **57**, 4189–4200.

Edens RM, Anand SC, Bolla RI, 1995. Enzymes of the phenylpropanoid pathway in soybean infected with *Meloidogyne incognita* or *Heterodera glycines*. *Journal of Nematology* **27**, 292–303.

Eigenbrode SD, Kabalo NN, Stoner KA, 1999. Predation, behaviour, and attachment by *Chrysoperla plarabunda* larvae on *Brassica oleracea* with different surface waxblooms. *Entomologia Experimentalis et Applicata* **90**, 225–235.

Eisenberg JF, 1980. The density and biomass of tropical mammals. In: Soulé ME, Wilcox BA, eds. *Conservation Biology: An Evolutionary-Ecological Perspective*. Sunderland: Sinauer, pp. 35–55.

Fahn A, 1979. *Secretory Tissues in Plants*. London: Academic Press.

Fiala B, Maschwitz U, Pong TY, 1991. The association between *Macaranga* trees and ants in S.E. Asia. In: Huxley CR, Cutler DF, eds. *Ant–Plant Interactions*. Oxford: Oxford University Press, pp. 263–270.

Field B, Jordan F, Osbourn A, 2006. First encounters—deployment of defence-related natural products by plants. *New Phytologist* **172**, 193–207.

Filip V, Dirzo R, Maass JM, Sarukhan J, 1995. Within- and among-year variation in the levels of herbivory on the foliage of trees from a Mexican tropical deciduous forest. *Biotropica* **27**, 78–86.

Fink LS, Brower LP, 1981. Birds can overcome the cardenolide defense of Monarch butterflies in Mexico. *Nature* **291**, 67–70.

Fraenkel GS, 1959. The *raison d'etre* of secondary plant substances. *Science* **129**, 1466–1470.

Freeman JL, Zhang LH, Marcus MA, Fakra S, Pilon-Smits EAH, 2006. Spatial imaging, speciation and quantiWcation of selenium in the hyperaccumulator plants *Astragalus bisulcatus* and *Stanleya pinnata*. *Plant Physiology* **142**, 124–134.

Frischnecht PM, Dufek JV, Baumann TW, 1986. Purine alkaloid formation in buds and developing leaflets of *Coffea arabica:* an expression of an optimal defense strategy. *Phytochemistry* **25**, 613–616.

Fuller VL, Lilley CJ, Unwin PE, 2008. Nematode resistance. *New Phytologist* **180**, 27–44.

Garcion C, Lamotte O, Métraux J-P, 2007. Mechanisms of defence to pathogens: biochemistry and physiology. In: Walters D, Newton A, Lyon G, eds. *Induced Resistance for Plant Defence: A Sustainable Approach to Crop Protection.* Oxford: Blackwell Publishing Ltd., pp. 109–132.

Giebel J, 1982. Mechanism of resistance to plant nematodes. *Annual Review of Phytopathology* **20**, 257–279.

Gomez JM, Zamora R, 2002. Thorns as induced mechanical defense in a long-lived shrub (*Hormathophylla spinosa*). *Ecology* **83**, 885–890.

Grubb PJ, 1992. A positive distrust in simplicity—lessons from plant defences and from competition among plants and among animals. *Journal of Ecology* **80**, 585–610.

Grubb PJ, Jackson RV, 2007. The adaptive value of young leaves being tightly folded or rolled on monocotyledons in tropical lowland rain forest: an hypothesis in two parts. *Plant Ecology* **192**, 317–327.

Grubb PJ, Jackson RV, Barberis IM, Bee JN, Coomes DA, Dominy NJ, De La Fuente MAS, Lucas PW, Metcalfe DJ, Svenning J-C, Turner IM, Vargas O, 2008. Monocot leaves are eaten less than dicot leaves in tropical lowland rain forests: correlations with toughness and leaf presentation. *Annals of Botany* **101**, 1379–1389.

Gurney AL, Grimanelli D, Kanampiu F, Hoisington D, Scholes JD, Press MC, 2003. Novel sources of resistance to *Striga hermonthica* in *Tripsacum dactyloides*, a wild relative of maize. *New Phytologist* **160**, 557–568.

Hain R, Reif HJ, Krausse E, Langebartels R, Kindl H, Vornam B, Wiese W, Schmelzer E, Schreier PH, Stocker SK, 1993. Disease resistance results from foreign phytoalexin expression in a novel plant. *Nature* **361**, 153–156.

Halpern M, Raats D, Lev-Yadun S, 2007. Plant biological warfare: thorns inject pathogenic bacteria into herbivores. *Environmental Microbiology* **9**, 584–592.

Hanson B, Garifullina GF, Lindbloom SD, Wangeline A, Ackley A, Kramer K, Norton AP, Lawrence CB, Pilon Smits EAH, 2003. Selenium accumulation protects *Brassica juncea* from invertebrate herbivory and fungal infection. *New Phytologist* **159**, 461–469.

Hanson B, Lindblom SD, LoeZer ML, Pilon-Smits EAH, 2004. Selenium protects plants from phloem feeding aphids due to both deterrence and toxicity. *New Phytologist* **162**, 655–662.

Harborne JB, 1979. Flavonoid pigments. In: Rosenthal GA, Janzen DH, eds. *Herbivores: Their Interaction with Secondary Plant Metabolites.* New York: Academic Press, pp. 619–655.

Harborne JB, 1991. The chemical basis of plant defence. In: Palo RT, Robbins CT, eds. *Plant Defences against Mammalian Herbivory.* Boca Raton, FL: CRC Press, pp. 45–59.

Harborne JB, 1993. *Introduction to Ecological Biochemistry*, fourth edition. London: Elsevier Academic Press.

Harborne JB, 1999. The comparative biochemistry of phytoalexin induction in plants. *Biochemical Systematics and Ecology* **27**, 335–367.

Harrison NA, Beckman CH, 1982. Time/space relationships of colonization and host response in wilt-resistant and wilt-susceptible cotton (*Gossypium*) cultivars inoculated with *Verticillium dahliae* and *Fusarium oxysporum* f. sp. *vasinfectum*. *Physiological Plant Pathology* **21**, 193–207.

Hartmann T, Ehmke A, Sonder H, Borstel KV, Adolph R, Toppel G, 1989. Metabolic integration of the pyrrolizidine alkaloids in respect of their function in chemical protection of *Senecio vulgaris*. *Planta Medica* **55**, 218–219.

Haussmann BIG, Hess DE, Welz HG, Geiger HH, 2000. Improved methodologies for breeding *Striga*-resistant sorghums. *Field Crops Research* **66**, 195–211.

He X-Z, Dixon RA, 2000. Genetic manipulation of isoflavone 7-*O*-methyltransferase enhances biosynthesis of 4'-*O*-methylated isoflavonoid phytoalexins and disease resistance in alfalfa. *The Plant Cell* **12**, 1689–1702.

Hedin PA, Jenkins JN, Collum DH, White WH, Parrott WL, 1983. Multiple factors in cotton contributing to resistance to the tobacco budworm, *Heliothis virescens*. In: Hedin PA, ed. *Plant Resistance to Insects*. Washington, DC: American Chemical Society Symposium 208, pp. 347–365.

Hedrick SA, Bell JN, Boller T, Lamb CJ, 1988. Chitinase cDNA cloning and mRNA induction by fungal elicitor, wounding and infection. *Plant Physiology* **86**, 182–186.

Hostettmann K, Marston A, 1995. *Saponins (chemistry and pharmacology of natural products)*. Cambridge: Cambridge University Press.

Huang JS, Barker KR, 1991. Glyceollin 1 in soybean-cyst nematode interactions: spatial and temporal distribution in roots of resistant and susceptible soybeans. *Plant Physiology* **96**, 1302–1307.

Hückelhoven R, Kogel K-H, 1998. Tissue-specific superoxide generation at interaction sites in resistant and susceptible near-isogenic barley lines attacked by the powdery mildew fungus (*Erysiphe graminis* f.sp. *hordei*). *Molecular Plant-Microbe Interactions* **11**, 292–300.

Isman MB, Duffey SS, 1982. Toxicity of tomato phenolic compounds to the fruitworm, *Heliothis zea*. *Entomologia Experimentalis et Applicata* **31**, 370–376.

Jasmer DP, Goverse A, Smart G, 2003. Parasitic nematode interactions with mammals and plants. *Annual Review of Phytopathology* **41**, 245–270.

Jassbi AR, Gase K, Hettenhausen C, Schmidt A, Baldwin IT, 2008. Silencing geranyl diphosphate synthase in *Nicotiana attenuata* dramatically impairs resistance to tobacco hormworm. *Plant Physiology* **146**, 974–986.

Johnson MB, 1992. The genus *Bursera* (Bursaraceae) in Sonora, Mexico and Arizona, U.S.A. *Desert Plants* **10**, 126–143.

Jones DA, 1988. Cyanogenesis in animal-plant interactions. In: Evered D, Harnett S, eds. *Cyanide Compounds in Biology*. Chichester: John Wiley & Sons, Ltd., pp. 151–176.

Karban R, Baldwin IT, 1997. *Induced Responses to Herbivory*. Chicago, IL: University of Chicago Press.

Kato T, Ishida K, Sato H, 2008. The evolution of nettle resistance to heavy deer browsing. *Ecological Research* **23**, 339–345.

Kistler HC, Van Etten HD, 1984. Regulation of pisatin demethylation in *Nectria haematococca* and its influence on pisatin tolerance and virulence. *Journal of General Microbiology* **130**, 2605–2613.

Kodan A, Kuroda H, Sakai F, 2002. A stilbene synthase from Japanese red pine (*Pinus densiflora*): implication for phytoalexin accumulation and downregulation of flavonoid biosynthesis. *Proceedings of the National Academy of Sciences, USA* **99**, 3335–3339.

Konno K, Hirayama C, Nakamura M, Tateishi K, Tamura Y, Hattori M, Kohno K, 2004. Papain protects papaya trees from herbivorous insects: role of cysteine proteases in latex. *The Plant Journal* **37**, 370–378.

Kubo I, Klacke J, Asano S, 1983. Effects of ingested phytoecdysteroids on the growth and development of two lepidopterous larvae. *Journal of Insect Physiology* **29**, 307–316.

Kuć J, 1995. Phytoalexins, stress metabolism, and disease resistance in plants. *Annual Review of Phytopathology* **33**, 275–297.

Kusnierczyk A, Winge P, Jorstad TS, Troczynska J, Rossiter JT, Bones AM, 2008. Towards global understanding of plant defence against aphids—timing and dynamics of early Arabidopsis defence responses to cabbage aphid (*Brevicoryne brassicae*) attack. *Plant Cell and Environment* **31**, 1097–1115.

Lamb RJ, 1980. Hairs protect pods of mustard (*Brassica hirta* "gisilba") from flea beetle feeding damage. *Canadian Journal of Plant Science* **60**, 1439–1440.

Lane GA, Biggs DR, Sutherland OWR, Williams EM, Maindonald JM, Donnell DJ, 1985. Isoflavonoid feeding deterrents for *Costelytra zealandica:* structure-activity relationships. *Journal of Chemical Ecology* **11**, 1713–1735.

Lauvergeat V, Lacomme C, Lacombe E, Lasserre E, Roby D, Grima-Pettenati J, 2001. Two cinnamoyl-CoA reductase (CCR) genes from *Arabidopsis thaliana* are differentially expressed during development and in response to infection with pathogenic bacteria. *Phytochemistry* **57**, 1187–1195.

Lee Y-I, 1983. The potato leafhopper, *Empoasca fabae*, soybean pubescence, and hopperburn resistance, Ph.D dissertation. University of Illinois Urbana-Champaign, IL.

Lev-Yadun S, 2001. Aposematic (warning) coloration associated with thorns in higher plants. *Journal of Theoretical Biology* **210**, 385–388.

Lichtenthaler HK, 1999. The 1-deoxy-D-xylulose-5-phosphate pathway of isoprenoid biosynthesis in plants. *Annual Review of Plant Physiology and Plant Molecular Biology* **50**, 47–65.

Lo S-CC, de Verdier K, Nicholson RL, 1999. Accumulation of 3-deoxyanthocyanidin phytoalexins and resistance to *Colletotrichum sublineolum* in sorghum. *Physiological and Molecular Plant Pathology* **55**, 263–273.

Lozano-Baena MD, Moreno MT, Rubiales D, Pérez-de-Luque A, 2007. *Medicago truncatula* as a model for non-host resistance in legume-parasitic plant interactions. *Plant Physiology* **145**, 437–449.

Lucas JA, 1998. *Plant Pathology and Plant Pathogens*, third edition. Oxford: Blackwell Publishing Ltd.

Lucas PW, Turner IM, Dominy NJ, Yamashita N, 2000. Mechanical defences to herbivory. *Annals of Botany* **86**, 913–920.

Mace ME, 1978. Contributions of tyloses and terpenoid aldehyde phytoalexins to *Verticillium dahliae*. *Physiological Plant Pathology* **12**, 1–12.

Martin JT, Juniper BE, 1970. *The Cuticles of Plants*. London: Edward Arnold.

Massey AB, 1925. Antagonism of the walnuts (*Juglans nigra* L. and. *J. cinerea* L.) in certain plant associations. *Phytopathology* **15**, 773–784.

Matsui K, 2006. Green leaf volatiles: hydroperoxide lyase pathway of oxylipin metabolism. *Current Opinion in Plant Biology* **9**, 274–280.

Melillo MT, Leonetti P, Bongiovanni M, Castagnone-Sereno P, Bleve-Zacheo T, 2006. Modulation of reactive oxygen species activities and H_2O_2 accumulation during compatible and incompatible tomato–root-knot nematode interactions. *New Phytologist* **170**, 501–512.

Melotto M, Underwood W, Koczan J, Nomura K, He SY, 2006. Plant stomata function in innate immunity against bacterial invasion. *Cell* **126**, 969–980.

Melotto M, Underwood W, He SY, 2008. Role of stomata in plant innate immunity and foliar bacterial diseases. *Annual Review of Phytopathology* **46**, 101–122.

Mondor EB, Addicott JF, 2003. Conspicious extra-floral nectarines are inducible in *Vicia faba*. *Ecology Letters* **6**, 495–497.

Mooney HA, Emboden WA, 1968. The relationship of terpene composition, morphology, and distribution of populations of *Bursera microphylla* (Burseraceae). *Brittonia* **20**, 44–51.

Mordue AJ, Blackwell A, 1993. Azadirachtin: an update. *Journal of Insect Physiology* **39**, 903–924.

Morrissey JP, Osbourn AE, 1999. Fungal resistance to plant antibiotics as a mechanism of pathogenesis. *Microbiological Molecular Biological Reviews* **63**, 708–724.

Muller CH, 1970. Phytotoxins as plant habitat variables. *Recent Advances in Phytochemistry* **3**, 106–121.

Nicholson RL, Hammerschmidt R, 1992. Phenolic compounds and their role in disease resistance. *Annual Review of Phytopathology* **30**, 369–389.

Nomura K, Melotto M, He S-Y, 2005. Suppression of host defense in compatible plant-*Pseudomonas syringae* interactions. *Current Opinion in Plant Biology* **8**, 361–368.

Oates J, Swain T, Zantovska J, 1977. Secondary compounds and food selection by Colobus monkeys. *Biochemical Systematics and Ecology* **5**, 317–321.

Osbourn AE, Qi X, Townsend B, Qin B, 2003. Secondary metabolism and plant defence. *New Phytologist* **159**, 101–108.

Paxton JD, 1981. Phytoalexins—a working redefinition. *Phytopathologische Zeitschrift* **101**, 106–109.

Pérez-de-Luque A, Rubiales D, Cubero JI, Press MC, Scholes J, Yoneyama K, Takeuchi Y, Plakhine D, Joel DM, 2005. Interaction between *Orobanche crenata* and its host legumes: unsuccessful haustorial penetration and necrosis of the developing parasite. *Annals of Botany* **95**, 935–942.

Pérez-de-Luque A, Lozano MD, Moreno MT, Testillano PS, Rubiales D, 2007. Resistance to broomrape (*Orobanche crenata*) in faba bean (*Vicia faba*): cell wall changes associated with prehaustorial defensive mechanisms. *Annals of Applied Biology* **151**, 89–98.

Pérez-de-Luque A, Moreno MT, Rubiales D, 2008. Host plant resistance against broomrapes (*Orobanche* spp.): defence reactions and mechanisms of resistance. *Annals of Applied Biology* **152**, 131–141.

Pieterse CMJ, Dicke M, 2007. Plant interactions with microbes and insects: from molecular mechanisms to ecology. *Trends in Plant Science* **12**, 564–569.

Prithiviraj B, Perry LG, Badri DV, Vivanco JM, 2007. Chemical facilitation and induced pathogen resistance mediated by a root-secreted phytotoxin. *New Phytologist* **173**, 852–860.

Prusky D, Dinoor A, Jacoby B, 1980. The sequence of death of haustoria and host cells during the hypersensitive reaction of oat to crown rust. *Physiological Plant Pathology* **17**, 33–40.

Quinn CF, Freeman JL, Galeas ML, Klamper EM, Pilon-Smits EAH, 2008. The role of selenium in protecting plants against prairie dog herbivory: implications for the evolution of selenium hyperaccumulation. *Oecologia* **155**, 267–275.

Ratzka A, Vogel H, Kliebenstein DJ, Mitchell-Olds T, Kroymann J, 2002. Disarming the mustard oil bomb. *Proceedings of the National Academy of Sciences, USA* **99**, 11223–11228.

Rees SB, Harborne JB, 1985. The role of sesquiterpene lactones and phenolics in the chemical defence of the chicory plant. *Phytochemistry* **24**, 2225–2231.

Reichardt PB, Bryant JP, Clausen TP, Wieland GD, 1984. Defense of winter-dormant Alaska paper birch against snowshoe hares. *Oecologia* **65**, 58–69.

Renwick JAA, 2002. The chemical world of crucivores: lures, treats and traps. *Entomologia Expereimentalis et Applicata* **104**, 35–42.

Rich PJ, Grenier U, Ejeta G, 2004. *Striga* resistance in the wild relatives of sorghum. *Crop Science* **44**, 2221–2229.

Root RB, Cappuccino N, 1992. Patterns in population change and the organization of the insect community associated with goldenrod. *Ecological Monographs* **62**, 393–420.

Scheirs J, de Bruyn L, Verhagen R, 2001. A test of the C_3-C_4 hypothesis with two grass miners. *Ecology* **82**, 410–421.

Schoonhoven LM, Van Loon JJA, Dicke M, 2005. *Insect–Plant Biology*. Oxford: Oxford University Press.

Schulze-Lefert P, Vogel J, 2000. Closing the ranks to attack by powdery mildew. *Trends in Plant Science* **5**, 343–348.

Scott P, 2008. *Physiology and Behaviour of Plants*. Chichester: John Wiley & Sons, Ltd.

Seiber JN, Lee SM, Benson JM, 1984. Chemical characteristics and ecological significance of cardenolides of *Asclepias* species. In: Nes WD, Fuller G, Tsai LS, eds. *Isopentenoids in Plants: Biochemistry and Function*. New York: Marcel Dekker, pp. 563–588.

Serghini K, Pérez-de-Luque A, Castejón-Muñoz M, García-Torres L, Jorrín JV, 2001. Sunflower (*Helianthus annuus* L.) response to broomrape (*Orobanche crenata* Loefl.) parasitism: induced synthesis and excretion of 7-hydroxylated simple coumarins. *Journal of Experimental Botany* **52**, 2227–2234.

Shi J, Mueller WC, Beckman CH, 1992. Vessel occlusion and secretory activities of vessel contact cells in resistant or susceptible cotton plants infected with *Fusarium oxysporum* f. sp. *vasinfectum*. *Physiological and Molecular Plant Pathology* **40**, 133–147.

Showalter AM, Bell JN, Cramer CL, Bailey JA, Varner JA, Lamb CJ, 1985. Accumulation of hydroxyproline-rich glycoprotein mRNAs in response to fungal elicitor and infection. *Proceedings of the National Academy of Sciences USA* **82**, 6551–6555.

Shroff R, Vergara F, Muck A, Svatos A, Gershenzon J, 2008. Nonuniform distribution of glucosinolates in *Arabidopsis thaliana* leaves has important consequences for plant defense. *Proceedings of the National Academy of Sciences, USA* **105**, 6196–6201.

Shutt DA, 1976. The effects of plant oestrogens on animal production. *Endeavour* **35**, 110–113.

Silva MC, Nicole M, Guerra-GuimarÃes L, Rodrigues, Jr., CJ, 2002. Hypersensitive cell death and post-haustorial defence responses arrest the orange rust (*Hemileia vastatrix*) growth in resistant coffee leaves. *Physiological and Molecular Plant Pathology* **60**, 169–183.

Simmonds MSJ, 2001. Importance of flavonoids in insect–plant interactions: feeding and oviposition. *Phytochemistry* **56**, 245–252.

Snook ME, Johnson AW, Severson RF, Teng Q, White RA, Sisson VA, Jackson DM, 1997. Hydroxygeranyllinalool glycosides from tobacco exhibit antibiosis activity against the tobacco budworm (*Heliothis virescens*). *Journal of Agricultural and Food Chemistry* **45**, 2299–2308.

Sobczak M, Avrova A, Jupowicz J, Phillips MS, Ernst K, Kumar A, 2005. Characterization of susceptibility and resistance responses to potato cyst nematode (*Globodera* spp.) infection of tomato lines in the absence and presence of the broad-spectrum nematode resistance Hero gene. *Molecular Plant-Microbe Interactions* **18**, 158–168.

Soriano IR, Asenstorfer RE, Schmidt O, Riley IT, 2004. Inducible flavone in oats (*Avenae sativa*) is a novel defense against plant parasitic nematodes. *Phytopathology* **94**, 1207–1214.

Stange, Jr., RR, Ralph J, Peng JP, Sims JJ, Midland SL, McDonald RE, 2001. Acidolysis and hot water extraction provide new insights into the composition of the induced "lignin-like" material from squash fruit. *Phytochemistry* **57**, 1005–1011.

Stanton ML, Palmer TM, Young TP, 2002. Competition-colonization trade-offs in a guild of African *Acacia* ants. *Ecological Monographs* **72**, 347–363.

Strauss SY, Zangerl AR, 2002. Plant-insect interactions in terrestrial ecosystems. In: Herrera CM, Pellmyr O, eds. *Plant-Animal Interactions: An Evolutionary Approach*. Oxford: Blackwell Publishing Ltd., pp. 77–106.

Tahvanainen J, Helle E, Julkunen-Tiito R, Lavola A, 1985. Phenolic compounds of willow bark as deterrents against feeding by mountain hare. *Oecologia* **65**, 319–323.

Tallamy DW, Stull J, Ehresman NP, Gorski PM, Mason CE, 1997. Cucurbitacins as feeding and oviposition deterrents to insects. *Environmental Entomology* **26**, 678–683.

Tegos G, Stermitz FR, Lomovskaya O, Lewis K, 2002. Multidrug inhibitors uncover remarkable activity of plant antimicrobials. *Antimicrobial Agents and Chemotherapy* **46**, 796–815.

Thomma B, Nelissen I, Eggermont K, Broekaert WF, 1999. Deficiency in phytoalexin production causes enhanced susceptibility of *Arabidopsis thaliana* to the fungus *Alternaria brassicicola*. *The Plant Journal* **19**, 163–171.

Thordal-Christensen H, Zhang Z, Wei Y, Collinge DB, 1997. Subcellular localization of H_2O_2 in plants. H_2O_2 accumulation in papillae and hypersensitive response during the barley powdery mildew interaction. *The Plant Journal* **11**, 1187–1194.

Tomiyama K, 1971. Cytological and biochemical studies of the hypersensitive response of potato cells to *P. infestans*. In: Akai S, Ouchi S, eds. *Morphological and Biochemical Events in Plant/Parasite Interactions*. Tokyo, Japan: Phytopathological Society of Japan, pp. 387–401.

Towers GHN, 1980. Photosensitizers from plants and their photodynamic action. *Progress in Phytochemistry* **6**, 183–202.

Traw MB, Dawson TE, 2002. Differential induction of trichomes by three herbivores of black mustard. *Oecologia* **131**, 526–532.

Trudgill DL, 1991. Resistance to and tolerance of plant parasitic nematodes in plants. *Annual Review of Phytopathology* **29**, 167–192.

Turlings TCJ, Tumlinson JH, 1992. Systemic release of chemical signals by herbivore-injured corn. *Proceedings of the National Academy of Sciences, USA* **89**, 8399–8402.

Urban M, Bhargava T, Hamer JE, 1999. An ATP-driven efflux pump is a novel pathogenicity factor in rice blast disease. *EMBO Journal* **18**, 512–521.

Valette C, Andary C, Geiger JP, Sarah JL, Nicole M, 1998. Histochemical and cytochemical investigations of phenols in roots of banana *Radopholus similis*. *Phytopathology* **88**, 1141–1148.

Van Baarlen P, Staats M, Van Kan JAL, 2004. Induction of programmed cell death in lily by the fungal pathogen *Botrytis elliptica*. *Molecular Plant Pathology* **5**, 559–574.

Van Lenteren JC, Hua LZ, Kamerman JW, Xu R, 1995. The parasite–host relationship between *Encarsia formosa* (Hym., Aphelinidae) and *Trialeurodes vaporariorum* (Hom., Aleyrodidae). XXVI. Leaf hairs reduce the capacity of *Encarsia* to control greenhouse whitefly on cucumber. *Journal of Applied Entomology* **119**, 553–559.

Van Loon LC, Rep M, Pieterse CMJ, 2006. Significance of inducible defense-related proteins in infected plants. *Annual Review of Phytopathology* **44**, 135–162.

Wäckers FL, Zuber D, Wunderlin R, Keller F, 2001. The effect of herbivory on temporal and spatial dynamics of foliar nectar production in cotton and castor. *Annals of Botany* **87**, 365–370.

Wang P, Turner NE, 2000. Virus resistance mediated by ribosome inactivating proteins. *Advances in Virus Research* **55**, 325–355.

Wharton PS, Julian AM, O'Connell RJ, 2001. Ultrastructure of the infection of *Sorghum bicolor* by *Colletotrichum sublineolum*. *Phytopathology* **91**, 149–158.

Whittaker RH, 1972. The biochemical ecology of higher plants. In: Sondheimer E, Simeone JB, eds. *Chemical Ecology*. New York: Academic Press, pp. 43–70.

Williamson VM, 1998. Root-knot nematode resistance genes in tomato and their potential for future use. *Annual Review of Phytopathology* **36**, 277–293.

Williamson VM, Kumar A, 2006. Nematode resistance in plants: the battle underground. *Trends in Genetics* **22**, 396–403.

Wink M, 1999. *Functions of Plant Secondary Metabolites and Their Exploitation in Biotechnology*. Sheffield, UK: Sheffield Academic Press.

Wittstock U, Gerschenzon J, 2002. Constitutive plant toxins and their role in defense against herbivores and pathogens. *Current Opinion in Plant Biology* **5**, 300–307.

Wittstock U, Halkier BA, 2002. Glucosinolate research in the *Arabidopsis* era. *Trends in Plant Science* **7**, 263–270.

Wittstock U, Agerbirk N, Stauber EJ, Olsen CE, Hippler M, Mitchell-Olds T, Gershenzon J, Vogel H, 2004. Successful herbivore attack due to metabolic diversion of a plant chemical defense. *Proceedings of the National Academy of Sciences, USA* **101**, 4859–4864.

Wróbel-Kwiatkowska M, Lorenc-Kukula K, Starzycki M, Oszmianski J, Kepczyńska E, Szopa J, 2004. Expression of β-1,3-glucanase in flax causes increased resistance to fungi. *Physiological and Molecular Plant Pathology* **65**, 245–256.

Wuyts N, Lognay G, Swennen R, De Waele D, 2006. Nematode infection and reproduction in transgenic and mutant *Arabidopsis* and tobacco with an altered phenylpropanoid metabolism. *Journal of Experimental Botany* **57**, 2825–2835.

Wuyts N, Lognay G, Verscheure M, Marlier M, De Waele D, Swennen R, 2007. Potential physical and chemical barriers to infection by the burrowing nematode *Radopholus similis* in roots of susceptible and resistant banana (*Musa* spp.). *Plant Pathology* **56**, 878–890.

Wyss U, 1997. Root parasitic nematodes: an overview. In: Fenoll C, Grundler FMW, Ohl SA, eds. *Cellular and Molecular Aspects of Plant–Nematode Interactions*. Dordrecht: Kluwer, pp. 5–22.

Zalucki MP, Malcolm SB, Paine TD, Hanlon CC, Brower LP, Clark AR, 2001. It's the first bite that counts: survival of first-instar monarchs on milkweeds. *Austral Ecology* **26**, 547–555.

Zhu YJ, Agbayani R, Jackson MC, Tang CS, Moore PH, 2004. Expression of the grapevine stilbene synthase gene VST1 in papaya provides increased resistance against diseases caused by *Phytophthora palmivora*. *Planta* **220**, 241–250.

Zhu-Salzman K, Luthe DS, Felton GW, 2008. Arthropod-inducible proteins: broad spectrum defenses against multiple herbivores. *Plant Physiology* **146**, 852–858.

Zipfel C, 2008. Pattern-recognition receptors in plant innate immunity. *Current Opinion in Immunology* **20**, 10–16.

Chapter 3
Sounding the Alarm: Signaling and Communication in Plant Defense

3.1   Introduction

Each of the approximately 300,000 plant species present on Earth is attacked by a multitude of other organisms. As we saw in the previous chapter, plants have evolved an arsenal of physical and chemical weapons and barriers to tackle this assault. The question for plants is, how and when to unleash these defenses? In other words, how do plants perceive these threats, and further, how do they translate that perception into the appropriate action? In the sections below, we look at the signaling pathways responsible for regulating plant defenses to the range of different attackers encountered by plants. We examine the signals involved in regulating defenses within the tissue being attacked, and the signals involved in alerting other plant tissues not yet under attack. We also look at the evidence for communicating these threats to neighboring plants.

3.2   Signaling in plant–pathogen interactions

3.2.1   Introduction

As described in Chapter 2 (Section 2.2.3), plants sense microbial pathogens via the perception of pathogen-associated molecular patterns (PAMPs) by pattern-recognition receptors, which are located on the cell surface. This represents the first level of recognition and is referred to as PAMP-triggered immunity (PTI). This is a basal resistance and acts as the first barrier to pathogen infection in plants. A variety of intracellular responses are associated with PTI or basal resistance, including rapid ion fluxes across the plasma membrane, activation of calcium-dependent mitogen-activated protein (MAP) kinases, production of reactive oxygen species (ROS), rapid expression of defense-related genes, and cell wall reinforcement (Thomma et al., 2001). Successful pathogens have evolved to overcome basal resistance, in many cases by suppressing PTI using virulence effectors. Plants, in turn, have evolved resistance (R) proteins that can recognize these pathogen effectors either directly or indirectly. This represents the second line of defense and is called effector-triggered immunity (ETI) (Zipfel, 2008). ETI is often accompanied by the hypersensitive response (HR;

Plant Defense, First Edition, by Dale Walters   © 2011 by Blackwell Publishing Ltd.

see Section 2.2.3.2) and is associated with additional locally induced defense responses that block further pathogen growth (De Wit, 2007). The signals involved in activating these defense responses are examined next.

3.2.2   *Local signaling and basal resistance*

The main molecules signaling the activation of defense-related genes have repeatedly been shown to be salicylic acid (SA), jasmonic acid (JA), and ethylene (ET). These plant hormones have multiple roles in plants and have important functions in the regulation of plant growth and development. Thus, SA is involved in the regulation of cell growth (Vanacker *et al.*, 2001), flowering, and thermogenesis (Raskin, 1992; Shah & Klessig, 1999), while JA plays an important role in a range of developmental processes, including seed germination, tuber formation, senescence, and flower development (Wasternack, 2007). Finally, ET plays an essential role in flowering, fruit ripening and abscission, and leaf senescence (Dugardeyn & Van der Straeten, 2008). In many cases, pathogen infection is associated with enhanced production of these hormones and a concomitant activation of defense-related genes (Maleck *et al.*, 2000; Schenk *et al.*, 2000). Further, application of these compounds to plants often leads to an enhancement of resistance to pathogens (Reignault & Walters, 2007). The involvement of SA, JA, and ET in basal resistance depends on the host–pathogen interaction, and indeed, it has been proposed that the signaling pathway induced is dependent on the mode of attack of the pathogen, that is, whether it is a biotroph or a necrotroph (see below).

3.2.2.1   *SA signaling*

In plants, SA can be made from phenylalanine via the enzyme phenylalanine ammonia lyase (PAL), although the main pathway for its biosynthesis is now known to be through chorismate via isochorismate synthase (Figure 3.1) (Wildermuth *et al.*, 2001; Ogawa *et al.*, 2005). The SA pathway is linked mainly to resistance to biotrophic pathogens (Glazebrook, 2005) and is involved in the expression of *PR* genes encoding proteins with antimicrobial or other defensive functions during basal resistance, as well as in the production of ROS and the induction of the HR in gene-for-gene resistance (Ton *et al.*, 2006). The importance of SA in plant defense became apparent with the use of tobacco and *Arabidopsis* transformed to constitutively express the bacterial gene *NahG*. This gene codes for the enzyme salicylate hydroxylase, which converts SA into inactive catechol. NahG plants show enhanced susceptibility to a broad range of oomycete, fungal, bacterial, and viral pathogens (Delaney *et al.*, 1994; Kachroo *et al.*, 2000). In addition, the *Arabidopsis* mutants *sid1*, *sid2*, and *pad4* are unable to accumulate SA following pathogen infection and as a result show enhanced susceptibility to the bacterial pathogen *Pseudomonas syringae* pv. *tomato* DC3000 and the oomycete pathogen *Hyaloperonospora parasitica* (Zhou *et al.*, 1998; Nawrath & Métraux, 1999; Wildermuth *et al.*, 2001). These results confirm the importance of SA in basal resistance against these different types of pathogens.

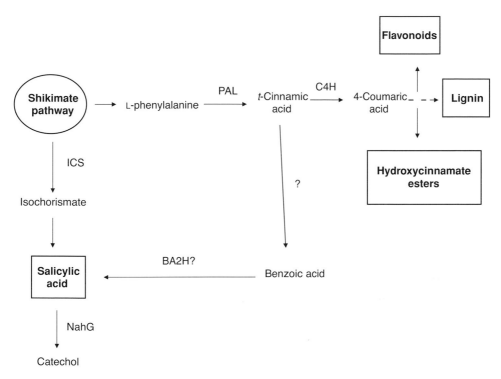

Figure 3.1 Pathways for the biosynthetic origin of the salicylic acid involved in plant defense signaling. PAL, L-phenylalanine ammonia-lyase; C4H, cinnamate 4-hydroxylase; BA2H, benzoate 2-hydroxylase; ICS, isochorismate synthase. (Reproduced from Suzuki *et al.* (2004), with permission of Oxford University Press.)

3.2.2.2 *JA signaling*

JA belongs to a family of active molecules, the oxylipins, which play diverse roles in plant biology as signal molecules and antimicrobial compounds (Weber, 2002). Many oxylipins are generated by the action of enzymes known as lipoxygenases (LOX). These enzymes add oxygen to pentadiene fatty acids, such as linoleic and linolenic acids. The resulting products, fatty acid hydroperoxides, are, in turn, subject to a diverse array of modifications, leading to the formation of large numbers of oxylipins (Figure 3.2) (Blée, 2002). Two carbon atoms in the polyunsaturated 18-carbon fatty acids, linoleic and linolenic acids, are targets for the addition of oxygen—C_9 and C_{13}. The addition of oxygen to C_{13} on these substrates by 13-LOX leads to the formation of the jasmonate family, including JA. Work using *Arabidopsis thaliana* mutants affected in the biosynthesis of, or responsiveness to, JA has shown that JA and ET are involved in resistance to necrotrophic pathogens (Glazebrook, 2005). JA-dependent defenses contribute to basal resistance to a range of pathogens, including *A. brassicicola*, *Botrytis cinerea*, and *Erwinia carotovora* pv. *carotovora* (Ton *et al.*, 2006).

Terrestrial plants use derivatives of both C_{18} (known as octadecanoids) and C_{16} (known as hexadecanoids) fatty acids as developmental or defense hormones (Weber, 2002). In contrast, animals use derivatives of C_{20} fatty acids (known as eicosanoids) to regulate

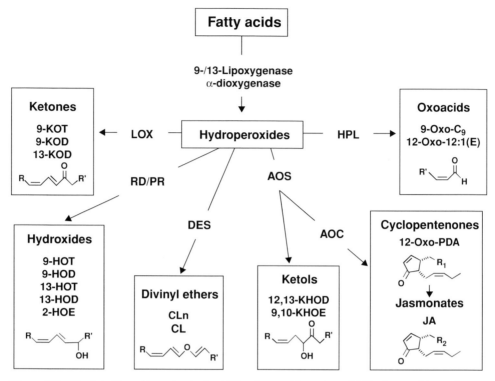

Figure 3.2 Oxylipin biosynthesis in plants. AOC, allene oxide cyclase; AOS, allene oxide synthase; DES, divinyl ether synthase; HPL, hydroperoxide lyase; LOX, lipoxygenase; PR, peroxygenase; RD, reductase. (Modified from Vellosillo *et al.* (2007), with permission.)

cell differentiation and immune responses (Funk, 2001). Interestingly, marine red algae, which emerged as an independent lineage early in the evolution of eukaryotes (Baldauf *et al.*, 2000), contain both octadecanoids and eicosanoids. Moreover, when challenged by pathogen extracts, these algae accumulated both octadecanoids and eicosanoids (Bouarab *et al.*, 2004). These results suggest that in early eukaryotes both animal-like (eicosanoids) and plant-like (octadecanoids) oxylipins are essential components of innate immunity mechanisms.

3.2.2.3 *ET signaling*

The first committed and in most cases the rate-limiting step in ET biosynthesis involves the conversion of *S*-adenosylmethionine (AdoMet) to 1-aminocyclopropane-1-carboxylic acid (ACC) by the enzyme ACC synthase (Figure 3.3). Compared to SA and JA, the role of ET in plant resistance to pathogens seems more ambiguous. Although there are cases where ET is clearly involved in disease resistance, there are others where it is associated with symptom development. Nevertheless, mutants of *Arabidopsis* have been used to demonstrate that ET-dependent defenses contribute to basal resistance to, for example, *P. syringae* pv. *syringae* and *Xanthomonas campestris* pv. *vesicatoria* (Ton *et al.*, 2006). Similarly, soybean mutants

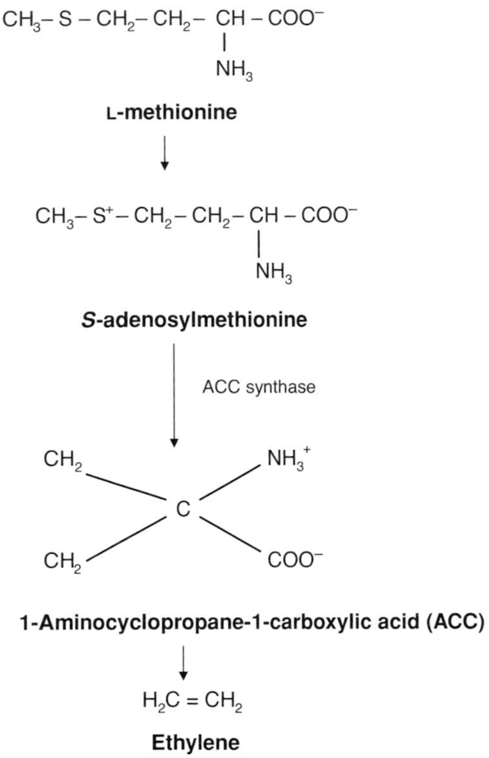

$$CH_3 - S - CH_2 - CH_2 - CH - COO^-$$
$$|$$
$$NH_3$$

L-methionine

↓

$$CH_3 - S^+ - CH_2 - CH_2 - CH - COO^-$$
$$|$$
$$NH_3$$

S-adenosylmethionine

ACC synthase

$$CH_2 \qquad\qquad NH_3^+$$
$$\diagdown\qquad\diagup$$
$$C$$
$$\diagup\qquad\diagdown$$
$$CH_2 \qquad\qquad COO^-$$

1-Aminocyclopropane-1-carboxylic acid (ACC)

↓

$$H_2C = CH_2$$

Ethylene

Figure 3.3   Ethylene biosynthesis in plants.

with reduced sensitivity to ET were found to develop more severe symptoms in response to infection with *Septoria glycines* and *Rhizoctonia solani* (Hoffman *et al.*, 1999). Recent work has suggested that ET, together with SA and nitric oxide (NO), is involved in the development of the HR in some plants (see Box 3.1).

Box 3.1   Signaling during the HR—involvement of ET, SA, and NO

As we saw in Chapter 2, the HR is a cell death phenomenon associated with localized resistance to pathogens. It is initiated following host recognition of the pathogen-encoded avirulence (*avr*) gene product by a plant resistance (R) gene. Regulation of the HR involves ROS (mainly the superoxide anion and hydrogen peroxide, H_2O_2) and NO (Lamb & Dixon, 1997; Delledonne *et al.*, 1998). NO production has been demonstrated during the HR in a number of systems. For example, a transient but significant burst of NO was observed in epidermal cells of barley challenged with the powdery mildew fungus, *Blumeria graminis* f.sp. *hordei* just before their HR-associated collapse (Prats *et al.*, 2005). In this work, the use of an NO scavenger delayed cell death, suggesting that NO was involved in the HR. NO has also been shown to activate proteases that are known to contribute to HR-type cell death (e.g., Belenghi *et al.*, 2003) and to induce the expression of defense-associated genes (Grun *et al.*, 2006). There is also evidence that NO can affect the generation of ROS, which is closely associated with the HR (Zeier *et al.*, 2004).

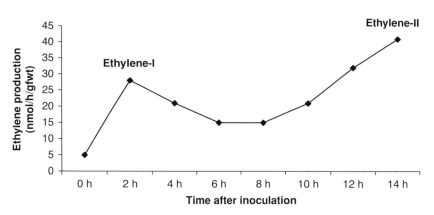

Figure A   Ethylene production in tobacco following challenge with *P. syringae* pv. *phaseolicola*. Note the two peaks in ethylene production, designated "ethylene-I" and "ethylene-II." (Redrawn from Mur *et al.* (2008), with permission.)

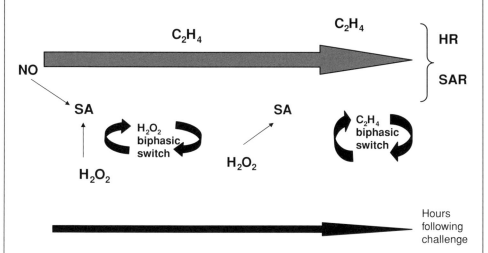

Figure B   A possible model for biphasic signal regulation during the HR elicited in tobacco by challenge with *P. syringae* pv. *phaseolicola*. During an initial phase, NO and H_2O_2 act to initiate SA biosynthesis, where SA and NO act to initiate an "H_2O_2 biphasic switch." This could initially suppress both SA and H_2O_2 generation, but subsequently acts to potentiate a second phase of H_2O_2 generation. This, in turn, increases SA biosynthesis, which could act with NO to initiate the "C_2H_4 biphasic switch" to potentiate ethylene production. These (and other) signals contribute to initiation of the HR and SAR. (From Mur *et al.* (2009), with permission.)

One feature of the generation of H_2O_2 during the HR is that it follows a biphasic pattern. The initial increase in H_2O_2 is induced by PAMPs (see Section 2.2.3 in Chapter 2), while the second, more persistent rise results from the interaction between a pathogen *avr* gene product and a plant *R* gene. The kinetics and amplitude of the second rise in H_2O_2 influence the rate of cell death during the HR and, in turn, the effectiveness of the associated defenses (Mur *et al.*, 2000). An important aspect of the initial rise in H_2O_2 is the initiation of SA synthesis that potentiates the second increase in H_2O_2, thereby increasing the effectiveness of the HR (Lamb & Dixon, 1997). Some workers consider that this potentiation mechanism is similar to priming for induced resistance (Mur *et al.*,

2009) (see Section 3.2.3.1 below). Recently, Mur *et al.* (2008) demonstrated biphasic ET production during the HR in tobacco elicited by the bacterial pathogen *P. syringae* pv. *phaseolicola* (Figure A). As with the biphasic pattern of H_2O_2 production, the pattern of ET generation reflected PAMP- and AVR-dependent elicitation events. They also found that NO was a crucial component in the SA-potentiation mechanism. On the basis of these changes in H_2O_2, SA, ET, and NO, Mur *et al.* (2009) proposed a model for biphasic signal regulation during the HR in the tobacco *P. syringae.* pv. *phaseolicola* system (Figure B). Here, in the first phase, NO and H_2O_2 act to initiate SA biosynthesis, while SA and NO initiate an "H_2O_2 biphasic switch." Initially, this switch could suppress both SA and the generation of H_2O_2, but thereafter, it would potentiate a second phase of H_2O_2 generation. This would then lead to increased SA formation, which, together with NO, would initiate an "ET biphasic switch" to potentiate ET production. These signals, along with others, would contribute to the development of the HR and systemic acquired resistance (SAR; see Section 3.2.3.1 below).

3.2.2.4   Signaling involving other plant hormones

It is well known that plant growth and development are regulated through signaling pathways governed by hormones such as auxins, cytokinins, and abscisic acid (ABA), and it should be no surprise therefore that the regulation of these pathways can help to determine the outcome of a plant–pathogen interaction. For example, evidence suggests that part of the invading strategy of pathogens might be the stimulation of auxin signaling, since increased auxin levels could suppress plant defenses (Chen *et al.*, 2007), and at the same time alter other aspects of host physiology to favor pathogen growth and establishment (Lopez *et al.*, 2008). However, plants can repress auxin signaling as a component of basal resistance. Thus, Navarro *et al.* (2006) have shown that down-regulation of auxin responses is the result, in part, of activation of microRNAs that repress auxin signaling.

Cytokinins have also been implicated in the outcome of host-pathogen interactions, for example, in the establishment of biotrophic fungal pathogens and the formation of green islands (Walters & McRoberts, 2006; Walters *et al.*, 2008). Less information is available on cytokinins and defense against pathogens. Nevertheless, there are reports suggesting that cytokinins can increase disease resistance. Thus, cytokinins have been shown to induce programmed cell death and expression of *PR1*, contributing to resistance against biotrophic pathogens (Sano *et al.*, 1994; Mlejnek & Prochazka, 2002; Carimi *et al.*, 2003).

There have also been reports of altered ABA levels in plant–pathogen interactions, and, depending on the pathogen, ABA can enhance resistance or enhance susceptibility (Mauch-Mani & Mauch, 2005). For example, ABA enhances resistance of *Arabidopsis* to the oomycete *Pythium irregulare*, while ABA-deficient and ABA-insensitive mutants are more susceptible to the pathogen (Adie *et al.*, 2007). ABA has also been shown to protect *Arabidopsis* against the necrotrophic fungi *A. brassicicola* and *Plectosphaerella cucumerina*, while the ABA-deficient and ABA-insensitive mutants display enhanced susceptibility to these pathogens (Ton & Mauch-Mani, 2004; Adie *et al.*, 2007). More recently, ABA signaling was shown to regulate resistance to the powdery mildew pathogen *Golovinomyces cichoracearum* in *edr1 Arabidopsis* mutants (Wawrzynska *et al.*, 2008). However, as indicated above, ABA can also enhance susceptibility. For example, ABA application enhances susceptibility of *Arabidopsis* to *Fusarium oxysporum* (Anderson *et al.*, 2004) and susceptibility of tomato (*Solanum lycopersicum*) to *Botrytis cineria* (Audenaert *et al.*, 2002). Similarly, ABA-deficient *Arabidopsis* mutants were more resistant to *F. oxysporum* (Anderson *et al.*, 2004) and *B. cinerea* (Adie *et al.*, 2007). A number of mechanisms have been proposed to underpin these actions of ABA on plant susceptibility/resistance, including stimulation of callose deposition (Ton & Mauch-Mani, 2004), induction of stomatal closure (Melotto *et al.*, 2006), suppression of ROS generation (Asselbergh *et al.*, 2007, 2008b), and suppression of SA- and JA/ET-dependent basal defenses (Asselbergh *et al.*, 2008a).

From the above, it seems that the outcome of any changes in ABA levels or signaling in plants is dependent on the particular plant–pathogen interaction, rather than pathogen lifestyle (in other words, whether it is a biotroph or a necrotroph), although the timing of infection appears to be crucial in its role in defense against pathogens (Asselbergh *et al.*, 2008a). Although the multifarious regulatory effects of ABA on plant defense preclude any clear-cut model of its effects on resistance, it has been suggested that ABA might be involved in controlling a global shift between plant responses to abiotic and biotic stresses. It is suggested that the increase in ABA levels in response to, for example, water stress, shifts the priority away from pathogen resistance toward abiotic stress tolerance. This would be consistent with ABA acting as a pathogen virulence factor, suppressing SA- and JA/ET-dependent defense responses. Equally, a reduction in ABA levels, as in ABA deficiency, would lead to reduced tolerance of abiotic stress and greatly increased disease resistance (Asselbergh *et al.*, 2008a).

3.2.3   *Systemic signaling and induced resistance*

3.2.3.1   *Induced resistance*

When a plant is attacked by a pathogen, in addition to reacting locally, it can also mount a systemic response, establishing an enhanced defensive capacity in parts distant from the site of primary attack. This phenomenon, known as induced resistance, was first described more than 100 years ago and in the intervening years has burgeoned into a major area of plant biology (Hammerschmidt, 2007). A classic example of this systemically induced resistance is activated following primary infection with a necrotizing pathogen (one which causes cell death), and renders distant, uninfected parts of the plant more resistant to a broad spectrum of pathogens, for example, viruses, bacteria, and fungi (Kuć, 1982). This type of induced resistance is known as systemic acquired resistance (SAR) and is characterized by a restriction of pathogen growth and a suppression of disease symptoms compared to noninduced plants infected by the same pathogen. Another, phenotypically similar form of induced resistance, is activated on colonization of plant roots by selected strains of non-pathogenic rhizobacteria and is referred to as induced systemic resistance (ISR) (Van Loon *et al.*, 1998). The signaling involved in SAR and ISR is complex, with pathogen-induced SAR requiring SA and rhizobacteria-mediated ISR requiring, in most cases, JA and ET.

3.2.3.2   *Signaling during SAR*

The onset of SAR is associated with an early increase in SA levels both locally at the site of infection and systemically in distant tissues. It is also associated with the immediate expression of a specific set of so-called *SAR* genes, some of which encode PR proteins (Pieterse & Van Loon, 2007). If SA is applied exogenously to plants, SAR is induced and is accompanied by activation of the same set of genes. Compelling evidence that SA plays an important role in the onset of SAR comes from work using transgenic NahG plants (see Section 3.2.2.1 above). These plants, which are unable to accumulate SA, cannot establish SAR and do not show *PR* gene activation following pathogen attack (Gaffney *et al.*, 1993; Delaney *et al.*, 1994). Although SA can be transported in the plant, it is not the translocated signal (Vernooij *et al.*, 1994). So if SA, although crucial for the establishment of SAR, is not the long-distance signal, what is? It appears that SA is transported in plants mostly as its methyl ester, methyl salicylic acid (MeSA) (Heil & Ton, 2008). Two enzymes are known to control the balance between SA and MeSA in tobacco: SA methyltransferase1 (SAMT1), which catalyzes the formation of MeSA from SA, and SA-binding protein2 (SABP2), which converts biologically inactive MeSA into active SA. In some revealing experiments using tobacco in which the *SABP2* gene was silenced, there was a reduced level of basal resistance and plants were unable to express SAR, suggesting that SABP2 might generate SA from MeSA during SAR (Kumar & Klessig, 2003). The final piece in the jigsaw, at least in tobacco, came from some elegant work demonstrating that the long-distance SAR signal is MeSA (Figure 3.4) (Park *et al.*, 2007). Whether MeSA is the long-distance signal for SAR in other plant species is not known. Indeed, work on *Arabidopsis* suggests that long-distance SAR signaling is mediated by jasmonates. JA, and not SA, was found to accumulate rapidly in phloem exudates of leaves challenged with an avirulent strain of *P. syringae*, while in systemically responding leaves, transcripts associated with JA biosynthesis were up-regulated within 4 hours and JA levels increased transiently. Moreover, foliar application of JA induced SAR, and *Arabidopsis* mutants defective in JA biosynthesis or response were unable to establish SAR (Truman *et al.*, 2007). These workers suggested that, in *Arabidopsis*, JA might be involved in the early initiation of systemic resistance, with SA contributing to subsequent events in establishment of systemic immunity.

Irrespective of the nature of the mobile signal, SA in the distant tissues induces defense proteins locally. The mechanisms responsible for bringing about SAR-related gene expression are complex and are examined in Box 3.2.

3.2.3.3 Signaling during ISR

A number of strains of bacteria in the rhizosphere possess the ability to stimulate plant growth and as such are known as plant growth-promoting rhizobacteria (PGPR). Although growth-promoting effects result mainly from the suppression of soilborne pathogens and other deleterious microbes, some strains of PGPR are capable of reducing disease on shoots via ISR (Van Loon *et al.*, 1998). In *Arabidopsis*, ISR activated by the rhizobacteria *Pseudomonas fluorescens* WCS417r and *Pseudomonas putida* WCS358r has been shown to function independently of SA and *PR* gene expression (e.g., Pieterse *et al.*, 1996). Importantly, SA-independent ISR has also been demonstrated in tobacco and tomato (Yan *et al.*, 2002; Zhang *et al.*, 2002). Instead of SA, ISR signaling requires JA and ET, because *Arabidopsis* mutants impaired in their ability to respond to either of these two plant hormones are unable to express ISR (Pieterse *et al.*, 1998; Ton *et al.*, 2001). Rather surprisingly, ISR, like SA-dependent SAR (Box 3.2), was also found to be dependent on NPR1 (Pieterse *et al.*, 1998). It might be expected that because SAR is associated with NPR1-dependent *PR* gene expression and ISR is not, ISR and SAR would be mutually exclusive. However, simultaneous activation of ISR and SAR leads to enhanced defensive activity compared

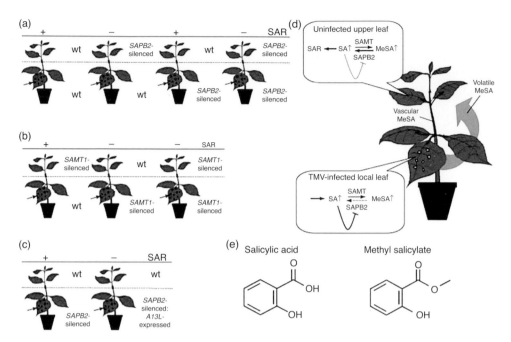

Figure 3.4 The role of MeSA as a transported signal in SAR. (a) Grafting experiments with plants in which the *SA-binding protein 2* (*SABP2*) gene is silenced demonstrated that an *SABP2*-silenced rootstock, which cannot convert MeSA into SA, was still capable of generating a long-distance signal to activate SAR in an upper wild-type scion. (b) Conversely, an *SA methyltransferase1* (*SAMT1*)-silenced rootstock, in which MeSA production is impaired, was not capable of generating this long-distance signal. (c) Furthermore, a recombinant SABP2 protein was constructed that is affected in SA-induced feedback inhibition of MeSA esterase activity. When this so-called A13L protein was expressed in *SABP2*-silenced rootstocks, no SAR signal was transported to the upper scion. Hence, SA-induced feedback inhibition of SABP2 is necessary to generate enough MeSA for long-distance transport to distal plant parts, where it is converted by SABP2 into active SA that triggers SAR. Together with the observation that exposure of lower plant parts to MeSA in gas-tight chambers can trigger SAR in nonexposed, upper leaves, these findings demonstrate that MeSA functions as the crucial long-distance signal in tobacco. +, expression of SAR; −, no expression of SAR. (d) The interaction of *SAMT1* with *SABP2* (which is inhibited by free SA) regulates the synthesis of MeSA from SA in the infected tissue, and the release of SA from the transported MeSA in the systemic tissue. Although MeSA can be transported within the plant, it is possible that under natural circumstances a combination of airborne and vascular transport of MeSA is responsible for long-distance regulation of SAR. (e) Structures of SA and methyl salicylate (MeSA). (Reproduced from Heil & Ton (2008), with permission of Elsevier.)

to either type of induced resistance alone (Van Wees *et al.*, 2000). This suggests an important role for NPR1 in regulating and coordinating different hormone-dependent defense pathways (Pieterse & Van Loon, 2007).

3.2.3.4 Priming

In plants expressing induced resistance, defense mechanisms can be activated directly following the resistance-inducing treatment (i.e., infection by a necrotizing pathogen or

Box 3.2   A closer look at SAR signaling

Following infection by a necrotizing pathogen, SA accumulates and the cellular redox environment changes, with cells attaining a more reducing environment. Transduction of SA to activate *PR* gene expression and SAR requires the function of a regulatory protein known as NPR1 (NON-EXPRESSOR OF PATHOGENESIS-RELATED1, also known as NIM1—NONIMMUNITY1). During induction of SAR and the associated redox changes, NPR1 oligomers are reduced to an active monomeric state and the monomeric NPR1 is then translocated into the nucleus. Once in the nucleus, monomeric NPR1 interacts with members of the TGACG motif binding (TGA) class of transcription factors such as TGA2. This binding of monomeric NPR1 to TGAs increases the ability of these transcription factors to bind to the cognate *cis*-element on DNA, leading to activation of *PR-1* gene expression (Figure C) (Loake & Grant, 2007; Pieterse & Van Loon, 2007). A further factor implicated in the regulation of SAR is the SNI1 protein (SUPPRESSOR OF npr1-1, INDUCIBLE 1), which appears to be a negative regulator in the establishment of SAR. SNI1 probably acts as a transcriptional repressor of SAR that can be counteracted by NPR1 after activation of the SA-dependent SAR pathway. Thereafter, the transcription factors of the TGA family would be allowed to activate the expression of *PR-1* and other genes involved in establishing SAR (Ton *et al.*, 2006).

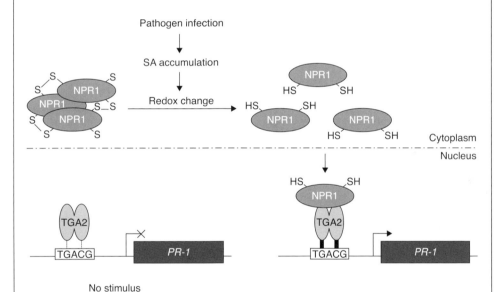

Figure C   A model for SAR illustrating the role of SA-mediated redox changes, NPR1, and TGA transcription factors in SAR-related gene expression. In noninduced cells, oxidized NPR1 is present as inactive oligomers that remain in the cytosol. Binding of TGAs to the cognate SA-responsive promoter elements (TGACG) does not activate *PR-1* gene expression (indicated by "no stimulus"). Upon infection by a necrotizing pathogen, SA accumulates and plant cells attain a more reducing environment, possibly due to the accumulation of antioxidants. Under these conditions, NPR1 oligomers are reduced to an active monomeric state through reduction of intermolecular disulfide bonds. Monomeric NPR1 is translocated into the nucleus, where it interacts with TGAs, such as TGA2. The binding of NPR1 to TGAs increases the DNA-binding activity of these transcription factors to the cognate *cis*-element (black boxes), resulting in the activation of *PR-1* gene expression. (Reproduced from Pieterse & van Loon (2007), with permission of Blackwell Publishing Ltd.)

application of a chemical), or following challenge of the induced tissues by a pathogen. The latter is known as priming and occurs in both SAR and ISR.

The defense mechanisms triggered during priming also operate in noninduced plants, albeit at a lower level or at a later stage of pathogen attack (Hammerschmidt, 1999). Thus, cucumber plants attacked by the fungal pathogen *Colletotrichum lagenarium* formed papillae containing callose and lignin at sites of attempted pathogen penetration. In cucumber plants expressing induced resistance, more papillae were formed at sites of attempted penetration, papilla formation was more rapid, and the papillae contained more callose and lignin (Hammerschmidt & Kuć, 1982). Similarly, in carnation plants expressing ISR against *F. oxysporum* f.sp. *dianthi*, no change in phytoalexin levels could be detected in induced plants before pathogen challenge. However, on subsequent inoculation, phytoalexin levels in ISR-expressing plants rose significantly faster than on challenge of noninduced plants (Van Peer *et al.*, 1991). It appears therefore that the greater defensive capacity of induced resistance is based on enhanced expression of basal defense mechanisms (Ton *et al.*, 2006). If this is so, it suggests that genotypes differing in genetically determined basal resistance could differ in the extent to which they express induced resistance. Indeed, there are reports of cultivars of the same crop plant differing in their expression of induced resistance (see Walters *et al.*, 2005).

So, what about signaling during priming for induced resistance? Well, since priming can occur in both SAR and ISR, it is perhaps no surprise that SA, JA, and ET have all been implicated in regulating the priming of defense responses. Thus, pretreatment of parsley cells with SA, JA, or their functional analogs resulted in priming of a range of defenses, including accumulation of ROS, secretion of cell wall phenolics, phytoalexin accumulation, and accumulation of transcripts for various defense-associated genes (e.g., Kauss *et al.*, 1992, 1994; Katz *et al.*, 1998). Interestingly, *Arabidopsis* plants pretreated with ET were primed to SA-induced *PR-1* gene expression, suggesting that ET potentiates defense mechanisms that contribute to SAR (Lawton *et al.*, 1994). In fact, transgenic tobacco plants with an insensitivity to ET displayed a reduced SAR response (Knoester *et al.*, 2001).

As we have seen above, ISR requires an intact response to both JA and ET (Pieterse *et al.*, 1998). However, what became clear subsequently was that the production of these hormones did not change during ISR, suggesting that ISR is based on an enhanced sensitivity to JA and ET, rather than changes in their levels (Pieterse *et al.*, 2000). This being so, then plants expressing ISR would be primed to respond more quickly and intensely to JA and ET produced following pathogen attack (Pieterse *et al.*, 2006).

As part of a screening process to identify *Arabidopsis* mutants with enhanced resistance to the bacterial pathogen *P. syringae* pv. *tomato* (strain DC3000), a mutant with constitutively enhanced resistance to this pathogen and the fungus *Erysiphe cichoracearum* was discovered (Frye & Innes, 1998). This mutant, *edr1* (enhanced disease resistance 1), differed from other disease-resistant mutants of *Arabidopsis* in not exhibiting constitutive expression of *PR-1* and *PR-2*, although expression of both of these genes was increased following pathogen challenge. The *edr1* mutant also showed stronger expression of defense responses such as the HR and callose deposition following pathogen attack. These results suggest that the EDR1 protein is involved in priming in *Arabidopsis*. Interestingly, EDR1 appears to be a mitogen-activated protein kinase kinase kinase (MAPKKK), which mediates disease resistance via SA-inducible defenses (Frye *et al.*, 2001).

We have already seen that induced resistance seems to be based on the enhanced expression of defense mechanisms used in basal resistance. However, the defenses used might differ in plants primed for SAR or ISR. For example, when *Arabidopsis* plants expressing SAR were challenged with *P. syringae* pv. *tomato*, there was primed expression of SA-inducible *PR* genes, while *Arabidopsis* plants expressing ISR displayed primed expression of the JA-inducible *AtVSP* gene (Van Wees *et al.*, 1999). One might speculate therefore that SAR is achieved through primed expression of SA-dependent basal defenses, while ISR (e.g., that mediated by WCS417r) is achieved through primed expression of JA/ET-dependent basal resistance (Pieterse *et al.*, 2006) (Figure 3.5). If this is the case, then as with basal resistance, where SA-dependent defenses appear to be effective against biotrophic pathogens, while JA/ET-dependent defenses appear to be effective against necrotrophic

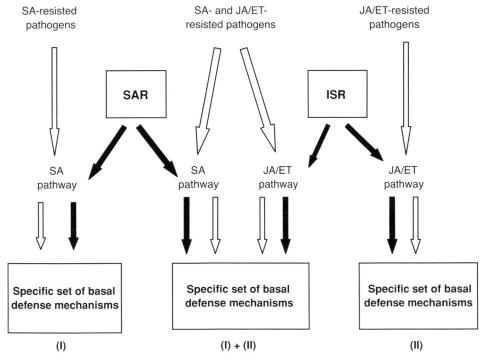

Figure 3.5 A model explaining pathogen-induced SAR and PGPR-mediated ISR as a primed expression of basal defense mechanisms. SA-dependent basal defenses (I) are primed in SAR-induced plants. As a result, infection of tissues expressing SAR triggers a more rapid and stronger activation of SA-dependent defense mechanisms, resulting in an effective protection against pathogens that are resisted through SA-dependent basal resistance, for example, *P. parasitica* and turnip crinkle virus. On the other hand, pathogen infection of plants pretreated with ISR-inducing WCS417r bacteria results in priming of JA/ET-dependent basal defenses (II). Here, ISR-expressing tissues exhibit a more rapid and stronger expression of JA/ET-dependent defenses on infection, resulting in effective protection against pathogens that are resisted through JA/ET-dependent basal resistance, for example, *A. brassicicola*. Pathogens that are resisted through a combination of SA- and JA/ET-dependent basal resistance, for example, *P. syringae* and *X. campestris*, are sensitive to both SAR and ISR (I) + (II). (From Ton *et al.* (2006), with kind permission of Springer Science and Business Media.)

pathogens, SAR and ISR might also be effective against different pathogens. Interestingly, the biotrophs *Peronospora parasitica* and turnip crinkle virus, which are both resisted mainly via SA-dependent basal defenses, are strongly inhibited by expression of SAR, but weakly resisted by expression of ISR. In contrast, the necrotroph *A. brassicicola*, which is resisted through JA/ET-dependent basal defenses, was inhibited in plants expressing ISR, but not in plants expressing SAR (Ton *et al.*, 2002).

3.2.4   *Volatile signaling*

Although most research on long-distance signaling has concentrated on movement of putative signals within the plant vascular system, the fact that plants can produce volatile organic compounds (VOCs), including methylated forms of SA and JA, suggests that volatile signaling could be important in plant defense. Indeed, methyl salicylic acid (MeSA) and methyl jasmonic acid (MeJA) are not only volatile but also potent inducers of defense responses in plants (Shulaev *et al.*, 1997; Park *et al.*, 2007). Such volatile signals might move both within the plant and between plants.

We have already seen (Chapter 2, Section 2.5.3.4) that plants produce green leaf volatiles (GLVs) following damage. Relatively little information is available on GLVs and plant–pathogen interactions. Nevertheless, work on *Arabidopsis* with genetically modified GLV biosynthesis showed that GLV production was significantly increased following inoculation with *B. cinerea*, compared to wild-type controls (Shiojiri *et al.*, 2006). Further, genetically modified plants were less susceptible to *B. cinerea* than wild-type plants, suggesting that the increased GLV production by these plants might be at least partly responsible for the observed protection. Although the mechanisms by which the GLVs might bring about such protection remain to be elucidated, it is known that GLVs can up-regulate defense-related genes in *Arabidopsis* (Kishimoto *et al.*, 2005), and can alter phytoalexin accumulation in both cotton and *Arabidopsis* (Zeringue, 1992; Kishimoto *et al.*, 2006). In addition, GLVs are known to possess antifungal properties and could possibly exert a direct effect on the fungus *in planta* (Matsui, 2006). GLVs can also exert bactericidal activity. For example, lipid-derived volatiles, including (Z)-3-hexenol and (E)-2-hexenal, were released from *Phaseolus vulgaris* reacting hypersensitively to the bacterial pathogen *P. syringae* pv. *phaseolicola*, and both volatiles were found to be bactericidal (Croft *et al.*, 1993).

As mentioned above, MeSA is a powerful inducer of plant defenses. There is also evidence that plants attacked by pathogens emit MeSA. For example, peanut plants infected with the white mold fungus *Sclerotium rolfsii* emitted volatile MeSA. Moreover, volatile MeSA was shown to inhibit growth of the fungus on solid culture, hinting at the possibility that it might play a defensive role against this pathogen *in planta* (Cardoza *et al.*, 2002). In incompatible interactions between tobacco and avirulent strains of *P. syringae* (*P. syringae* pv. *maculicola* ES4326 and *P. syringae* pv. *tomato* DC3000), MeSA was released rapidly within 12 hours, just prior to the formation of necrotic lesions, and emission started to decline once necrotic lesions had fully formed, 36 hours after inoculation (Figure 3.6a) (Huang *et al.*, 2003). In contrast, less MeSA was emitted during the compatible interaction of tobacco with *P. syringae* pv. *tabaci*, with peak emission occurring 60 hours after inoculation, much later than in the incompatible interactions (Figure 3.6b). Interestingly, the volatile blend released in the incompatible interactions was complex, containing a number of different compounds, including (E)-β-ocimene, linalool, two unidentified sesquiterpenes,

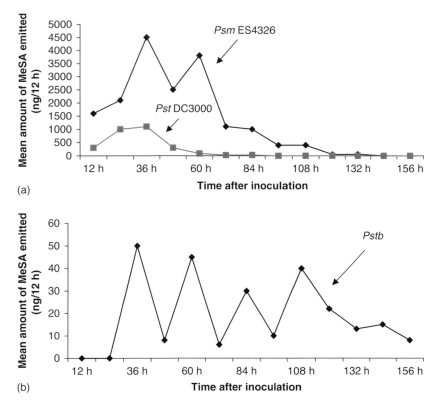

Figure 3.6 Emissions of methyl salicylate (MeSA) from tobacco plants inoculated with *P. syringae* pv. *maculicola* (*Psm*) ES4326 or *P. syringae* pv. *tomato* (*Pst*) DC3000 (a), or with *P. syringae* pv. *tabaci* (*Pstb*) (b). Volatiles were collected continuously in two samples, 12-hour light and 12-hour dark periods, over the collection period of 7 days. (Redrawn from Huang *et al.* (2003), with permission of Springer Science + Business Media.)

and MeSA. The volatile blend released during the compatible interaction was much less complex, comprising just MeSA and two unidentified sesquiterpenes. The correlation of volatile emissions with formation of the HR lesions suggests that either they play a direct role in defense against the pathogen or they are by-products of defenses elicited against *P. syringae* (Huang *et al.*, 2003).

3.3   Signaling in plant–nematode interactions

3.3.1   Introduction

As described in Chapter 2, sedentary endoparasitic nematodes, such as cyst nematodes and root-knot nematodes, are obligate biotrophs that induce the formation of complex feeding sites within the roots of their plant host. Because a living host cell is required to act as an initial feeding cell, plant parasitic nematodes, in common with many other biotrophic pathogens, have evolved mechanisms to avoid or negate both constitutive and inducible plant defenses in susceptible hosts.

A number of nematode resistance genes (Nem-*R*) have been isolated from plants and all confer resistance to sedentary endoparasites. The tomato genes *Mi-1* and *Hero A* confer broad-spectrum resistance to several root-knot nematode species and several pathotypes of two potato cyst nematode species, while the potato genes *Gpa2* and *Gro1-4* confer resistance to a narrow range of pathotypes of a single potato cyst nematode species (Williamson & Kumar, 2006). Most *R*-genes are expressed constitutively in the plant at low levels, consistent with the hypothesis that they code for surveillance proteins responsible for detecting pathogen effector molecules and initiating an effective defensive response. The Nem-*R* genes *Mi-1* and *Hero A* are expressed constitutively, although *Hero A* is up-regulated in roots in response to potato cyst nematode infection (Martinez de Ilarduya & Kaloshian, 2001; Sobczak *et al.*, 2005). Information on the signaling involved in activation of defenses against nematodes is limited, although some data are available on the role of SA and JA in signaling in plant–nematode interactions.

3.3.2   SA signaling

In order to examine defense signaling in resistance to root-knot nematodes mediated by *Mi-1*, tomato was transformed with the bacterial salicylate hydroxylase gene, *NahG*. Transformed plants cannot accumulate SA, and in roots of such plants, resistance to the root-knot nematode *Meloidogyne javanica* was greatly decreased. Further, resistance could be restored by application of the SA analog, acibenzolar-*S*-methyl (ASM), indicating that the resistance conferred by *Mi-1* is, at least in part, mediated via SA signaling (Branch *et al.*, 2004). Later work using the beet cyst nematode, *Heterodera schachtii*, found that a number of SA-deficient *Arabidopsis* mutants were more susceptible to infection than the wild-type (Wubben *et al.*, 2008). A similar result was obtained using SA-insensitive mutants impaired in the SA signaling component NPR1, while the *npr1*-suppressor mutation, *sni1*, showed decreased susceptibility. These results led the authors to suggest that SA acts via NPR1 to inhibit nematode parasitism and, in turn, this is negatively regulated by SNI1. Intriguingly, cyst nematode infection of wild-type *Arabidopsis* led to increased levels of SA in shoots, but not in roots. This was paralleled by induction of the SAR marker gene *PR-1* in shoots, indicating that cyst nematode infection induces SAR in *Arabidopsis*. Moreover, the inverse correlation between basal expression of *PR-1* in roots and susceptibility to cyst nematode suggests that successful cyst nematode infection might involve local suppression of SA signaling in roots (Wubben *et al.*, 2008).

3.3.3   JA signaling

In contrast to the situation for SA, the involvement of JA in defense signaling in plant–nematode interactions is, at present, less clear. Thus, although exogenous application of MeJA to various plants was found to increase resistance to nematodes (Soriano *et al.*, 2004a, 2004b; Cooper *et al.*, 2005), JA signaling was shown not to be involved in *Mi-1*-mediated resistance (Bhattarai *et al.*, 2008). On the other hand, roots of the tomato mutant *jai1*, which is defective in the JA-Ile receptor component COI1, exhibited increased resistance to root-knot nematodes, suggesting that COI1-mediated JA signaling increases susceptibility to root-knot nematodes. Interestingly, Bhattarai *et al.* (2008) suggest that, analogous to the situation with the bacterial pathogen *P. syringae*, susceptibility to root-knot nematodes mediated by COI1 might not be caused by JA-Ile produced by the plant, but by a structurally similar molecule produced by the nematodes to suppress SA signaling. In contrast to this, work on the interaction between soybean roots and cyst nematodes has suggested that JA signaling is suppressed during fully established compatible interactions (Ithal *et al.*, 2007a, 2007b).

3.4   Signaling in plant–insect herbivore interactions

3.4.1   Introduction

Wounding following insect attack leads to the rapid (minutes to several hours) induction of a series of events, including the generation and release of specific signals, the subsequent perception and transduction of those signals, and, finally, activation of wound-related defenses (León *et al.*, 2001). Insect damage will activate defenses both locally in the damaged tissue and systemically in distant tissues. Following on from the early events that take place within the first few minutes after attack, secondary signals are generated, and these, in turn, lead to the further propagation of defense responses both locally and systemically. Before we look at the signaling involved in triggering plant responses to insect herbivory, let us first examine how a plant manages to distinguish between wounding caused by insect damage and by abiotic agents such as hail or wind.

Much evidence suggests that the oral secretions (OS) of insect herbivores contain fatty acid–amino acid conjugates (FACs), which are capable to eliciting various defense responses. The first FAC identified was *N*-17-hydroxylinolenoyl-L-glutamine (volicitin) (Figure 3.7). This was found in the OS of the army beetworm, *Spodoptera exigua*, and was shown to induce the release of volatiles in maize seedlings (Alborn *et al.*, 1997). Other examples of OS that contain FACs include the OS from *Manduca sexta*, which, when applied to mechanical wounds on *Nicotiana attenuata*, elicited bursts of JA and ET, which were greater than those elicited by mechanical wounding alone (Kahl *et al.*, 2000), and also elicited accumulation of trypsin proteinase inhibitor (TPI) (Zavala *et al.*, 2004). Moreover, if the FACs were removed from the OS of *M. sexta* by ion-exchange chromatography, the defense responses of *N. attenuata* were not elicited, while adding synthetic FACs to the FAC-free OS restored the responses (Alborn *et al.*, 1997; Halitschke *et al.*, 2003). These findings show that perception by the plant of FACs in an insect herbivore's OS activates antiherbivore defenses. However, not all plants respond to FACs, and indeed, other elicitors are present in the OS from insect herbivores (Howe & Jander, 2008). OS from the fall armyworm, *Spodoptera frugiperda*, contains peptides known as inceptins (Figure 3.7). These are produced in the insect gut by proteolytic digestion of chloroplast ATP synthase (Schmelz *et al.*, 2006, 2007). Other work has shown that OS from orthopteran insects such as the American bird grasshopper, *Shistocera americana*, contains sulfated fatty acids called caeliferins (Figure 3.7), which are capable of eliciting the release of volatile terpenes from maize (Alborn *et al.*, 2007).

Plants can also respond to oviposition, when herbivorous insects lay eggs on plants. These responses include formation of necrotic tissue, production of ovicidal substances,

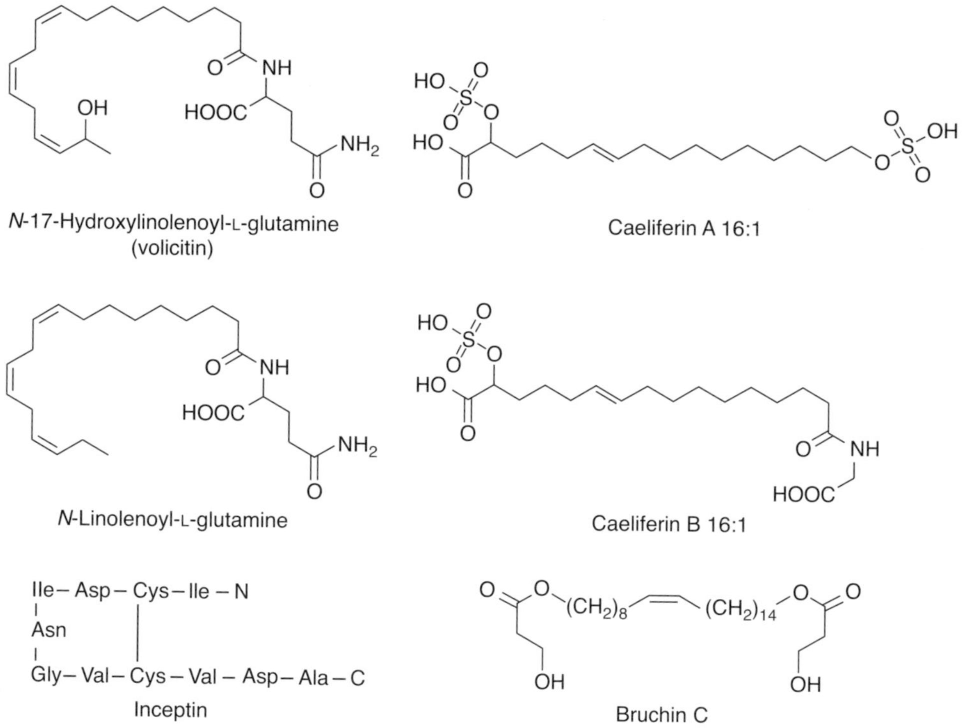

N-17-Hydroxylinolenoyl-L-glutamine
(volicitin)

Caeliferin A 16:1

N-Linolenoyl-L-glutamine

Caeliferin B 16:1

Ile—Asp—Cys—Ile—N
|
Asn
|
Gly—Val—Cys—Val—Asp—Ala—C
Inceptin

Bruchin C

Figure 3.7   Insect-derived elicitors of host plant defense responses. Volicitin and
N-linolenoyl-L-glutamine belong to the family of FACs found in oral secretions of lepidopteran larvae.
The fatty acid and amino acid moieties of FACs are derived from the insect and host plant,
respectively. Inceptin, which was also isolated from oral secretions of lepidopteran larvae, is produced
by proteolytic degradation of chloroplast ATP synthase in the insect gut. FACs and inceptin therefore
represent examples of elicitors that are produced by modification of plant compounds within the
insect. Caeliferins were isolated from the oral secretions of the grasshopper *Schistocerca americana*.
Bruchins, which are produced by pea weevils and related bruchids, stimulate neoplastic growth at the
site of weevil oviposition. (Reproduced from Howe & Jander (2008), with permission.)

and emission of volatile signals (Hilker & Meiners, 2006). Interestingly, oviposition fluid
is known to contain compounds capable of eliciting such plant responses. For example, the
oviposition fluid of the pea weevil, *Bruchus pisorum*, was found to contain long-chain diols
called bruchins (Figure 3.7), which, following application to pea pods, elicit tumor-like
growth beneath the egg, preventing larvae from entering the pod (Doss *et al.*, 2000).

Recently, it was suggested that, in line with the classification of PAMPs and MAMPs (see
Section 2.2.3), these herbivore-derived elicitors should be known as herbivore-associated
molecular patterns (HAMPs) (Mithöfer & Boland, 2008). Yet, little is known about how
these elicitors are perceived by plants. Nevertheless, work using radiolabeled volicitin in
maize leaves has revealed the existence of a receptor-like binding site for this elicitor (Truitt
et al., 2004).

In addition to the recognition of chemical elicitors, plants might also be able to discriminate between insect herbivory and mechanical damage by gauging the quantity and

quality of tissue damage, and the pattern of feeding, for example, frequency and period of feeding (Howe & Jander, 2008; Wu & Baldwin, 2009). Thus, by using a mechanical device that mimics tissue damage caused by caterpillars, wounding of lima bean leaves elicited a pattern of volatile emission that was qualitatively similar to that induced by caterpillar herbivory (Mithöfer *et al.*, 2005).

3.4.2   Local signaling

Following perception of HAMPs or insect damage by the plant, a series of events are set in motion, involving local and systemic signaling. Some of these changes can be detected within minutes of insect attack. For example, feeding by larvae of *Spodoptera littoralis* on lima bean leaves led to a large and rapid membrane depolarization in cells next to damaged cells (Maffei *et al.*, 2004). These changes in membrane depolarization were associated with a rapid influx of calcium into cells close to the site of insect feeding. Other work showed that treating lima bean leaves with a calcium chelator to block the transient increase in cytosolic calcium prevented the induction of defense genes following attack by the two-spotted spider mite, *Tetranychus urticae* (Arimura *et al.*, 2000b). These changes in intracellular calcium are significant, given the role of calcium as a second messenger capable of mediating the transduction of external and internal signals into physiological changes (Lecourieux *et al.*, 2006).

Plants subject to insect attack also produce ROS within the first few hours. This includes production of the superoxide anion in the damaged tissue and H_2O_2 both locally and systemically (Orozco-Cardenas & Ryan, 1999; Maffei *et al.*, 2007). The superoxide anion is generated rapidly, with maximal production within minutes, while production of H_2O_2 peaks at between 4 and 6 hours and declines thereafter. As with ROS elicited by pathogen attack, nicotinamide adenine nucleotide phosphate (reduced) (NADPH) oxidases appear to be the main source of ROS following insect herbivory and wounding (Orozco-Cardenas & Ryan, 1999; Sagi *et al.*, 2004). Interestingly therefore, inhibiting NADPH oxidases in tomato was found to block the increased expression of several herbivore defense genes following various treatments, including wounding (Orozco-Cardenas & Ryan, 1999).

Although much is known about the involvement of NO in plant resistance to pathogens (see Box 3.1 above and Hong *et al.*, 2008), little information is available on NO in wounding, and as yet, there have been no reports of NO induced by insect herbivory (Wu & Baldwin, 2009).

3.4.2.1   JA signaling

There is much information demonstrating the central role played by the jasmonate family of signaling compounds in regulating plant defense against herbivorous arthropods (Howe & Jander, 2008). Wounding caused by chewing insects or mechanical damage is associated with rapid accumulation of JA at the wound site. Concentrations of JA and its lipid precursor linolenic acid increase in tomato leaves after wounding and JA concentrations are also increased by application of various elicitors of defense responses (Pena-Cortés *et al.*, 1993, 1995). Moreover, application of exogenous JA to plants induces synthesis of proteinase inhibitors (PIs) (Farmer & Ryan, 1992). Very strong evidence for JA as a positive regulator of defense is that mutants defective in the synthesis or perception of JA are compromised in their resistance to insect herbivores. For example, a tomato mutant

Figure 3.8   Death and protection of *Arabidopsis* mutants from *Bradysia* larvae attack. (a) A mixed population of wild-type and *fad3-2 fad7-2 fad8* plants were grown in a net enclosure populated with 20–25 adult *Bradysia* flies. Each day, eight pots were sprayed with 0.8 mL of water and seven pots were sprayed with 0.8 mL of a dilute aqueous solution of MeJA. Data from two experiments are shown in which the concentration of MeJA used was either 0.001% (*open symbols*) or 0.01% (*solid symbols*). The graph shows the percentage survival of 117 wild-type plants (○ and ● both treatments), 73 mutant plants treated with water (□, ■), and 73 mutant plants treated with MeJA (△, ▲). (b and c) For clarity, wild-type and mutant seeds were sown in pots in two rows but were otherwise treated as described above. The photographed plants correspond to day 50 in (a). (b) Compared with wild-type controls (on the left), mutant plants (on the right) sprayed with water show extensive damage 20 days after the introduction of adult *Bradysia* flies. Some leaves on mutant plants have been almost completely eaten. Wilting of other leaves was attributed to damage to the petiole or to the roots of the plants. (c) Mutant plants sprayed with 0.01% methyl jasmonate (on the right) remained healthy and vigorous within the same environment. (From McConn *et al.* (1997), with permission of the National Academy of Sciences, USA.)

deficient in JA synthesis failed to accumulate PI in response to wounding and displayed increased susceptibility to attack by tobacco hornworm larvae than wild-type plants (Howe *et al.*, 1996). Similarly, an *Arabidopsis* mutant, deficient in the JA precursor linolenic acid, showed extremely high mortality (~80%) from attack by larvae of the fungal gnat *Bradysia impatiens*, while application of exogenous MeJA reduced mortality to approximately 12% (Figure 3.8) (McConn *et al.*, 1997). Interestingly, *B. impatiens* is not a serious pest of wild-type *Arabidopsis*, and these results demonstrate quite clearly that genetic removal of the JA pathway can transform a nonhost plant into a suitable host for insect herbivores (Browse & Howe, 2008). Conversely, constitutive activation of the JA pathway in plants results in enhanced resistance to insect attackers. Thus, constitutive activation of JA signaling in an *Arabidopsis* mutant led to activation of JA-dependent defense genes and enhanced resistance to the green peach aphid, *Myzus persicae* (Ellis *et al.*, 2002).

Precursors of JA and JA derivatives are also capable of eliciting antiherbivore defenses. For example, the *Arabidopsis* mutant *opr3* cannot convert the JA precursor

12-oxo-phytodienoic acid (OPDA) to JA. Nevertheless, resistance to *B. impatiens* is not compromised in this mutant, suggesting a signaling function for OPDA or earlier intermediates in the JA biosynthetic pathway (Stintzi *et al.*, 2001). JA can also be conjugated to amino acids, especially isoleucine (Ile), to form JA-isoleucine (JA-Ile). Silencing the enzyme responsible for synthesis of JA-Ile in *Arabidopsis* impaired resistance to *M. sexta*, while exogenous application of JA-Ile to such plants restored resistance (Kang *et al.*, 2006), suggesting that JA-Ile is an important oxylipin signal in plant–herbivore interactions.

The work described briefly above highlights the central role played by jasmonates in signaling in defense against herbivorous arthropods. But how are jasmonates perceived by the plant, and how do they exert their effects on plant defense? Work in the past few years has begun to provide answers to these questions, and a picture of considerable complexity is emerging (see Box 3.3).

3.4.2.2   *ET signaling*

ET emission following attack by herbivorous insects is a widespread occurrence (von Dahl & Baldwin, 2007). In *N. attenuata*, ET emission following attack by *M. sexta* was elicited by FACs in the insect's OS (Halitschke *et al.*, 2001), while in cowpea, *Vigna unguiculata*, ET emission was elicited by inceptins in the OS of *S. frugiperda* (Schmelz *et al.*, 2006). This ET can affect the expression of defensive proteins and secondary metabolites, and in maize, ET synthesis and perception are required for effective defense against *S. frugiperda* (Harfouche *et al.*, 2006). However, it appears that the defenses regulated by ET are also highly dependent on JA. Indeed, in many cases, ET also requires JA induction for the effective expression of defenses (von Dahl & Baldwin, 2007). Thus, the elicitation of *PI* gene expression in tomato requires the synergistic action of both ET and JA (O'Donnell *et al.*, 1996), while a similar synergism between ET and JA was observed for the elicitation of terpenoid emission in the interaction between maize and *S. exigua* (Schmelz *et al.*, 2003).

3.4.2.3   *SA signaling*

A number of studies have shown that phloem-feeding insects, such as aphids and silverleaf whitefly, induce SA-dependent responses (e.g., De Vos *et al.*, 2005; Zarate *et al.*, 2007). Nevertheless, the role of SA in defense against aphids is controversial. Thus, although it has been shown to be important for defense against aphids in tomato (Li *et al.*, 2006), it was found to exert neutral or negative effects on the growth of aphids and whiteflies (Pegadaraju *et al.*, 2005; Zarate *et al.*, 2007). Little information exists on SA in interactions between plants and chewing insects, although weak accumulation of SA was detected in *N. attenuata* attacked by *M. sexta* (cited in Wu & Baldwin, 2009). Whether SA plays any role in defense against chewing insects is unclear.

3.4.2.4   *Specificity and regulation of jasmonate-based defenses*

As we have seen in Chapter 2, insects use different types of feeding behavior (e.g., chewing or piercing-sucking) to obtain food from the host plant. In other words, they belong to different feeding guilds. A reasonable question to ask, therefore, is whether jasmonate-regulated responses are specific for different insect attackers. The answer has much to do with the

---

Box 3.3   Jasmonate perception by plants

Coronatine is a phytotoxin produced by the plant pathogenic bacterium *P. syringae*. It has a structure similar to JA-Ile and has been shown to mimic the effects of MeJA in plants. Mutant plants that are defective in the *CORONATINE INSENSITIVE 1 (COI1)* gene have impaired jasmonate signaling and show a greatly enhanced susceptibility to insect attack (e.g., Mewis *et al.*, 2005; Paschold *et al.*, 2007). *COI1* codes for an F-box protein—a protein motif of approximately 50 amino acids that functions as a site of protein–protein interaction. In fact, COI1 is the F-box component of a multiprotein complex called SCF^COI1, an ubiquitin ligase complex named after the main components, Skp I, Cullin, and F-box protein. SCF^COI1 binds substrates for ubiquitin-mediated proteolysis. This suggests that COI1 negatively regulates jasmonate-induced responses by degrading repressors of jasmonate responses (Devoto & Turner, 2005). These repressors have recently been identified as jasmonate ZIM-domain (JAZ) proteins (Chini *et al.*, 2007; Thines *et al.*, 2007). So, how does all of this fit together? Well, it appears that following insect attack or wounding, JA is converted to JA-Ile, the latter specifically promoting the interaction between COI1 and JAZ proteins. This, in turn, results in the degradation of JAZ proteins via the ubiquitin/26S proteasome pathway (Figure D).

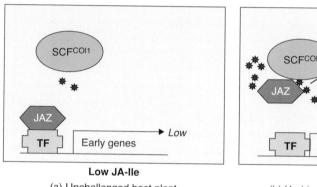

Low JA-Ile

(a) Unchallenged host plant

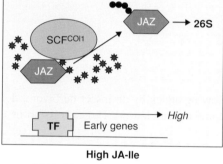

High JA-Ile

(b) Herbivore-challenged host plant

Figure D   The JAZ repressor model of jasmonate signaling. (a) Low intracellular levels of jasmonoyl-isoleucine (JA-Ile) (*stars*) favor the accumulation of jasmonate ZIM domain (JAZ) proteins, which bind to and repress the activity of transcription factors (TF) such as MYC2 that positively regulate jasmonate-responsive early genes. (b) Tissue injury, such as that caused by chewing insects, results in rapid accumulation of JA-Ile. These high levels of JA-Ile promote SCF^COI1-mediated ubiquitination (*black circles*) and subsequent degradation of JAZ repressor proteins via the 26S proteasome (26S), resulting in the derepression of transcription factors and the expression of early response genes. (Reproduced from Howe & Jander (2008), with permission.)

---

feeding guilds to which the insects belong. In general, insects from different feeding guilds tend to elicit distinct, but overlapping, patterns of gene expression, whereas those from the same feeding guild evoke similar responses (Heidel & Baldwin, 2004; Reymond *et al.*, 2004; De Vos *et al.*, 2005). For example, whereas caterpillars of the crucifer specialist *Pieris rapae* and generalist *S. littoralis* elicited nearly identical patterns of gene expression in *Arabidopsis* (Reymond *et al.*, 2004), 61% of the genes elicited by *P. rapae* (chewing) and *Frankliniella occidentalis* (piercing-sucking) exhibited an expression pattern specific to the attacker (De Vos *et al.*, 2005). Interestingly, gene expression profiles elicited by phloem-feeding insects are very different from those induced by herbivores from other feeding guilds. They tend to cause little damage as they access the host's vascular tissue to obtain amino acids and carbohydrates. Aphids feeding on *Arabidopsis* have been shown to elicit the expression of SA-regulated genes, but only weak expression of JA/ET-regulated genes (Moran & Thompson, 2001; De Vos *et al.*, 2005). Another phloem-feeding insect, the silver-leaf whitefly, *Bemisia tabaci*, did not activate JA-responsive genes in *Arabidopsis*, probably reflecting the fact that they cause even less damage to the host than to aphids (Zarate *et al.*, 2007). *B. tabaci* also induced the expression of SA-responsive genes, and since SA is known to antagonize the action of JA (see Chapter 4), it is possible that in the *B. tabaci*/*Arabidopsis* interaction the expression of JA-responsive genes was suppressed by SA.

JA synthesis and JA-regulated defenses can be activated not just by insect herbivory but also by a range of other inductive signals, including abiotic stress, developmental signals, and infection by necrotrophic pathogens (see Section 3.2.2.2). However, the activity of the jasmonate pathway is likely to be greatly amplified during plant–insect interactions, for example, via FACs and other substances in the insect's OS (Kessler & Baldwin, 2002; Schmelz *et al.*, 2006). Ultimately, because most insect herbivores trigger JA synthesis in plants, it is possible that JA-regulated defenses might have evolved as a means of deterring a wide range of different herbivores (Howe & Jander, 2008).

3.4.3   Systemic signaling

Following attack by an insect herbivore, defenses are expressed rapidly both in damaged and undamaged leaves. The expression of defenses in as yet undamaged leaves requires the transmission of a signal from damaged leaves. The nature of this signal has been the topic of considerable research over the past three decades, and a number of candidates were proposed for this role, including electric and hydraulic signals (Malone *et al.*, 1994; Stankovic & Davies, 1997) and a polypeptide called systemin.

3.4.3.1 Systemin

Work on tomato in the early 1970s demonstrated that wounding following insect attack led to induction of PIs in both damaged and undamaged leaves (Green & Ryan, 1972). Subsequent work showed that application of extracts from wounded leaves to excised, but otherwise undamaged leaves, also resulted in PI induction, hinting at the existence of a mobile signal (Ryan, 1974). Indeed, detailed studies identified an 18-amino acid peptide, systemin, which is released at wound sites following damage by chewing insects. In fact, following wounding, systemin is formed from a 200-amino acid polypeptide precursor, prosystemin, which is present constitutively in leaves at low levels.

The importance of systemin in the wound response can be gauged from the fact that transgenic plants in which the expression of prosystemin is blocked not only show severe disruption to their systemic wound responses but also exhibit increased susceptibility to attack by the tobacco hornworm (McGurl *et al.*, 1992; Orozco-Cardenas *et al.*, 1993). Moreover, transgenic plants overexpressing the *prosystemin* gene produced PIs constitutively at high levels (McGurl *et al.*, 1994). Subsequent research identified a systemin receptor in tomato, a 160-kDa plasma membrane-bound protein (Scheer & Ryan, 1999, 2002). The binding of systemin to this receptor on the cell surface is likely to lead to a signal transduction event that activates a series of processes within the cell. However, although systemin plays a central role in triggering the wound response in tomato, peptides with sequence similarity to systemin and prosystemin are present in very few plant species, including potato, black nightshade (*Solanum nigrum*), and black pepper (*Capsicum annum*) (Constabel & Ryan, 1998). Perhaps surprisingly, the systemin homolog in black nightshade does not mediate PI expression or other direct defense responses (Schmidt & Baldwin, 2006). It is possible, therefore, that different plants might use distinct mechanisms to regulate JA synthesis and systemic responses to herbivory (Howe, 2004).

Systemin also possesses phloem mobility and was thought to be the mobile signal in the wound response of tomato. However, some elegant grafting experiments revealed that although systemin is involved in propagating the local wound-induced JA signal, it is not the long-distance signal. Rather, JA or a JA-elicited signal moves to systemic leaves and induces PI expression (Li *et al.*, 2002; Lee & Howe, 2003).

3.4.3.2   JA signaling

Given the importance of JA in local signaling in response to insect attack, it is not unreasonable to ask whether JA is also involved in systemic signaling. Indeed, there is evidence to support such a role for JA. For example, application of JA to single tomato leaf elicited PI gene expression in untreated, distal leaves on the same plant (Farmer *et al.*, 1992) (Figure 3.9). Further, radiolabeled JA applied to a single leaf moved to roots and young leaves, but not into mature leaves, suggesting the transport of exogenous JA occurs via the phloem (Zhang & Baldwin, 1997). What about evidence that JA levels increase in undamaged leaves on wounded plants? Such evidence exists, and, for example, mechanical wounding of first leaves of *Vicia faba*, resulted in significant increases in JA levels in undamaged second leaves (Walters *et al.*, 2006). However, in many cases, wounding leads to small increases in

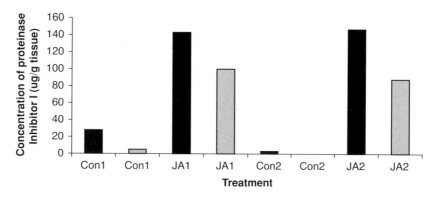

Figure 3.9 Systemic induction of proteinase inhibitor I in young tomato plants in response to treatment with JA. JA was applied to a single lower leaf of young, two-leaf stage tomato plants. Inhibitor I accumulation was assayed 24 hours later in both the lower (treated; black bars) and upper (untreated; grey bars) leaves. Controls were treated with water. Con1 and JA1, treatments applied via pin pricks; Con2 and JA2, treatments applied to adaxial surface of lower leaf. (Data reproduced from Farmer *et al.* (1992), with permission.)

JA levels in distal, undamaged tissues, and whether any increase can be detected seems to depend on the quantity and quality of the leaf damage inflicted. Thus, although mechanical wounding of tobacco did not result in a systemic increase in JA, application of caterpillar regurgitant to wounds did (von Dahl & Baldwin, 2004).

The role of JA in the systemic wound response was examined in various grafting experiments in tomato (Li *et al.*, 2002). For example, wild-type scions of tomato grafted onto damaged rootstocks of a jasmonate-insensitive mutant (*jasmonate-insensitive 1; jai1*) were still capable of expressing PIs, while *jai1* scions grafted onto damaged rootstocks of the wild-type failed to activate PI expression. These results demonstrated that responsiveness to JA was not strictly required for the production of the systemic signal in damaged leaves, instead, was required for the perception of that signal in systemic leaves. Importantly, rootstocks of plants defective in systemin perception were unable to generate the transported signal, while scions of such plants were able to perceive the signal from wild-type rootstocks and activate PI expression. Finally, grafts between wild-type and JA-deficient mutants showed that JA synthesis was required for production of the systemic signal in wounded leaves, but was not required in undamaged systemic leaves. Taken together, these data suggest that JA or a related compound, and not systemin, acts as a mobile signal in the wound response (Li *et al.*, 2002). Indeed, a number of characteristics make jasmonates ideal candidates for systemic, long-distance signals. Thus, they induce the transient expression of genes necessary for their own synthesis, leading to amplification of the signal during transport, and they have also been demonstrated to promote their own transport (Stenzel *et al.*, 2003; Thorpe *et al.*, 2007).

3.4.3.3   *Within leaf signaling*

So far in this chapter, we have looked at local signaling in the wounded leaf and systemic signaling to distal, as yet undamaged, plant parts. But what about signaling between damaged and undamaged regions within the attacked leaf? Using tomato, Howe *et al.* (1996) found that if the leaf tip was wounded, greater PI gene expression was observed, not in the wounded tip region, but at the base of the leaf. Later work using *N. attenuata* showed that if the insect's OS was applied on the wounded region of a leaf, there was the rapid movement of a short-distance signal to undamaged parts of the leaf, where it activated MAPK signaling and JA biosynthesis (Wu *et al.*, 2007). It appears therefore that, during herbivory, FACs in the insect's OS bind to a receptor, thereby inducing the production of a short-distance signal, which, in turn, generates responses in nonwounded parts of the leaf. This is examined in more detail in Box 3.4.

Box 3.4   Signaling events leading to the activation of local and systemic defense responses to herbivory

Mitogen-activated protein kinase (MAPK) signaling plays a central role in transducing extracellular stimuli into intracellular responses in all eukaryotes. Two kinases, salicylic acid-induced protein kinase (SIPK) and wound-induced protein kinase (WIPK), have been shown to be activated following wounding. Thus, an MAPK was activated both locally and systemically in wounded tomato (Stratmann & Ryan, 1997), and MMK4, a homolog of WIPK in alfalfa, was rapidly activated following wounding (Bogre *et al.*, 1997). Further, WIPK was found to be necessary for JA production after wounding and for the accumulation of PIs in tobacco (Seo *et al.*, 1999). Subsequent work by Wu *et al.* (2007) showed that application of OS from *M. sexta* to wounds in *N. attenuata* led to rapid and marked activation of MAPKs, much more so than wounding alone. This effect was shown to be due to FACs present in the *M. sexta* OS. By silencing the *SIPK* and *WIPK* genes, Wu *et al.* (2007) were also able to demonstrate the importance of these kinases in mediating wound and OS-elicited hormonal responses and transcriptional regulation of defense-related genes. Moreover, they found that after application of OS to wounds created in one part of the leaf, SIPK was activated in both wounded and unwounded parts of the leaf. This led the authors to suggest that, following herbivory by *M. sexta*, a mobile signal moves quickly to undamaged areas of the attacked leaf, activating MAPK signaling and downstream responses. It seems that a distinct signal travels to systemic leaves and triggers defense-related responses without activating MAPKs (Figure E).

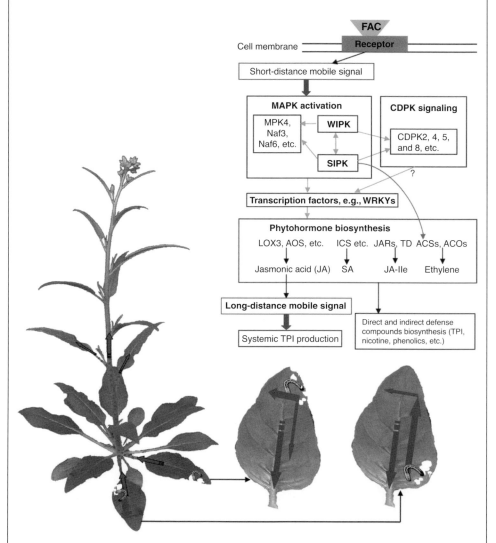

Figure E   A model summarizing how OS-elicited responses activate defenses in local and systemic leaves of *N. attenuata*. After attack from *M. sexta* larvae, FACs in the larvae's OS bind to hypothetical receptors in the cell membranes and activate a short-distance mobile signal that enhances SIPK and WIPK activity in both wounded regions and, in particular, nonwounded adjacent regions in the leaf. Afterward, activating SIPK and WIPK leads to the transcriptional regulation of other MAPKs, calcium-dependent protein kinases (CDPKs), and transcription factors, such as WRKYs. Through WRKY and other transcription factors, both kinases subsequently enhance transcript levels of genes involved in JA, SA, JA-Ile, and ethylene biosynthesis, which in turn enhance levels of JA, SA, JA-Ile, and ethylene. SIPK may also directly phosphorylate some of the protein products of these genes and thus enhance their activity. A long-distance mobile signal, such as JA or a JA-elicited substance, moves through the vascular system to distal leaves and enhances both local and systemic levels of TPI activity. Arrows within the MAPK activation box, between the MAPK activation box and both the CDPK signaling box and the transcription factors box, and between the transcription factors box and the phytohormone biosynthesis box, all represent regulation at the level of transcription. The arrow from SIPK to ACSs represents direct phosphorylation. Arrows pointing out of leaves via the petiole, and those pointing up stems, represent long-distance mobile signals. Other arrows in leaves represent short-distance mobile signals. (From Wu *et al.* (2007), with permission.)

3.4.4 *Volatile signaling*

The induction of defenses in distal leaves that lack a direct vascular connection with the attacked leaf suggests other mechanisms for signaling between leaves. As we saw in Chapter 2 (Section 2.5.3.4), plants can emit a cocktail of volatile chemicals in response to attack, including terpenoids and GLVs. In addition, both JA and SA can be methylated, and the resulting methyl-SA (MeSA) and methyl-JA (MeJA) are volatile. Plants also produce the gaseous hormone ET, which can increase the plant's response to GLVs, but is unlikely to act as a primary signal. Collectively, the volatiles emitted by plants in response to herbivore attack are known as herbivore-induced plant volatiles (HIPVs). They can be induced by elicitors present in the OS of herbivores (e.g., Schmelz *et al.*, 2006; Carroll *et al.*, 2008), and their induction is mediated by JA, ET, and SA (Boland *et al.*, 1995; Ozawa *et al.*, 2000; Horiuchi *et al.*, 2001).

Why should plants use volatile signaling when they are equipped with an efficient vascular transportation system? Well, the plant's vascular system might be sophisticated, but it does impose various restrictions. For example, directly adjacent leaves on a plant usually lack a direct vascular connection, and yet, insect herbivores often move between adjacent leaves. In trees, spatially neighboring leaves often originate from different branches, with different vascular connections. Moreover, signals produced in an attacked leaf are unlikely to move into more mature leaves, which have become net exporters of assimilate. In contrast, volatile signals can overcome such restrictions and are likely to reach distal leaves more quickly than signals moving through the vascular system (Heil & Ton, 2008).

Release of HIPV can provide a direct defensive benefit to the plant, for example, by deterring further oviposition by the insect, or an indirect benefit by attracting predators (De Moraes *et al.*, 2001; Kessler & Baldwin, 2001). Many carnivorous species are attracted by HIPV, including parasitoids, predatory mites, and beetles, and many can discriminate between volatiles induced by different species of herbivores (Dicke, 1999). Indeed, the composition of HIPV can be specific to the herbivore species, and moreover, many carnivorous arthropods can learn to discriminate between different HIPV (Smid *et al.*, 2007; Rasmann & Turlings, 2008). Applying JA to lima bean plants elicits a volatile cocktail that is similar to that induced by spider mite feeding, and amazingly, predatory mites that feed on these spider mites can discriminate between the mite-induced and JA-induced volatiles (Dicke *et al.*, 1999). An important difference in the composition of the two volatile cocktails is the lack of MeSA in that induced by JA, and it appears that this is largely responsible for the differential attraction of the predatory mites (De Boer & Dicke, 2004).

Various laboratory studies have shown that plant perception of volatile signals is associated with changes in expression of defense-related genes (Arimura *et al.*, 2000a, 2000b; Paschold *et al.*, 2006). Exposure to HIPV has also been shown to result in increased production of defense-related chemicals such as terpenoids, phenolics, and PIs (Farmer & Ryan, 1990; Tscharntke *et al.*, 2001; Ruther & Kleier, 2005). Work on sagebush under field conditions showed that clipped plants emitted a pulse of an epimer of MeJA, which induced the defense-related enzyme polyphenol oxidase in neighboring plants. Such plants suffered less damage from herbivorous arthropods than plants situated next to unclipped control plants (Karban *et al.*, 2000). Work using lima bean plants showed that herbivore damage induced the release of a volatile cocktail, which included the GLV (3Z)-hex-3-enyl acetate. These volatiles elicited the secretion of extrafloral nectar (EFN) in neighboring plants and also led to the attraction of greater numbers of predatory insects (ants and wasps). Such plants suffered less herbivore damage and exhibited increased production of leaves and flowers than control plants (Kost & Heil, 2006). However, since neighboring plants are likely to be in competition for resources, giving one's competitors an advantage would seem to be a problem. This aspect of volatile signaling is discussed further in Chapter 5.

Interestingly, HIPVs display diurnal or nocturnal rhythms that seem to be coordinated with the habits of the parasitoids that are attracted to them. Thus, daytime exposure of caterpillars of *Mythimna separata* to volatiles, emitted from host plants in the dark, caused the insects to behave as if they were in the dark, while if they were exposed to volatiles produced by plants during the day, the caterpillars behaved as if they were in the light, irrespective of the amount of light available (Shiojiri *et al.*, 2006). Work on lima beans (*Phaseolus lunatus*) damaged using MecWorm, a robotic device designed to reproduce tissue damage caused by herbivore attack, showed that leaves damaged during the day emitted maximal levels of the volatiles β-ocimene and (Z)-3-hexenyl acetate. In contrast, leaves damaged during the dark period produced (Z)-3-hexenyl acetate, but only small amounts of β-ocimene, although this was followed by a burst of β-ocimene release at the onset of the light period (Figure 3.10) (Arimura *et al.*, 2008). This light-dependent emission of β-ocimene was found to be regulated by the availability of terpenoid precursors, which, in turn, was dependent on photosynthesis (Figure 3.11). The authors suggested that the early morning terpenoid emission burst, following damage suffered during the night, might serve to attract predators that are present during the early part of the day.

It appears that volatile emission can be differentially affected depending on the feeding guild of the insect attacking the plant. For example, potato plants attacked by the aphid *M. persicae* elicited the emission of a greater number of volatiles than attack by the Colorado potato beetle (*Leptinotarsa decemlineata*) (Gosset *et al.*, 2009). In aphid-attacked plants, the changes in volatile emission were associated with the accumulation of products of 9-LOX activity, which are usually associated with defense against pathogens (Howe & Schilmiller, 2002; Weber, 2002).

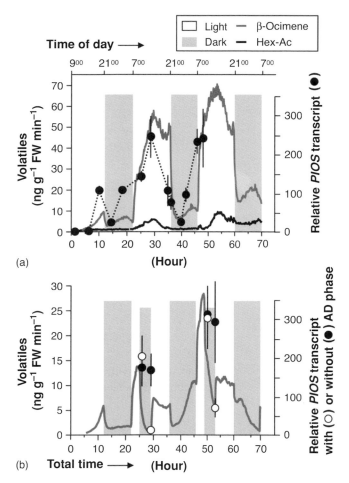

Figure 3.10 Emission of the volatiles β-ocimene and Hex-Ac and transcript levels of the β-ocimene synthase gene, *PIOS*, in damaged lima bean leaves. Damage conditions: larvae of *S. littoralis* under a 14:10 light/dark cycle (a) and under a light/dark cycle set with 4 hours of additional darkness (AD) during the second (25–29 hours) and third (49–53 hours) photoperiods (b). Time of day is shown across the top x-axis, and total elapsed time is shown on the bottom x-axis. (Reproduced from Arimura *et al.* (2008), with permission.)

Another volatile compound produced by plants is isoprene. Indeed, isoprene is the most abundant volatile compound produced by vegetation (Guenther *et al.*, 2006), although its effects on plant–insect interactions were little studied until fairly recently. Using isoprene-emitting transgenic tobacco and nonemitting azygous control plants, isoprene was shown to deter feeding by caterpillars of *M. sexta* (Laothawornkitkul *et al.*, 2008). In contrast, the performance of two lepidopteran herbivores (*P. rapae* and *Plutella xylostella*) on *Arabidopsis* was found to be unaffected by isoprene (Loivamäki *et al.*, 2008). In this work, isoprene was shown to interfere with the attraction of the parasitic wasp *Diadegma semiclausum* to HIPV, although the behavior of another parasitic wasp, *Cotesia rubecula*, was not affected.

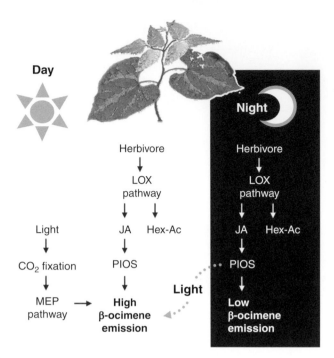

Figure 3.11   Schematic representation of the signaling and metabolic pathways required for herbivore-induced β-ocimene and Hex-Ac emissions in lima bean leaves in a daily cycle. LOX, lipoxygenase. (Reproduced from Arimura *et al.* (2008), with permission.)

3.4.5   Priming

Plants have a remarkable ability to adapt to the complex environment in which they live. One important adaptation is the ability of plants, once they have experienced a biotic stress, to enter a state of readiness for any future recurrence of the biotic encounter. This is called priming and it does not confer resistance per se, but allows the plant to mount defenses quickly when attacked. Although much attention has been focused on priming for defense against pathogens (see Section 3.2.3.4), priming also occurs in plant–herbivore interactions. For example, Engelberth *et al.* (2004) found that maize plants were primed by GLVs released from damaged plants. Here, exposure to the GLVs caused yet undamaged plants to produce JA and terpenes more intensely in response to damage by caterpillars than plants that were damaged without GLV pretreatment. Subsequent work by Ton *et al.* (2007) identified 10 defense-related genes in maize that were responsive to wounding, JA or caterpillar regurgitant. When maize was exposed to volatiles from caterpillar-infested plants, these genes were not activated directly, but six of them were primed for earlier and/or stronger induction on subsequent defense elicitation. This priming for defense-related gene expression was correlated with reduced caterpillar feeding and development (Ton *et al.*, 2007). Priming has also been shown to affect indirect defenses. Thus, EFN production in wild lima bean plants was primed by both naturally produced HIPVs and a synthetic HIPV cocktail (Figure 3.12) (Kost & Heil, 2006; Heil & Silva Bueno, 2007). The physical

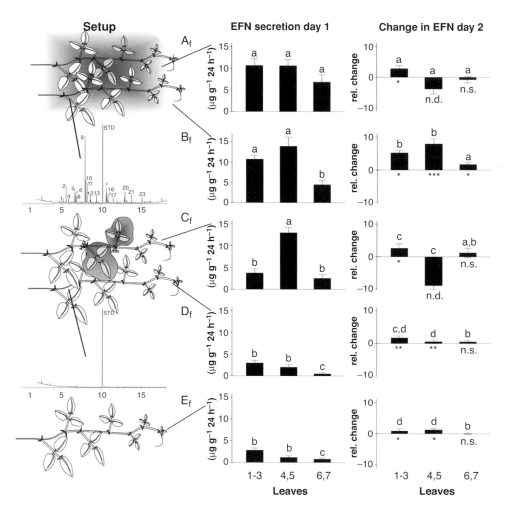

Figure 3.12 Induction and priming of EFN secretion by volatiles. (Left) The experimental setup with receiver tendrils being exposed (B_f) or not exposed (D_f) to VOCs of artificially induced emitter tendrils (A_f and C_f) and the respective GC profiles are displayed. (Center) EFN secretion of different leaf age classes (leaves 1–3, leaves 4 and 5, and leaves 6 and 7) on day 1. (Right) Change in EFN secretion on day 2 relative to day 1 (a value of $+2$ indicating a twofold higher secretion on day 2 than on day 1). Asterisks indicate significant differences in EFN secretion on day 2 as compared with day 1 (n.s., not significant; n.d., not determined). Bars marked by different letters within the same leaf age group are significantly different. Identity of volatile compounds: 1, *cis*-hexenyl acetate; 2, 2-ethylhexanol; 3, *cis*-β-ocimene; 4, *trans*-β-ocimene; 5, linalool; 6, nonanal; 8, C_{11} homoterpene; 9, *cis*-3-hexen-1-yl-butyrat; 10, MeSA; 12, unidentified; 17, β-caryophyllene; 18, *trans*-geranylacetone; 20, 4,8,12-trimethyltrideca,1,3,7,11-tetraene; 22, unidentified. (From Heil & Silva Bueno (2007), with permission of the National Academy of Sciences, USA.)

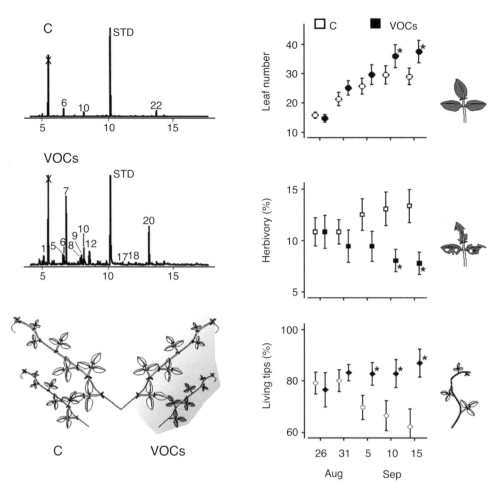

Figure 3.13   Protection of volatile-exposed plants in the field. (Left) Representative gas chromatographic profiles of headspaces of tendrils exposed to herbivore-induced emitter tendrils (VOCs) and to undamaged emitters (C). (Right) Development of leaf number, herbivory (percent missing leaf area), and percentage of living shoot tips (means ± SE) during the experiment. Asterisks indicate significant differences between C and VOC tendrils. Identity of volatile compounds compounds: 1, *cis*-hexenyl acetate; 2, 2-ethylhexanol; 3, *cis*-β-ocimene; 4, *trans*-β-ocimene; 5, linalool; 6, nonanal; 8, C_{11} homoterpene; 9, *cis*-3-hexen-1-yl-butyrat; 10, MeSA; 12, unidentified; 17, β-caryophyllene; 18, *trans*-geranylacetone; 20, 4,8,12-trimethyltrideca,1,3,7,11-tetraene; 22, unidentified. (From Heil & Silva Bueno (2007), with permission of the National Academy of Sciences, USA.)

protection provided by predatory ants attracted to the EFN is an effective defense and reduced herbivore damage significantly (Figure 3.13) (Heil & Silva Bueno, 2007).

Heil & Ton (2008) have proposed a two-step regulatory system in which airborne and vascular signals interact to yield an optimal systemic defense response. They suggest that airborne signals prime distal leaves to respond more efficiently to vascular signals or direct attack (Figure 3.14). In this system, self-priming by volatile signals prepares distal tissues

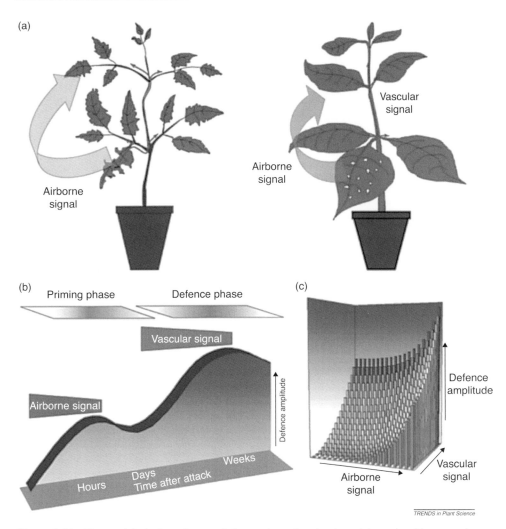

Figure 3.14 The model of a two-step regulation system of systemic resistance by airborne and vascular long-distance signals. (a) Upon local attack by insects (left) or pathogens (right), plants rapidly emit airborne signals, such as GLVs, MeJA, or MeSA, which can sensitize distal plant parts for a second vascular signal, such as JA or SA. (b) Kinetics of systemically expressed induced resistance. Rapid airborne signals (minutes—hours) trigger priming (priming phase) and relatively little defense expression. These signals are followed by a second vascular signal (days—weeks; defense phase) that boosts the induced defense expression. (c) Relationship between induced defense expression (defense amplitude) and dosage of airborne and vascular long-distance signals. Airborne signals can trigger induced defense at relatively high concentrations, but at lower concentrations, they prime for an enhanced defense induction by the vascular signal. (Reproduced from Heil & Ton (2008), with permission of Elsevier.)

for a rapid response, only allowing full activation of defenses following confirmation by the long-distance vascular signal.

Because priming involves rapid induction of defenses only on attack, it is assumed to be less costly to the plant in terms of energy and resources than constitutive defenses, which operate all the time. This is an important topic in plant defense, and this, together with the evolution of plant defenses, is covered fully in Chapter 5.

3.5   Signaling in interactions between plants and vertebrate herbivores

As discussed in Chapter 2, some physical and chemical defenses can be induced in plants following mammalian herbivory. Thus, herbivory can induce formation of spines in various species, such as acacias, and can lead to increased concentrations of terpenes and phenolics in, for example, deciduous trees and shrubs in Alaska (Bryant, 1981; Danell & Bergström, 2002). Research on induced defense responses to mammalian herbivory, and vertebrate herbivory in general, continues to lag behind similar work on pathogens and insect herbivores. Not surprisingly, therefore, little is known about signaling in plant defense against mammalian herbivores. Although it might be assumed that any leaf damage, irrespective of the cause, should lead to similar plant signaling and defense responses, such an assumption might not be justified. For example, as we have seen above, although tissue damage can elicit a defense response, this can be modified by the insect's OS. Whether saliva from mammalian herbivores alters plant defense responses is not known. Further, the type of tissue damage and its duration and timing can influence the plant defense response. Indeed, the design and development of MecWorm reflects the importance of more closely simulating the damage caused by chewing insects. The speed and pattern of leaf area removal by a mammalian herbivore might not lead to appreciable defense activation (Baldwin, 1990; Cipollini & Sipe, 2001).

3.6   Signaling in interactions between plants and parasitic plants

As discussed in Chapter 2, plant defenses against parasitic plants can be divided into preattachment, prehaustorial, and posthaustorial mechanisms. Indeed, information on resistance mechanisms has increased over the past few years, as has our understanding of the mechanisms used by parasitic plants in locating an appropriate host (e.g., Runyon et al., 2006). In contrast, very little is known about the signaling involved in host defense against parasitic plants. In fact, what little is known, based on work using different systems, presents a confusing picture. Thus, work on the compatible interaction between Arabidopsis and Orobanche ramosa found no activation of SA-dependent defenses, although the ET and JA pathways were activated (Vieira Dos Santos et al., 2003a, 2003b; Die et al., 2007). Similarly, up-regulation of JA-responsive genes, and down-regulation of SA-responsive genes, was found in compatible interactions between sorghum and Striga hermonthica (Hiraoka & Sugimoto, 2008). Interestingly, JA-responsive gene expression was weak in a more resistant variety of S. hermonthica. Work on sunflower parasitized by Orobanche cumana reported strong activation of both the SA and JA pathways in a resistant genotype, but not the ET pathway, based on expression of marker genes (Letousey *et al.*, 2007). However, these data were at odds with biochemical data, which found no changes in levels of SA and JA in the plants. This led the authors to conclude that the resistance of this genotype was regulated independently of SA, JA, and ET (Letousey *et al.*, 2007). In terms of SA signaling, a number of studies have shown that application of SA or its functional analog, ASM, to monocotyledonous and dicotyledonous hosts reduced parasitism by *Oronanche* and *Striga* spp. (e.g., Pérez-de-Luque *et al.*, 2004; Kusumoto *et al.*, 2007; Hiraoka & Sugimoto, 2008). These studies are supported by work showing the up-regulation of several SA-responsive genes in an incompatible interaction between rice and *S. hermonthica*, as well as in the interaction between *S. hermonthica* and a partially resistant sorghum variety (Hiraoka & Sugimoto, 2008; Swarbrick *et al.*, 2008). Finally, in a study of the interaction between tomato and *Cuscuta pentagona*, Runyon *et al.* (2008) obtained evidence to suggest that parasitism by *C. pentagona* induced volatile release, thereby eliciting an SA-mediated defense response in the host. This work was part of a more comprehensive study of the effects of parasitic plant infection on host defense against insect herbivores, which is dealt with in Chapter 4.

At present, it is impossible to say whether the increased expression of JA-responsive genes in some of the interactions mentioned above is a side effect of wounding caused by penetration of host tissue by the parasite, or whether JA signaling mediates susceptibility to parasitic plants, by suppressing SA signaling, for example (Gutjahr & Paszkowski, 2009). These diverse reports demonstrate that research on signaling in plants interacting with parasitic plants is in its infancy and highlights the need for future, detailed work in this area.

3.7   Conclusions

Plants face a continual dilemma during their lives. They need to defend themselves against attackers, but they also need to accommodate beneficial organisms. The problem is how to differentiate friends from foes. In order to do this, they have evolved sophisticated mechanisms to perceive interactions with their biotic environment and to translate what they perceive into an appropriate response—vigorous defense or relaxation of border controls. As we have seen in this chapter, the signaling networks that have evolved to bring about the appropriate plant response to its biotic environment are highly complex. The plant hormones SA, JA, and ET are seen as key players in the regulation of signaling networks, and signaling pathways that are dependent on these hormones are differentially effective against different types of attackers. So, generally speaking, biotrophs are more sensitive to SA-dependent defense, while necrotrophs and herbivorous insects are affected by JA- and ET-dependent defenses. There is considerable overlap in the signaling pathways that are activated following attack by the different types of attacker. This raises the possibility that activation of one type of defense might be effective against more than one type of attacker. Of course, the opposite might occur; that is, activation of one signaling pathway might compromise defense mediated by another signaling pathway, leading to signaling conflicts. In this chapter, we have considered the signaling pathways activated by different types of attackers separately. In reality, plants are likely to encounter more than one attacker at any one time. The question arises, therefore, as to how plants deal with multiple attackers. These topics are dealt with in the next chapter.

Recommended reading

De Wit PJGM, 2007. How plants recognise pathogens and defend themselves. *Cellular and Molecular Life Sciences* **64**, 2726–2732.

Glazebrook J, 2005. Contrasting mechanisms of defense against biotrophic and necrotrophic pathogens. *Annual Review of Phytopathology* **43**, 205–227.

Heil M, Ton J, 2008. Long-distance signalling in plant defence. *Trends in Plant Science* **13**, 264–272.

Howe GA, Jander G, 2008. Plant immunity to insect herbivores. *Annual Review of Plant Biology* **59**, 41–66.

Lamb C, Dixon RA, 1997. The oxidative burst in plant disease resistance. *Annual Review of Plant Physiology and Plant Molecular Biology* **48**, 251–275.

León J, Rojo E, Sánchez-Serrano JJ, 2001. Wound signalling in plants. *Journal of Experimental Botany* **52**, 1–9.

Loake G, Grant M, 2007. Salicylic acid in plant defence: the players and protagonists. *Current Opinion in Plant Biology* **10**, 466–472.

Lopez MA, Bannenberg G, Castresana C, 2008. Controlling hormone signaling is a plant and pathogen challenge for growth and survival. *Current Opinion in Plant Biology* **11**, 420–427.

Mauch-Mani B, Mauch F, 2005. The role of abscisic acid in plant–pathogen interactions. *Current Opinion in Plant Biology* **8**, 409–414.

Pieterse CMJ, Van Loon LC, 2007. Signalling cascades involved in induced resistance. In: Walters D, Newton A, Lyon G, eds. *Induced Resistance for Plant Disease Control: A Sustainable Approach to Crop Protection*. Oxford: Blackwell Publishing Ltd., pp. 65–88.

Weber H, 2002. Fatty acid derived signals in plants. *Trends in Plant Science* **7**, 217–224.

Wu J, Baldwin IT, 2009. Herbivory-induced signalling in plants: perception and action. *Plant Cell and Environment* **32**, 1161–1174.

References

Adie BA, Perez-Perez J, Perez-Perez MM, Godoy M, Sanchez-Serrano JJ, Schmelz EA, Solano R, 2007. ABA is an essential signal for plant resistance to pathogens affecting JA biosynthesis and the activation of defences in *Arabidopsis*. *The Plant Cell* **19**, 1665–1681.

Alborn HT, Turlings TCJ, Jones TH, Stenhagen G, Loughrin JH, Tumlinson JH, 1997. An elicitor of plant volatiles from beet armyworm oral secretion. *Science* **276**, 945–949.

Alborn HT, Hansen TV, Jones TH, Bennett DC, Tumlinson JH, Schmelz EA, Teal PEA, 2007. Disulfooxy fatty acids from the American grasshopper *Shistocerca americana*, elicitors of plant volatiles. *Proceedings of the National Academy of Sciences, USA* **104**, 12976–12981.

Anderson JP, Badruzsaufari E, Schenk PM, Manners JM, Desmond OJ, Ehlert C, Maclean DJ, Ebert PR, Kazan K, 2004. Antagonistic interaction between absciscic acid and jasmonate-ethylene signaling pathways modulates defense gene expression and disease resistance in *Arabidopsis*. *The Plant Cell* **16**, 3460–3479.

Arimura G, Ozawa R, Shimoda T, Nishioka T, Boland W, Takabayashi J, 2000a. Herbivory-induced volatiles elicit defence genes in lima bean leaves. *Nature* **406**, 512–515.

Arimura G, Tashiro K, Kuhara S, Nishioka T, Ozawa R, Takabayashi J, 2000b. Gene responses in bean leaves induced by herbivory and by herbivore-induced volatiles. *Biochemical and Biophysical Research Communications* **277**, 305–310.

Arimura G, Köpke S, Kunert M, Volpe V, David A, Brand P, Dabrowska P, Maffei ME, Boland W, 2008. Effects of feeding *Spodoptera littoralis* on lima bean leaves: IV. Diurnal and noctural damage differentially initiate plant volatile emission. *Plant Physiology* **146**, 965–973.

Asselbergh B, Curvers K, França SC, Audenaert K, Vuylsteke M, Van Breusegem F, Höfte M, 2007. Resistance to *Botrytis cinerea* in *sitiens*, an abscisic acid-deficient tomato mutant, involves timely production of hydrogen peroxide and cell wall modifications in the epidermis. *Plant Physiology* **144**, 1863–1877.

Asselbergh B, De Vleesschauwer D, Höfte M, 2008a. Global switches and fine-tuning—ABA modulates plant pathogen defense. *Molecular Plant-Microbe Interactions* **21**, 709–719.

Asselbergh B, Achuo AE, Höfte M, Van Gijsegem F, 2008b. Abscisic acid deficiency leads to rapid activation of tomato defence responses upon infection with *Erwinia chrysanthemi*. *Molecular Plant Pathology* **9**, 11–24.

Audenaert K, De Meyer GB, Hofte MM, 2002. Abscisic acid determines basal susceptibility of tomato to *Botrytis cinerea* and suppresses salicylic acid dependent signalling mechanisms. *Plant Physiology* **128**, 491–501.

Baldauf SL, Roger AJ, Wenk-Siefert I, Doolittle WF, 2000. A kingdom-level phylogeny eukaryotes based on combined protein data. *Science* **290**, 972–977.

Baldwin IT, 1990. Herbivory simulations in ecological research. *Trends in Ecology and Evolution* **5**, 91–93.

Belenghi B, Acconcia F, Trovato M, Perazzolli M, Bocedi A, Polticelli F, Ascenzi P, Delledonne M, 2003. AtCYS1, a cystatin from *Arabidopsis thaliana*, suppresses hypersensitive cell death. *European Journal of Biochemistry* **270**, 2593–2604.

Bhattarai KK, Xie Q-G, Mantelin S, Bishnoi U, Girke T, Navarre DA, Kaloshian I, 2008. Tomato susceptibility to root-knot nematodes requires an intact jasmonic acid signaling pathway. *Molecular Plant-Microbe Interactions* **21**, 1205–1214.

Blee E, 2002. Impact of phyto-oxylipins in plant defence. *Trends in Plant Science* **7**, 315–321.

Bogre L, Ligterink W, Meskiene I, Barker PJ, HeberleBors E, Huskisson NS, Hirt H, 1997. Wounding induces the rapid and transient activation of a specific MAP kinase pathway. *The Plant Cell* **9**, 75–83.

Boland W, Hopke J, Donath J, Nuske J, Bublitz F, 1995. Jasmonic acid and coronatin induce odor production in plants. *Angewandte Chemie* **34**, 1600–1602.

Bouarab K, Adas F, Gaquerel E, Kloareg B, Salaun JP, Pontin P, 2004. The innate immunity of a marine red alga involves oxylipins from both the eicosanoid and octadecanoid pathways. *Plant Physiology* **135**, 1838–1848.

Branch C, Hwang CF, Navarre DA, Williamson VM, 2004. Salicylic acid is part of the *Mi-1* mediated defense response to root-knot nematode in tomato. *Molecular Plant-Microbe Interactions* **17**, 351–357.

Browse J, Howe GA, 2008. New weapons and a rapid response against insect attack. *Plant Physiology* **146**, 832–838.

Bryant JP, 1981. Phytochemical deterrence of snowshoe hare browsing by adventitious shoots of four Alaskan trees. *Science* **213**, 889–890.

Cardoza YJ, Alborn HT, Tumlinson JH, 2002. *In vivo* volatile emissions from peanut plants induced simultaneous fungal infection and insect damage. *Journal of Chemical Ecology* **28**, 161–174.

Carimi F, Zottini M, Formentin E, Terzi M, Lo Schiavo F, 2003. Cytokinins: new apoptotic inducers in plants. *Planta* **216**, 413–421.

Carroll MJ, Schmelz EA, Teal PEA, 2008. The attraction of *Spodoptera frugiperda* neonates to cowpea seedlings is mediated by volatiles induced by conspecific herbivory and the elicitor inceptin. *Journal of Chemical Ecology* **34**, 291–300.

Chen Z, Agnew JL, Cohen JD, He P, Shan L, Sheen J, Kunkel BN, 2007. *Pseudomonas syringae* type III effector AvrRpt2 alters *Arabidopsis thaliana* auxin physiology. *Proceedings of the National Academy of Sciences, USA* **104**, 20131–20136.

Chini A, Fonseca S, Fernandez G, Adie B, Chico JM, Lorenzo O, Garcia-Casado G, Lopez-Vidriero I, Lozano FM, Ponce MR, Micol JL, Solano R, 2007. The JAZ family of repressors is the missing link in jasmonate signalling. *Nature* **448**, 666–672.

Cipollini DF, Sipe ML, 2001. Jasmonic acid treatment and mammalian herbivory differentially affect chemical defenses and growth of wild mustard (*Brassica kaber*). *Chemecology* **11**, 137–143.

Constabel CP, Ryan CA, 1998. A survey of wound- and methyl jasmonate-induced leaf polyphenol oxidase in crop plants. *Phytochemistry* **47**, 507–511.

Cooper W, Jia L, Goggin L, 2005. Effects of jasmonate-induced defences on root-knot nematode infection of resistant and susceptible tomato cultivars. *Journal of Chemical Ecology* **31**, 1953–1967.

Croft K, Juttner F, Slusarenko AJ, 1993. Volatile products of the lipoxygenase pathway evolved from *Phaseolus vulgaris* L. leaves inoculated with *Pseudomonas syringae* pv. *phaseolicola*. *Plant Physiology* **101**, 13–24.

Danell K, Bergström R, 2002. Mammalian herbivory in terrestrial environments. In: Herrera CM, Pellmyr O, eds. *Plant–Animal Interactions: An Evolutionary Approach*. Malden, MA: Blackwell Publishing Ltd., pp. 107–131.

De Boer JG, Dicke M, 2004. The role of methyl salicylate in prey searching behaviour of the predatory mite *Phytoseiulus persimilis*. *Journal of Chemical Ecology* **30**, 255–271.

De Moraes CM, Mescher MC, Tumlinson JH, 2001. Caterpillar-induced nocturnal plant volatiles repel nonspecific females. *Nature* **393**, 570–573.

De Vos M, Van Oosten VR, Van Poecke RMP, Van Pelt JA, Pozo MJ, Mueller MJ, Buchala AJ, Métraux J.-P, Van Loon LC, Dicke M, Pieterse CMJ, 2005. Signal signature and transcriptome changes of Arabidopsis during pathogen and insect attack. *Molecular Plant-Microbe Interactions* **18**, 923–937.

De Wit PJGM, 2007. How plants recognise pathogens and defend themselves. *Cellular and Molecular Life Sciences* **64**, 2726–2732.

Delaney TP, Uknes S, Vernooij B, Friedrich L, Weymann K, Negrotto D, Gaffney T, Gut-Rella M, Kessman H, Ward E, Ryals J, 1994. A central role for salicylic acid in plant disease resistance. *Science* **266**, 1247–1250.

Delledonne M, Xia YJ, Dixon RA, Lamb C, 1998. Nitric oxide functions as a signal in plant disease resistance. *Nature* **394**, 585–588.

Devoto A, Turner JG, 2005. Jasmonate-regulated *Arabidopsis* stress signalling network. *Physiologia Plantarum* **123**, 161–172.

Dicke M, 1999. Are herbivore-induced plant volatiles reliable indicators of herbivore identity to foraging carnivorous arthropods? *Entomologia Experimentalis et Applicata* **92**, 131–142.

Dicke M, Gols R, Ludeking D, Posthumus MA, 1999. Jasmonic acid and herbivory differentially induce carnivore-attracting plant volatiles in lima bean plants. *Journal of Chemical Ecology* **25**, 1907–1922.

Die JV, Dita MA, Krajinski F, Gonzalez-Verdejo CI, Rubiales D, Moreno MT, Roman B, 2007. Identification by suppression subtractive hybridization and expression analysis of *Medicago truncatula* putative defense genes during *Orobanche crenata* infection. *Physiological and Molecular Plant Pathology* **70**, 49–59.

Doss RP, Oliver JE, Proebsting WM, Potter SW, Kuy SR, Clement SL, Williamson RT, Carney JR, DeVilbiss ED, 2000. Bruchins: insect-derived plant regulators that stimulate neoplasm formation. *Proceedings of the National Academy of Sciences, USA* **97**, 6218–6223.

Dugardeyn J, Van der Straeten D, 2008. Ethylene: fine-tuning plant growth and development by stimulation and inhibition of elongation. *Plant Science* **175**, 59–70.

Ellis C, Karafyllidis I, Turner JG, 2002. Constitutive activation of jasmonate signalling in an *Arabidopsis* mutant correlates with enhanced resistance to *Erysiphe cichoracearum*, *Pseudomonas syringae* and *Myzus persicae*. *Molecular Plant-Microbe Interactions* **15**, 1025–1030.

Engelberth J, Alborn HT, Schmelz EA, Tumlinson JH, 2004. Airborne signals prime plants against insect herbivore attack. *Proceedings of the National Academy of Sciences, USA* **101**, 1781–1785.

Farmer EE, Ryan CA, 1990. Interplant communication: airborne methyl jasmonate induces synthesis of proteinase inhibitors in plant leaves. *Proceedings of the National Academy of Sciences, USA* **87**, 7713–7716.

Farmer EE, Ryan CA, 1992. Octadecanoid precursors of jasmonic acid activate the synthesis of wound-inducible proteinase inhibitors. *The Plant Cell* **4**, 129–134.

Farmer EE, Johnson RR, Ryan CA, 1992. Regulation of expression of proteinase inhibitor genes by methyl jasmonate and jasmonic acid. *Plant Physiology* **98**, 995–1002.

Frye CA, Innes RW, 1998. An *Arabidopsis* mutant with enhanced resistance to powdery mildew. *The Plant Cell* **10**, 947–956.

Frye CA, Tang D, Innes RW, 2001. Negative regulation of defense responses in plants by a conserved MAPKK kinase. *Proceedings of the National Academy of Sciences, USA* **98**, 373–378.

Funk CD, 2001. Prostaglandins and leukotrienes: advances in eicosanoid biology. *Science* **294**, 1871–1875.

Gaffney T, Friedrich L, Vernooij B, Negrotto D, Nye G, Uknes S, Ward E, Kessman H, Ryals J, 1993. Requirement of salicylic acid for the induction of systemic acquired resistance. *Science* **261**, 754–756.

Glazebrook J, 2005. Contrasting mechanisms of defense against biotrophic and necrotrophic pathogens. *Annual Review of Phytopathology* **43**, 205–227.

Gosset V, Harmel N, Göbel C, Francis F, Haubruge E, Wathelet J-P, du Jardin P, Feussner I, Fauconnier M-L, 2009. Attacks by a piercing-sucking insect (*Myzus persicae* Sultzer) or a chewing insect (*Leptinotarsa decemlineata* Say) on potato plants (*Solanum tuberosum* L.) induce differential changes in volatile compound release and oxylipin synthesis. *Journal of Experimental Botany* **60**, 1231–1240.

Green TR, Ryan CA, 1972. Wound-induced proteinase inhibitor in plant leaves: a possible defense mechanism against insects. *Science* **175**, 776–777.

Grun S, Lindemayer C, Sell S, Durner J, 2006. Nitric oxide and gene regulation in plants. *Journal of Experimental Botany* **57**, 507–516.

Guenther A, Karl T, Harley P, Wiedenmyer C, Palmer PI, Geron C, 2006. Estimates of global terrestrial isoprene emissions using MEGAN (Model of Emissions of Gases and Aerosols from Nature). *Atmospheric Chemistry and Physics* **6**, 3181–3210.

Gutjahr C, Paszkowski U, 2009. Weights in the balance: jasmonic acid and salicylic acid signaling in root-biotroph interactions. *Molecular Plant-Microbe Interactions* **22**, 763–772.

Halitschke R, Schittko U, Pohnert G, Boland W, Baldwin IT, 2001. Molecular interactions between the specialist herbivore *Manduca sexta* (Lepidoptera: Sphingidae) and its natural host *Nicotiana attenuata*. III. Fatty acid-amino acid conjugates in herbivore oral secretions are necessary and sufficient for herbivore-specific plant responses. *Plant Physiology* **125**, 711–717.

Halitschke R, Gase K, Hui D, Schmidt DD, Baldwin IT, 2003. Molecular interactions between the specialist herbivore *Manduca sexta* (Lepidoptera: Sphingidae) and its natural host *Nicotiana attenuata*. VI. Microarray analysis reveals that most herbivore-specific transcriptional changes are necessary and sufficient for herbivore-specific plant responses. *Plant Physiology* **131**, 1894–1902.

Hammerschmidt R, 1999. Induced disease resistance: how do induced plants stop pathogens? *Physiological and Molecular Plant Pathology* **55**, 77–84.

Hammerschmidt R, 2007. Introduction: definitions and some history. In: Walters D, Newton A, Lyon G, eds. *Induced Resistance for Plant Disease Control: A Sustainable Approach to Crop Protection*. Oxford: Blackwell Publishing Ltd., pp. 1–8.

Hammerschmidt R, Kuć J, 1982. Lignification as a mechanism for induced systemic resistance in cucumber. *Physiological Plant Pathology* **20**, 61–71.

Harfouche AL, Shivaji R, Stocker R, Williams PW, Luthe DS, 2006. Ethylene signaling mediates a maize defense response to insect herbivory. *Molecular Plant-Microbe Interactions* **19**, 189–199.

Heidel AJ, Baldwin IT, 2004. Microarray analysis of salicylic acid- and jasmonic acid-signalling in responses of *Nicotiana attenuata* to attack by insects from multiple feeding guilds. *Plant, Cell and Environment* **27**, 1362–1373.

Heil M, Silva Bueno JC, 2007. Within-plant signaling by volatiles leads to induction and priming of an indirect plant defense in nature. *Proceedings of the National Academy of Sciences, USA* **104**, 5467–5472.

Heil M, Ton J, 2008. Long-distance signalling in plant defence. *Trends in Plant Science* **13**, 264–272.

Hilker M, Meiners T, 2006. Early herbivore alert: insect eggs induce plant defense. *Journal of Chemical Ecology* **32**, 1379–1397.

Hiraoka Y, Sugimoto Y, 2008. Molecular responses of sorghum to purple witchweed (*Striga hermonthica*) parasitism. *Weed Science* **56**, 356–363.

Hoffman T, Schmidt JS, Zheng X, Bent AF, 1999. Isolation of ethylene-insensitive soybean mutants that are altered in pathogen susceptibility and gene-for-gene disease resistance. *Plant Physiology* **119**, 935–949.

Hong JK, Yun BW, Kang JG, Raja MU, Kwon E, Sorhagen K, Chu C, Wang Y, Loake GJ, 2008. Nitric oxide function and signalling in plant disease resistance. *Journal of Experimental Botany* **59**, 147–154.

Horiuchi J, Arimura G, Ozawa R, Shimoda T, Takabayashi J, Nishioka T, 2001. Exogenous ACC enhances volatiles production mediated by jasmonic acid in lima bean leaves. *FEBS Letters* **509**, 332–336.

Howe GA, 2004. Jasmonates as signals in the wound response. *Journal of Plant Growth Regulation* **23**, 223–235.

Howe GA, Jander G, 2008. Plant immunity to insect herbivores. *Annual Review of Plant Biology* **59**, 41–66.

Howe GA, Schilmiller A, 2002. Oxylipin metabolism in response to stress. *Current Opinion in Plant Biology* **5**, 230–236.

Howe GA, Lightner J, Browse J, Ryan CA, 1996. An octadecanoid pathway mutant (JL5) of tomato is compromised in signaling for defense against insect attack. *The Plant Cell* **8**, 2067–2077.

Huang J, Cardoza YJ, Schmelz EA, Rainer R, Engelberth J, Tumlinson JH, 2003. Differential volatile emissions and salicylic acid levels from tobacco plants in response to different strains of *Pseudomonas syringae*. *Planta* **217**, 767–775.

Ithal N, Recknor J, Nettleton D, Hearne L, Maier T, Baum TJ, Mitchum MG, 2007a. Parallel genome-wide expression profiling of host and pathogen during soybean cyst nematode infection of soybean. *Molecular Plant-Microbe Interactions* **20**, 293–305.

Ithal N, Recknor J, Nettleton D, Hearne L, Maier T, Baum TJ, Mitchum MG, 2007b. Developmental transcript profiling of cyst nematode feeding cells in soybean roots. *Molecular Plant-Microbe Interactions* **20**, 510–525.

Kachroo P, Yoshioka K, Shah J, Dooner KD, Klessig DF, 2000. Resistance to turnip crinkle virus in *Arabidopsis* is regulated by two host genes and is salicylic acid dependent but NPR1, ethylene, and jasmonate independent. *The Plant Cell* **12**, 677–690.

Kahl J, Siemens DH, Aerts RJ, Gabler R, Kuhnemann F, Preston CA, Baldwin IT, 2000. Herbivore-induced ethylene suppresses a direct defense but not a putative indirect defense against an adapted herbivore. *Planta* **210**, 336–342.

Kang JH, Wang L, Giri A, Baldwin IT, 2006. Silencing threonine deaminase and JAR4 in *Nicotiana attenuata* impairs jasmonic acid-isoleucine-mediated defenses against *Manduca sexta*. *The Plant Cell* **18**, 3303–3320.

Karban R, Baldwin I, Baxter K, Laue G, Felton G, 2000. Communication between plants: induced resistance in wild tobacco plants following clipping of neighboring sagebrush. *Oecologia* **125**, 66–71.

Katz VA, Thulke OU, Conrath U, 1998. Benzothiadiazole primes parsley cells for augmented elicitation of defence responses. *Plant Physiology* **117**, 1333–1339.

Kauss H, Krause K, Jeblick W, 1992. Methyl jasmonate conditions parsley suspension cells for increased elicitation of phenylpropanoid defence responses. *Biochemical and Biophysical Research Communications* **189**, 304–308.

Kauss H, Jeblick W, Ziegler J, Krabler W, 1994. Pretreatment of parsley (*Petroselinum crispum* L.) suspension cultures with methyl jasmonate enhances elicitation of activated oxygen species. *Plant Physiology* **105**, 89–94.

Kessler A, Baldwin IT, 2001. Defensive function of herbivore-induced plant volatile emissions in nature. *Science* **291**, 2141–2144.

Kessler A, Baldwin IT, 2002. Plant responses to insect herbivory: the emerging molecular analysis. *Annual Review of Plant Biology* **53**, 299–328.

Kishimoto K, Matsui K, Ozawa R, Takabayashi J, 2005. Volatile C6-aldehydes and allo-ocimene activate defense genes and induce resistance against *Botrytis cinerea* in *Arabidopsis thaliana*. *Plant and Cell Physiology* **46**, 1093–1102.

Kishimoto K, Matsui K, Ozawa R, Takabayashi J, 2006. Analysis of defence responses activated by volatile allo-ocimene treatment in *Arabidopsis thaliana*. *Phytochemistry* **67**, 1520–1529.

Knoester M, Linthorst HJM, Bol JF, Van Loon LC, 2001. Involvement of ethylene in lesion development and systemic acquired resistance in tobacco during the hypersensitive reaction to tobacco mosaic virus. *Physiological and Molecular Plant Pathology* **59**, 45–57.

Kost C, Heil M, 2006. Herbivore-induced plant volatiles induce an indirect defence in neighbouring plants. *Journal of Ecology* **94**, 619–628.

Kuć J, 1982. Induced immunity to plant disease. *BioScience* **32**, 854–860.

Kumar D, Klessig DF, 2003. High-affinity salicylic acid-binding protein 2 is required for plant innate immunity and has salicylic acid-stimulated lipase activity. *Proceedings of the National Academy of Sciences, USA* 100, 16101–16106.

Kusumoto D, Goldwasser Y, Xie X, Yoneyama K, Takeuchi Y, Yoneyama K, 2007. Resistance of red clover (*Trifolium pratense*) to the root parasitic plant *Orobanche minor* is activated by salicylate but not by jasmonate. *Annals of Botany* **100**, 537–544.

Lamb C, Dixon RA, 1997. The oxidative burst in plant disease resistance. *Annual Review of Plant Physiology and Plant Molecular Biology* **48**, 251–275.

Laothawornkitkul J, Paul ND, Vickers CE, Possell M, Taylor JE, Mullineaux PM, Hewitt CN, 2008. Isoprene emissions influence herbivore feeding decisions. *Plant Cell and Environment* **31**, 1410–1415.

Lawton KA, Potter SL, Uknes S, Ryals J, 1994. Acquired resistance signal transduction in *Arabidopsis* is ethylene independent. *The Plant Cell* **6**, 581–588.

Lecourieux D, Ranjeva R, Pugin A, 2006. Calcium in plant defence-signalling pathways. *New Phytologist* **171**, 249–269.

Lee GI, Howe GA, 2003. The tomato mutant *spr1* is defective in systemin perception and the production of a systemic wound signal for defense gene expression. *Plant Journal* **33**, 567–576.

León J, Rojo E, Sánchez-Serrano JJ, 2001. Wound signalling in plants. *Journal of Experimental Botany* **52**, 1–9.

Letousey P, de Zélicourt A, Vieira Dos Santos C, Thoiron S, Monteau F, Simier P, Thalouran P, Delavault P, 2007. Molecular analysis of resistance mechanisms to *Orobanche cumana* in sunflower. *Plant Pathology* **56**, 536–546.

Li L, Li C, Lee GI, Howe GA, 2002. Distinct roles for jasmonate synthesis and action in the systemic wound response of tomato. *Proceedings of the National Academy of Sciences, USA* **99**, 6416–6421.

Li Q, Xie QG, Smith-Becker J, Navarre DA, Kaloshian I, 2006. Mi-1-mediated aphid resistance involves salicylic acid and mitogen activated protein kinase signaling cascades. *Molecular Plant-Microbe Interactions* **19**, 655–664.

Loake G, Grant M, 2007. Salicylic acid in plant defence: the players and protagonists. *Current Opinion in Plant Biology* **10**, 466–472.

Loivamäki M, Mumm R, Dicke M, Schnitzler J-P, 2008. Isoprene interferes with the attraction of bodyguards by herbaceous plants. *Proceedings of the National Academy of Sciences, USA* **105**, 17430–17435.

Lopez MA, Bannenberg G, Castresana C, 2008. Controlling hormone signaling is a plant and pathogen challenge for growth and survival. *Current Opinion in Plant Biology* **11**, 420–427.

Maffei ME, Bossi S, Spiteller D, Mithöfer A, Boland W, 2004. Effects of feeding *Spodoptera littoralis* on lima bean leaves. I. Membrane potentials, intracellular calcium variations, oral secretions, and regurgitate components. *Plant Physiology* **134**, 1752–1762.

Maffei ME, Mithöfer A, Boland W, 2007. Before gene expression: early events in plant-insect interaction. *Trends in Plant Science* **12**, 310–316.

Maleck K, Levine A, Eulgem T, Morgan A, Schmid J, Lawton KA, Dangl JL, Dietrich RA, 2000. The transcriptome of *Arabidopsis thaliana* during systemic acquired resistance. *Nature Genetics* **26**, 403–410.

Malone M, Alarcon JJ, Palumbo L, 1994. An hydraulic interpretation of rapid, long-distance wound signaling in the tomato. *Planta* **193**, 181–185.

Martinez de Ilarduya O, Kaloshian I, 2001. *Mi-1.2* transcripts accumulate ubiquitously in root-knot nematode resistant *Lycopersicon esculentum*. *Journal of Nematology* **33**, 116–120.

Matsui K, 2006. Green leaf volatiles: hydroperoxide lyase pathway of oxylipin metabolism. *Current Opinion in Plant Biology* **9**, 274–280.

Mauch-Mani B, Mauch F, 2005. The role of abscisic acid in plant-pathogen interactions. *Current Opinion in Plant Biology* **8**, 409–414.

McConn M, Creelman RA, Bell E, Mullet JE, Browse J, 1997. Jasmonate is essential for insect defense in *Arabidopsis*. *Proceedings of the National Academy of Sciences, USA* **94**, 5473–5477.

McGurl B, Pearce G, Orozco-Cardenas M, Ryan CA, 1992. Structure, expression, and antisense inhibition of the systemin precursor gene. *Science* **255**, 1570–1573.

McGurl B, Orozco-Cardenas M, Pearce G, Ryan CA, 1994. Overexpression of the prosystemin gene in transgenic tomato plants generates a systemic signal that constitutively induces proteinase inhibitor synthesis. *Proceedings of the National Academy of Sciences, USA* **91**, 9799–9802.

Melotto M, Underwood W, Koczan J, Nomura K, He SY, 2006. Plant stomata function in innate immunity against bacterial invasion. *Cell* **126**, 969–980.

Mewis I, Appel HM, Hom A, Raina R, Schultz JC, 2005. Major signaling pathways modulate *Arabidopsis* glucosinolate accumulation and response to both phloem-feeding and chewing insects. *Plant Physiology* **138**, 1149–1162.

Mithöfer A, Boland W, 2008. Recognition of herbivory-asociated molecular patterns. *Plant Physiology* **146**, 2029–2040.

Mithöfer A, Wanner G, Boland W, 2005. Effects of feeding *Spodoptera littoralis* on lima bean leaves. II. Continuous mechanical wounding resembling insect feeding is sufficient to elicit herbivory-related volatile emission. *Plant Physiology* **137**, 1160–1168.

Mlejnek P, Prochazka S, 2002. Activation of caspase-like proteases and induction of apoptosis by isopentenyladenosine in tobacco BY-2 cells. *Planta* **215**, 158–166.

Moran PJ, Thompson GA, 2001. Molecular responses to aphid feeding in *Arabidopsis* in relation to plant defense pathways. *Plant Physiology* **125**, 1074–1085.

Mur LAJ, Brown IR, Darby RM, Bestwick CS, Bi YM, Mansfield JW, Draper J, 2000. A loss of resistance to avirulent bacterial pathogens in tobacco is associated with the attenuation of a salicylic acid-potentiated oxidative burst. *Plant Journal* **23**, 609–621.

Mur LAJ, Laarhoven LJJ, Harren FJM, Hall MA, Smith AR, 2008. Nitric oxide interacts with salicylate to regulate biphasic ethylene production during the hypersensitive response. *Plant Physiology* **148**, 1537–1546.

Mur LAJ, Lloyd AJ, Cristescu SM, Harren FJM, Hall MA, Smith AR, 2009. Biphasic ethylene production during the hypersensitive response in *Arabidopsis:* a window into defense priming mechanisms? *Plant Signalling and Behaviour* **4**, 610–613.

Navarro L, Dunoyer P, Jay F, Arnold B, Dharmasiri N, Estelle M, Voinnet O, Jones JDG, 2006. A plant miRNA contributes to antibacterial resistance by repressing auxin signaling. *Science* **312**, 436–439.

Nawrath C, Métraux J-P, 1999. Salicylic acid induction-deficient mutants of *Arabidopsis* express PR-2 and PR-5 and accumulate high levels of camalexin after pathogen inoculation. *The Plant Cell* **11**, 1393–1404.

O'Donnell PJ, Calvert C, Atzorn R, Wasternack C, Leyser HMO, Bowles DJ, 1996. Ethylene as a signal mediating the wound response of tomato plants. *Science* **274**, 1914–1917.

Ogawa D, Nakajima N, Sano T, Tamaoki M, Aono M, Kubo A, Kamada H, Saji H, 2005. Regulation of salicylic acid synthesis in ozone-exposed tobacco and *Arabidopsis. Phyton-Annales Rei Botanicae* **45**, 169–175.

Orozco-Cardenas M, Ryan CA, 1999. Hydrogen peroxide is generated systemically in plant leaves by wounding and systemin via the octadecanoid pathway. *Proceedings of the National Academy of Sciences, USA* **96**, 6553–6557.

Orozco-Cardenas M, McGurl B, Ryan CA, 1993. Expression of an antisense prosystemin gene in tomato plants reduces resistance toward *Manduca sexta* larvae. *Proceedings of the National Academy of Sciences, USA* **90**, 8273–8276.

Ozawa R, Arimura G, Takabayashi J, Shimoda T, Nishioka T 2000. Involvement of jasmonate- and salicylate-related signaling pathway for the production of specific herbivore-induced volatiles in plants. *Plant Cell Physiology* **41**, 391–398.

Park SW, Kaimoyo E, Kumar D, Mosher S, Klessig DF, 2007. Methyl salicylate is a critical mobile signal for plant systemic acquired resistance. *Science* **318**, 113–116.

Paschold A, Halitschke R, Baldwin IT, 2006. Using "mute" plants to translate volatile signals. *Plant Journal* **45**, 275–291.

Paschold A, Halitschke R, Baldwin IT, 2007. Co(i)-ordinating defenses: NaCOI1 mediates herbivore-induced resistance in *Nicotiana attenuata* and reveals the role of herbivore movement in avoiding defenses. *Plant Journal* **51**, 79–91.

Pegadaraju V, Knepper C, Reese J, Shah J, 2005. Premature leaf senescence modulated by the *Arabidopsis* PHYTOALEXIN DEFICIENT4 gene is associated with defense against the phloem-feeding green peach aphid. *Plant Physiology* **139**, 1927–1934.

Pena-Cortés H, Albrecht T, Prat S, Weiler EW, Willmitzer L, 1993. Aspirin prevents wound-induced gene expression in tomato leaves by blocking jasmonic acid synthesis. *Planta* **191**, 123–128.

Pena-Cortés H, Fisahn J, Willmitzer L, 1995. Signals involved in wound-induced proteinase inhibitor Ii gene expression in tomato and potato plants. *Proceedings of the National Academy of Sciences, USA* **92**, 4106–4113.

Pieterse CMJ, Van Loon LC, 2007. Signalling cascades involved in induced resistance. In: Walters D, Newton A, Lyon G, eds. *Induced Resistance for Plant Disease Control: A Sustainable Approach to Crop Protection*. Oxford: Blackwell Publishing Ltd., pp. 65–88.

Pieterse CMJ, Van Wees SCM, Hoffland E, Van Pelt JA, Van Loon LC, 1996. Systemic resistance in *Arabidopsis* induced by biocontrol bacteria is independent of salicylic acid and pathogenesis-related gene expression. *Plant Cell* **8**, 1125–1237.

Pieterse CMJ, Van Wees SCM, Van Pelt JA, Knoester M, Laan R, Gerrits N, Weisbeek PJ, Van Loon LC, 1998. A novel signaling pathway controlling induced systemic resistance in *Arabidopsis*. *Plant Cell* **10**, 1571–1580.

Pieterse CMJ, Van Pelt JA, Ton J, Parchmann S, Mueller MJ, Buchala AJ, Métraux J-P, Van Loon LC, 2000. Rhizobacteria-mediated induced systemic resistance (ISR) in *Arabidopsis* requires sensitivity to jasmonate and ethylene but is not accompanied by an increase in their production. *Physiological and Molecular Plant Pathology* **57**, 123–134.

Pérez-de-Luque A, Jorrín J, Rubiales D, 2004. Crenate broomrape control in pea by foliar application of benzothiadiazole (BTH). *Phytoparasitica* **32**, 21–29.

Pieterse CMJ, Schaller A, Mauch-Mani B, Conrath U, 2006. Signaling in plant resistance responses: divergence and cross-talk of defense pathways. In: Tuzun S, Bent E, eds. *Multigenic and Induced Systemic Resistance in Plants*. New York: Springer, pp. 166–196.

Prats E, Mur LAJ, Sanderson R, Carver TLW, 2005. Nitric oxide contributes both to papilla-based resistance and the hypersensitive response in barley attacked by *Blumeria graminis* f.sp. *hordei*. *Molecular Plant Pathology* **6**, 65–78.

Raskin I, 1992. Role of salicylic acid in plants. *Annual Review of Plant Physiology and Plant Molecular Biology* **43**, 439–463.

Rasmann S, Turlings TCJ, 2008. First insights into specificity of belowground tritrophic interactions. *Oikos* **117**, 362–369.

Reignault P, Walters D, 2007. Topical application of inducers for disease control. In: Walters D, Newton A, Lyon G, eds. *Induced Resistance for Plant Disease Control: A Sustainable Approach to Crop Protection*. Oxford: Blackwell Publishing Ltd., pp. 179–200.

Reymond P, Bodenhausen N, Van Poecke RM, Krishnamurthy V, Dicke M, Farmer EE, 2004. A conserved transcript pattern in response to a specialist and a generalist herbivore. *The Plant Cell* **16**, 3132–3147.

Runyon JB, Mescher MC, De Moraes CM, 2006. Volatile chemical cues guide host location and host selection by parasitic plants. *Science* **313**, 1964–1967.

Runyon JB, Mescher MC, De Moraes CM, 2008. Parasitism by *Cuscuta pentagona* attenuates host plant defenses against insect herbivores. *Plant Physiology* **146**, 987–995.

Ruther J, Kleier S, 2005. Plant-plant signalling: ethylene synergizes volatile emission in *Zea mays* induced by exposure to (Z)-3-hexen-1-ol. *Journal of Chemical Ecology* **31**, 2217–2222.

Ryan CA, 1974. Assay and biochemical properties of proteinase inhibitor inducing factor wound hormone. *Plant Physiology* **54**, 328–332.

Sagi M, Davydov O, Orazova S, Yesbergenova Z, Ophir R, Stratmann JW, Fluhr R, 2004. Plant respiratory burst oxidase homologs impinge on wound responsiveness and development in *Lycopersicon esculentum*. *The Plant Cell* **16**, 616–628.

Sano H, Seo S, Orudgev E, Youssefian S, Ishizuka K, Ohashi Y, 1994. Expression of the gene for a small GTP binding protein in transgenic tobacco elevates endogenous cytokinin levels, abnormally induces salicylic acid in response to wounding, and increases resistance to tobacco mosaic virus infection. *Proceedings of the National Academy of Sciences, USA* **91**, 10556–10560.

Scheer JM, Ryan CA, 1999. A 160-kD systemin receptor on the surface of *Lycopersicon peruvianum* suspension cultured cells. *The Plant Cell* **11**, 1525–1535.

Scheer JM, Ryan CA, 2002. The systemin receptor SR160 from *Lycopersicon peruvianum* is a member of the LRR receptor kinase family. *Proceedings of the National Academy of Sciences, USA* **99**, 9585–9590.

Schenk PM, Kazan K, Wilson I, Anderson JP, Richmond T, Sommerville SC, Manners JM, 2000. Coordinated plant defense responses in *Arabidopsis* revealed by microarray analysis. *Proceedings of the National Academy of Sciences, USA* **97**, 11655–11660.

Schmelz EA, Alborn HT, Tumlinson JH, 2003. Synergistic interactions between volicitin, jasmonic acid and ethylene mediate insect-induced volatile emission in *Zea mays*. *Physiologia Plantarum* **117**, 403–412.

Schmelz EA, Carroll MJ, LeClere S, Phipps SM, Meredith J, Chourey PS, Alborn HT, Teal PEA, 2006. Fragments of ATP synthase mediate plant perception of insect attack. *Proceedings of the National Academy of Sciences, USA* **103**, 8894–8899.

Schmelz EA, LeClere S, Carroll MJ, Alborn HT, Teal PEA, 2007. Cowpea (*Vigna unguiculata*) chloroplastic ATP synthase is the source of multiple plant defense elicitors during insect herbivory. *Plant Physiology* **144**, 793–805.

Schmidt S, Baldwin IT, 2006. Systemin in *Solanum nigrum*. The tomato-homologous polypeptide does not mediate direct defense responses. *Plant Physiology* **142**, 1751–1758.

Seo S, Sano H, Ohashi Y, 1999. Jasmonate-based wound signal transduction requires activation of WIPK, a tobacco mitogen-activated protein kinase. *The Plant Cell* **11**, 289–298.

Shah J, Klessig DF, 1999. Salicylic acid: signal perception and transduction. In: Hooykaas PPJ, Hall MA, Libbenga KR, eds. *Biochemistry and Molecular Biology of Plant Hormones*. New York: Elsevier Science, pp. 513–541.

Shiojiri K, Kishimoto K, Ozawa R, Kugimiya S, Urashimo S, Arimura G, Horiuchi J, Nishioka T, Matsui K, Takabayashi J, 2006. Changing green leaf volatile biosynthesis in plants: an approach for improving plant resistance against both herbivores and pathogens. *Proceedings of the National Academy of Sciences, USA* **103**, 16672–16676.

Shulaev V, Silverman P, Raskin I, 1997. Airborne signalling by methyl salicylate in plant pathogen resistance. *Nature* **385**, 718–721.

Smid HM, Wang GH, Bukovinsky T, Steidle JLM, Bleeker MAK, van Loon JJA, Vet LEM, 2007. Species-specific acquisition and consolidation of long-term memory in parasitic wasps. *Proceedings of the Royal Society B—Biological Sciences* **274**, 1539–1546.

Sobczak M, Avrova A, Jupowicz J, Phillips MS, Ernst K, Kumar A, 2005. Characterization of susceptibility and resistance responses to potato cyst nematode (*Globodera* spp.) infection of tomato lines in the absence and presence of the broad-spectrum nematode resistance *Hero* gene. *Molecular Plant-Microbe Interactions* **18**, 158–168.

Soriano IR, Asenstorfer RE, Schmidt O, Riley IT, 2004a. Inducible flavone in oats (*Avenae sativa*) is a novel defense against plant parasitic nematodes. *Phytopathology* **94**, 1207–1214.

Soriano I, Ripley I, Potter M, Bowers W, 2004b. Phytoectysteroids: a novel defence against plant parasitic nematodes. *Journal of Chemical Ecology* **30**, 1885–1899.

Stankovic B, Davies E, 1997. Intercellular communication in plants: electrical stimulation of proteinase inhibitor gene expression in tomato. *Planta* **202**, 402–406.

Stenzel I, Hause B, Maucher H, Pitzschke A, Miersch O, Ziegler J, Ryan CA, Wasternack C, 2003. Allene oxide cyclase dependence of the wound response and vascular bundle specific generation of jasmonates in tomato—amplification in wound signalling. *Plant Journal* **33**, 577–589.

Stintzi A, Weber H, Reymond P, Browse J, Farmer EE, 2001. Plant defense in the absence of jasmonic acid: the role of cyclopentenones. *Proceedings of the National Academy of Sciences, USA* **98**, 12837–12842.

Stratmann JW, Ryan CA, 1997. Activation of a 52kDa MBP kinase by wounding and systemin. *Plant Physiology* **114**, 1442.

Suzuki H, Xia Y, Cameron R, Shadle D, Blount J, Lamb C, Dixon RA, 2004. Signals for local and systemic responses of plants to pathogen attack. *Journal of Experimental Botany* **55**, 169–179.

Swarbrick PJ, Huang K, Liu G, Slate J, Press MC, Scholes JD, 2008. Global patterns of gene expression in rice cultivars undergoing a susceptible or resistant interaction with the parasitic plant *Striga hermonthica*. *New Phytologist* **179**, 515–529.

Thines B, Katsir L, Melotto M, Niu Y, Mandaokar A, Liu GH, Nomura K, He SY, Howe GA, Browse J, 2007. JAZ repressor proteins are targets of the SCF^{COI1} complex during jasmonate signalling. *Nature* **448**, 661–666.

Thomma BPHJ, Penninckx IAMA, Broekaert WF, Cammue BPA, 2001. The complexity of disease signalling in *Arabidopsis*. *Current Opinion in Immunology* **13**, 63–68.

Thorpe MR, Ferrieri AP, Herth MM, Ferrieri RA, 2007. ^{11}C-imaging: methyl jasmonate moves in both phloem and xylem, promotes transport of jasmonate, and of photoassimilate even after proton transport is decoupled. *Planta* **226**, 541–551.

Ton J, Mauch-Mani B, 2004. β-Amino-butyric acid-induced resistance against necrotrophic pathogens is based on ABA-dependent priming for callose. *Plant Journal* **38**, 119–130.

Ton J, Davison S, Van Wees SCM, Van Loon LC, Pieterse CMJ, 2001. The *Arabidopsis* ISR1 locus controlling rhizobacteria-mediated induced systemic resistance is involved in ethylene signaling. *Plant Physiology* **125**, 652–661.

Ton J, Van Pelt JA, Van Loon LC, Pieterse CMJ, 2002. Differential effectiveness of salicylate-dependent and jasmonate/ethylene-dependent induced resistance in *Arabidopsis*. *Molecular Plant-Microbe Interactions* **15**, 27–34.

Ton J, Pieterse CMJ, Van Loon LC, 2006. The relationship between basal and induced resistance in *Arabidopsis*. In: Tuzun S, Bent E, eds. *Multigenic and Induced Systemic Resistance in Plants*. New York: Springer, pp. 197–224.

Ton J, D'Allesandro M, Jourdie V, Jakab G, Karlen D, Held M, Mauch-Mani B, Turlings TCJ, 2007. Priming by airborne signals boosts direct and indirect resistance in maize. *Plant Journal* **49**, 16–26.

Truitt CL, Wei HX, Pare PW, 2004. A plasma membrane protein from *Zea mays* binds with the herbivore elicitor volicitin. *The Plant Cell* **16**, 523–532.

Truman W, Bennett MH, Kubigsteltig I, Turnbull C, Grant M, 2007. *Arabidopsis* systemic immunity uses conserved defense signaling pathways and is mediated by jasmonates. *Proceedings of the National Academy of Sciences, USA* **104**, 1075–1080.

Tscharntke T, Thiessen S, Dolch R, Boland W, 2001. Herbivory, induced resistance, and interplant signal transfer in *Alnus glutinosa*. *Biochemical Systematics and Ecology* **29**, 1025–1047.

Van Loon LC, Bakker PAHM, Pieterse CMJ, 1998. Systemic resistance induced by rhizosphere bacteria. *Annual Review of Phytopathology* **36**, 453–483.

Van Peer R, Niemann GJ, Schippers B, 1991. Induced resistance and phytoalexin accumulation in biological control of fusarium wilt of carnation by *Pseudomonas* sp. strain WCS417r. *Phytopathology* **81**, 728–734.

Van Wees SCM, Luijendijk M, Smoorenburg I, Van Loon LC, Pieterse CMJ, 1999. Rhizobacteria-mediated induced systemic resistance (ISR) in *Arabidopsis* is not associated with a direct effect on expression of known defense-related genes but stimulates the expression of the jasmonate-inducible gene *Atvsp* upon challenge. *Plant Molecular Biology* **41**, 537–549.

Van Wees SCM, De Swart EAM, Van Pelt JA, Van Loon LC, Pieterse CMJ, 2000. Enhancement of induced disease resistance by simultaneous activation of salicylate- and jasmonate-dependent defense pathways in *Arabidopsis thaliana*. *Proceedings of the National Academy of Sciences, USA* **97**, 8711–8716.

Vanacker H, Lu H, Rate DN, Greenberg JT, 2001. A role for salicylic acid and NPR1 in regulating cell growth in *Arabidopsis*. *The Plant Journal* **28**, 209–216.

Vellosillo T, Martinez M, Lopez MA, Vicente J, Cascon T, Dolan L, Hamberg M, Castresana C, 2007. Oxylipins produced by the 9-lipoxygenase pathway in *Arabidopsis* regulate lateral root development and defense responses through a specific signaling cascade. *The Plant Cell* **19**, 831–846.

Vernooij B, Friedrich L, Morse A, Reist R, Kolditz-Jawhar R, Ward E, Uknes S, Kessmann H, Ryals J, 1994. Salicylic acid is not the translocated signal responsible for inducing systemic acquired resistance but is required in signal transduction. *Plant Cell* **6**, 959–965.

Vieira Dos Santos C, Delavault P, Letousy P, Thalouran P, 2003a. Defense gene expression analysis of *Arabidopsis thaliana* parasitized by *Orobanche ramosa*. *Phytopathology* **93**, 451–457.

Vieira Dos Santos C, Delavault P, Letousy P, Thalouran P, 2003b. Identification by suppression subtractive hybridization and expression analysis of *Arabidopsis thaliana* putative defense genes during *Orobanche ramosa* infection. *Physiological and Molecular Plant Pathology* **62**, 297–303.

von Dahl CC, Baldwin IT, 2004. Methyl jasmonate and cis-jasmone do not dispose of the herbivore-induced jasmonate burst in *Nicotiana attenuata*. *Physiologia Plantarum* **120**, 474–481.

von Dahl CC, Baldwin IT, 2007. Deciphering the role of ethylene in plant-herbivore interactions. *Journal of Plant Growth Regulation* **26**, 201–209.

Walters DR, McRoberts N, 2006. Plants and biotrophs: a pivotal role for cytokinins? *Trends in Plant Science* **11**, 581–586.

Walters D, Walsh D, Newton A, Lyon G, 2005. Induced resistance for plant disease control: maximising the efficacy of resistance elicitors. *Phytopathology* **95**, 1368–1373.

Walters DR, Cowley T, Weber H, 2006. Rapid accumulation of trihydroxy oxylipins and resistance to the bean rust pathogen *Uromyces fabae* following wounding in *Vicia faba*. *Annals of Botany* **97**, 779–784.

Walters DR, McRoberts N, Fitt BDL, 2008. Are green islands red herrings? Significance of green islands in plant interactions with pathogens and pests. *Biological Reviews* **83**, 79–102.

Wasternack C, 2007. Jasmonates: an update on biosynthesis, signal transduction and action in plant stress response, growth and development. *Annals of Botany* **100**, 681–697.

Wawrzynska A, Christiansen KM, Lan Y, Rodibaugh NL, Innes RW, 2008. Powdery mildew resistance conferred by loss of the ENHANCED DISEASE RESISTANCE 1 protein kinase is suppressed by a missense mutation in *KEEP ON GOING*, a regulator of abscisic acid signaling. *Plant Physiology* **148**, 1510–1522.

Weber H, 2002. Fatty acid derived signals in plants. *Trends in Plant Science* **7**, 217–224.

Wildermuth MC, Dewdney J, Wu G, Ausubel FM, 2001. Isochorismate synthase is required to synthesize salicylic acid for plant defense. *Nature* **414**, 562–565.

Williamson VM, Kumar A, 2006. Nematode resistance in plants: the battle underground. *Trends in Genetics* **22**, 396–403.

Wu J, Baldwin IT, 2009. Herbivory-induced signalling in plants: perception and action. *Plant Cell and Environment* **32**, 1161–1174.

Wu J, Hettenhausen C, Meldau S, Baldwin IT, 2007. Herbivory rapidly activates MAPK signaling in attacked and unattacked leaf regions but not between leaves of *Nicotiana attenuata*. *The Plant Cell* **19**, 1096–1122.

Wubben MJE, Jin J, Baum TJ, 2008. Cyst nematode parasitism of *Arabidopsis* thaliana is inhibited by salicylic acid (SA) and elicits uncoupled SA-independent pathogenesis-related gene expression in roots. *Molecular Plant-Microbe Interactions* **21**, 424–432.

Yan Z, Reddy MS, Ryu C-M, McInroy JA, Wilson M, Kloepper JW, 2002. Induced systemic protection against tomato late blight elicited by plant growth promoting rhizobacteria. *Phytopathology* **92**, 1329–1333.

Zarate SI, Kempema LA, Walling LL, 2007. Silverleaf whitefly induces salicylic acid defenses and suppresses effectual jasmonic acid defenses. *Plant Physiology* **143**, 866–875.

Zavala JA, Patankar AG, Gase K, Hui DQ, Baldwin IT, 2004. Manipulation of endogenous trypsin proteinase inhibitor production in *Nicotiana attenuata* demonstrates their function as antiherbivore defenses. *Plant Physiology* **134**, 1181–1190.

Zeier J, Delledonne M, Mishina T, Severi E, Sonoda M, Lamb C, 2004. Genetic elucidation of nitric oxide signalling in incompatible plant–pathogen interactions. *Plant Physiology* **136**, 2875–2886.

Zeringue HJ, 1992. Effects of C_6-C_{10} alkenals and alkanals on eliciting a defence response in the developing cotton boll. *Phytochemistry* **31**, 2305–2308.

Zhang S, Moyne A-L, Reddy MS, Kloepper JW, 2002. The role of salicylic acid in induced systemic resistance elicited by plant growth promoting rhizobacteria against blue mold of tobacco. *Biological Control* **25**, 288–296.

Zhang Z-P, Baldwin IT, 1997. Transport of [2-^{14}C] jasmonic acid from leaves to roots mimics wound-induced changes in endogenous jasmonic acid pools in *Nicotiana sylvestris*. *Planta* **203**, 436–441.

Zhou N, Tootle TL, Tsui F, Klessig DF, Glazebrook J, 1998. PAD4 functions upstream from salicylic acid to control defense responses in *Arabidopsis*. *The Plant Cell* **10**, 1021–1030.

Zipfel C, 2008. Pattern-recognition receptors in plant innate immunity. *Current Opinion in Immunology* **20**, 10–16.

Chapter 4

Plant Defense in the Real World: Multiple Attackers and Beneficial Interactions

4.1 Introduction

As we have seen in previous chapters, plants are equipped with an array of defense mechanisms to protect themselves against attack by pathogens, herbivores, and even other plants. Some of these defense mechanisms are preexisting or constitutive, while others are only activated upon attack. The primary immune response in plants evolved to recognize common features of organisms that interact with the plant and to convert this recognition into a defense response targeted specifically against the particular attacker (Jones & Dangl, 2006). But plants can also activate another line of defense, referred to as induced resistance (see Chapter 3). Depending on the type of organism attacking the plant, it can activate different types of induced resistance. It might, for example, activate systemic acquired resistance (SAR), induced systemic resistance (ISR), or wound-induced resistance. We saw in the previous chapter that salicylic acid (SA), jasmonic acid (JA), and ethylene (ET) are key players in the regulation of the signaling pathways involved in plant defense. When the plant is attacked, it responds by producing a specific blend of SA, JA, and ET—a sort of signal signature—which contributes to its primary induced defense response (De Vos *et al.*, 2005). The signaling pathways subsequently activated regulate different defense responses that are effective against different types of attackers. As indicated in Chapter 3, generally speaking, biotrophic pathogens tend to be sensitive to SA-mediated induced defenses, while necrotrophs and herbivorous insects tend to be more sensitive to defenses mediated by JA and ET (Glazebrook, 2005).

However, plants often need to deal with more than one type of attacker, sometimes simultaneously. Therefore, the regulatory and signaling mechanisms used by the plant in its defense must be able to adapt to this changing, hostile environment. It appears that the plant achieves this through cross-talk between the different signaling pathways. This cross-talk can have outcomes that are positive or negative, depending on whether the signal interactions are synergistic or mutually exclusive (Bostock, 2005). But activating defenses to ward off attackers also runs the risk of affecting interactions with beneficial microorganisms such as

Plant Defense, First Edition, by Dale Walters   © 2011 by Blackwell Publishing Ltd.

mycorrhizal fungi and nitrogen-fixing bacteria. This chapter examines, therefore, not just the mechanisms plants use to deal with multiple attackers, but also whether activation of defense compromises interactions with beneficial microbes.

4.2 Dealing with multiple attackers: cross-talk between signaling pathways

Cross-talk can have a negative functional outcome if there is an antagonistic interaction between two signaling pathways, such as might occur between SA- and JA-mediated defense response pathways. When this occurs, the negative outcome is manifested as a trade-off in defense against one or more types of attackers, for example between pathogens and pest resistance. However, as indicated above, cross-talk can also have a positive outcome. Below, we examine the trade-offs and positive outcomes that occur when individual defense signaling pathways are activated.

4.2.1 Trade-offs associated with triggering SA-mediated defenses

4.2.1.1 SA suppression of JA-induced defenses

There are many examples of negative cross-talk between the SA and JA signaling systems. As a result of such negative cross-talk, activation of the SA response should render the plant more susceptible to attackers that are resisted via JA-dependent defenses. Indeed, there are many examples of trade-offs between SA-dependent resistance against biotrophic pathogens and JA-dependent defenses against insects and necrotrophic pathogens. For example, tobacco plants expressing tobacco mosaic virus-induced SAR were more susceptible to grazing by the tobacco hornworm, *Manduca sexta*, than noninduced plants, while application of the chemical activator acibenzolar-*S*-methyl (ASM) to field-grown tomato plants reduced resistance to the beet armyworm *Spodoptera exigua* (Preston *et al.*, 1999; Thaler *et al.*, 1999). More recent studies showed that SA-mediated defenses that are triggered upon infection by a virulent strain of the biotrophic pathogen *Pseudomonas syringae* rendered infected tissues more susceptible to infection by the necrotrophic pathogen *Alternaria brassicicola* by suppressing the JA-signaling pathway (Figure 4.1) (Spoel *et al.*, 2007). SA-mediated suppression of JA-responsive genes is shown clearly in Figure 4.2. Application of SA to *Arabidopsis* plants leads to expression of the *PR-1* gene, while application of MeJA results in expression of the gene *PDF1.2*. However, when SA and MeJA are supplied together, although the *PR-1* gene is expressed, expression of *PDF1.2* is strongly suppressed (Koornneef & Pieterse, 2008).

Herbivory of tomato by the beet armyworm, *S. exigua*, leads to activation of JA-dependent responses, including production of JA and volatiles. Interestingly, tomato plants infested with the parasitic plant *Cuscuta pentagona* produced substantially less JA and volatiles, although they did contain elevated levels of SA (Figures 4.3 and 4.4) (Runyon *et al.*, 2008). When mutant tomato plants deficient in SA production were parasitized by *C. pentagona* and then infested with *S. exigua*, significantly more JA was produced. Taken together, these results hint at the possibility of SA suppression of JA responses in this tripartite interaction.

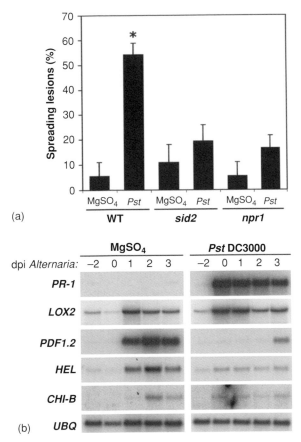

Figure 4.1 Biotroph infection locally suppresses JA-mediated defenses against necrotrophic *A. brassicicola* through SA and NPR1. (a) Percentage of spreading *A. brassicicola* lesions on WT, *sid2*, and *npr1* plants. Left halves of leaves were pressure-infiltrated with 10 mM $MgSO_4$ alone or with biotrophic virulent *P. syringae* pv. *tomato* (*Pst*) DC3000 (107 cfu/mL). After 2 days, the right halves of these leaves were challenge-inoculated with *A. brassicicola*. An asterisk indicates statistically significant differences compared with the control. Note: *sid2* plants cannot accumulate SA, while *npr1* plants are insensitive to SA. (b) RNA gel blot analysis of SA-responsive *PR-1* and JA/ethylene-responsive *PDF1.2*, *HEL*, *CHI-B*, and *LOX2* gene expression in *pad3* plants at different dpi (days postinoculation) with *A. brassicicola*. At −2 dpi, plants were infected with *Pst* DC3000, and at 0 dpi, challenge-inoculated with *A. brassicicola*. Only *A. brassicicola*-inoculated leaf halves were collected for RNA extraction. To check for equal loading, blots were stripped and hybridized for constitutively expressed ubiquitin (UBQ). Note: *pad3* plants are deficient in the phytoalexin camalexin. In WT *Arabidopsis*, resistance against *A. brassicicola* is established by two distinct mechanisms: (i) production of camalexin, and (ii) synthesis of JA and subsequent activation of a large set of defense genes. JA-dependent defenses against *A. brassicicola* are not affected by the *pad3* mutation. Therefore, by using *pad3* mutant plants, it is possible to examine the sole effect of JA-dependent defenses against *A. brassicicola*. (c) RNA gel blot analysis of JA/ethylene-responsive *PDF1.2*, *HEL*, and *CHI-B* gene expression in *pad3* and *pad3 npr1* plants at different dpi with *A. brassicicola*. To check for equal loading, rRNA was stained with ethidium bromide. (From Spoel *et al.* (2007), with permission of the National Academy of Sciences, USA.)

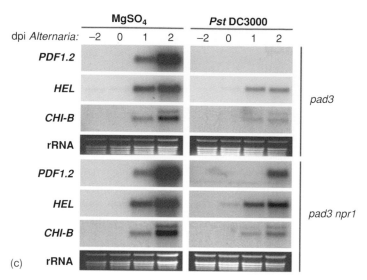

Figure 4.1 (*Continued*).

Given the examples described above of activation of SA-mediated defenses leading to suppression of JA-dependent defenses and increased susceptibility to herbivorous insects, it might be expected that plants compromised in their SA-regulated defenses might be more resistant to herbivorous insects. Indeed, there are reports supporting this supposition. For example, *Arabidopsis* mutants compromised in SA-dependent defense responses have been shown to exhibit enhanced resistance against feeding by the cabbage looper *Trichoplusia ni*, the Egyptian cotton worm *Spodoptera littoralis*, and the army beetworm *S. exigua* (Cui *et al.*, 2002; Stotz *et al.*, 2002; Cipollini *et al.*, 2004; Van Oosten *et al.*, 2008).

4.2.1.2   *Molecular basis of SA suppression of JA defenses*

So, what is the molecular basis for this SA suppression of JA-responsive genes? As we saw in Chapter 3, the regulatory protein NPR1 is required for transduction of the SA signal. Interestingly, SA-mediated suppression of JA-inducible gene expression is blocked in *npr1* mutants of *Arabidopsis*, demonstrating a crucial role for NPR1 in cross-talk between SA and JA signaling (Spoel *et al.*, 2003). A similar function for NPR1 in cross-talk was demonstrated in rice, where overexpression of *NPR1* suppressed JA-responsive transcription and enhanced susceptibility to insect herbivory (Yuan *et al.*, 2007). A further development in this story comes from work on wild tobacco in which NPR1 was silenced. Such plants accumulated increased levels of SA upon insect damage and were highly susceptible to herbivore attack (Rayapuram & Baldwin, 2007). These workers suggested that, in wild-type plants, NPR1 is needed to negatively regulate SA production during insect attack, thereby suppressing SA/JA cross-talk and thus allowing induction of JA-mediated defenses against herbivores.

Among the important regulators of SA-dependent defense responses are WRKY transcription factors, some of which have been shown to have a role in cross-talk between SA and JA. For example, overexpression of one of these transcription factors, *WRKY70*,

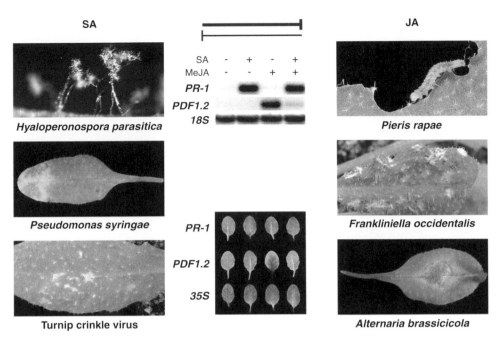

Figure 4.2   Schematic representation of mutually antagonistic cross-talk between SA- and JA-dependent defense-signaling pathways in *Arabidopsis*. SA-mediated defenses are predominantly effective against biotrophic pathogens, such as *Hyalopernospora parasitica*, *P. syringae*, and turnip crinkle virus, whereas JA-mediated defenses are primarily effective against herbivorous insects, such as *P. rapae* caterpillars and thrips (*Frankliniella occidentalis*), and against necrotrophic pathogens, such as *A. brassicicola*. Cross-talk between SA and JA signaling is typically visualized by monitoring the expression of SA-responsive genes, such as *PR-1*, and JA-responsive genes, such as *PDF1.2*. In the middle of the figure, a pharmacological experiment is shown in which exogenous application of 1 mM SA to the leaves of wild-type *Arabidopsis* Col-0 plants resulted in the accumulation of *PR-1* mRNA and activation of the *PR-1* promoter that is fused to the *GUS* reporter gene. Application of 100 µM MeJA resulted in the accumulation of *PDF1.2* mRNA and the activation of the *PDF1.2* promoter that is fused to the *GUS* reporter gene. The combined treatment with SA and MeJA resulted in strong SA-mediated suppression of the JA-responsive *PDF1.2* gene, thereby exemplifying the antagonistic effect of SA on JA signaling. Depending on the plant–attacker interaction, SA/JA cross-talk has been demonstrated in both directions. (Reproduced from Koorneef & Pieterse (2008), with permission.)

in *Arabidopsis*, led to enhanced expression of SA-responsive *PR* genes and suppression of MeJA-induced expression of *PDF1.2*. These results suggest that WRKY70 acts as a positive regulator of SA-mediated defenses, while suppressing JA responses (Li *et al.*, 2004).

Recent research suggests that SA/JA cross-talk is redox regulated. Glutaredoxins have been implicated in redox-dependent regulation of protein activities (Lemaire, 2004), and overexpression of one of these glutaredoxins, GRX480, completely abolished MeJA-induced *PDF1.2* gene expression (Ndamukong *et al.*, 2007). It was suggested that SA-activated NPR1 induces GRX480, and this, in turn, interacts with TGACG motif binding (TGA) transcription factors to suppress JA-responsive gene expression.

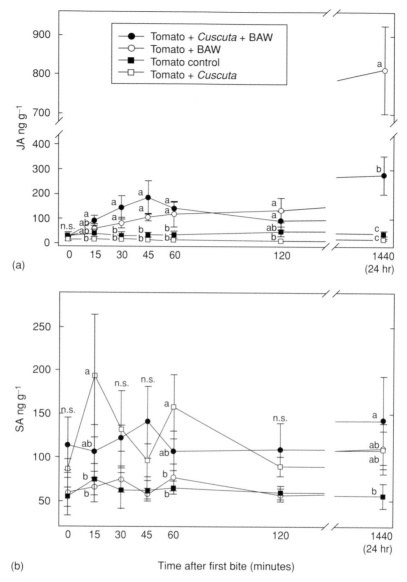

Figure 4.3   Time course of changes in JA (a) and SA (b) in unparasitized tomato plants and plants parasitized by *C. pentagona* in response to beet armyworm (BAW; *S. exigua*) feeding. Parasitized and unparasitized plants that did not receive insect feeding served as controls. Note breaks in the x-axis (a and b) and the y-axis (a). Different letters indicate significance differences within each time point; n.s., no significance between treatments. (Reproduced from Runyon *et al.* (2008), with permission.)

4.2.1.3   *Ecological costs of resistance to biotrophic versus necrotrophic pathogens*

Evidence suggests that necrotrophic interactions between plants and pathogens impart an ecological cost on plant resistance to biotrophic pathogens. This ecological cost is defined as resulting from any host mechanisms that provide resistance to biotrophic pathogens,

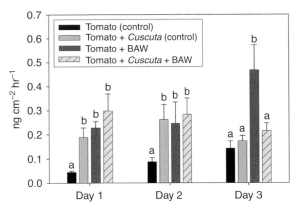

Figure 4.4   Total volatile production by unparasitized tomato plants and plants parasitized by *C. pentagona* on days 1 to 3 of beet armyworm (*S. exigua*) feeding. Parasitized and unparasitized plants that did not receive insect feeding served as controls. Different letters indicate significance differences within each day. (Reproduced from Runyon *et al.* (2008), with permission.)

but increase susceptibility to necrotrophs (Kliebenstein & Rowe, 2008). So, although a hypersensitive response (HR) is often associated with resistance to biotrophic pathogens, it can provide potential entry points for necrotrophic pathogens. For example, when leaves were pretreated with avirulent biotrophic bacteria capable of inducing an HR, and the necrotrophic pathogen *Botrytis cinerea* was placed in the middle of the pretreated zone, the resulting lesions were much larger than those produced by *B. cinerea* in the absence of bacterial pretreatment (Govrin & Levine, 2000). It is also suggested that there is a cost to the organism of possessing the machinery for programmed (or organized) cell death (Kliebenstein & Rowe, 2008). Thus, the very existence of plant signaling pathways that initiate an HR provides molecular targets that a necrotrophic pathogen could use to facilitate its infection and colonization by stimulating host cells to kill themselves. This is supported by evidence showing that the virulence of necrotrophic plant pathogens is increased by stimulation of the HR (Govrin & Levine, 2000; Van Baarlen *et al.*, 2007). These data suggest that plants that have evolved in an environment, biased toward resistance to biotrophic pathogens, may suffer from the inability to effectively resist necrotrophs.

4.2.1.4   Trade-offs with mutualistic symbioses

Induced resistance is a broad-spectrum resistance against microorganisms and it seems likely, therefore, that it will have an effect on interactions with mutualistic symbionts, such as mycorrhizal fungi and the nitrogen-fixing bacteria *Rhizobium* and *Bradyrhizobium*. Work in this area has, rather surprisingly, been limited to date. Nevertheless, some experiments have reported negative effects of SA application on nodule formation and functioning in legumes (Ramanujam *et al.*, 1998; Lian *et al.*, 2000). For example, following inoculation of alfalfa, with a compatible strain of *Rhizobium meliloti*, no accumulation of SA was detected. In contrast, when plants were inoculated with an incompatible strain of *R. meliloti* or a mutant of *Rhizobium* blocked in Nod factor synthesis, SA accumulated (Martínez-Abarca *et al.*, 1998). Further, application of SA to alfalfa plants prior to inoculation, with a compatible

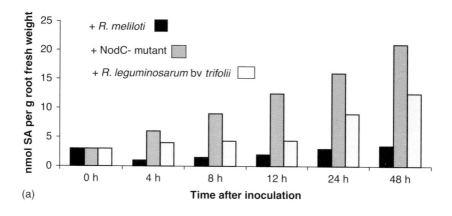

(a)

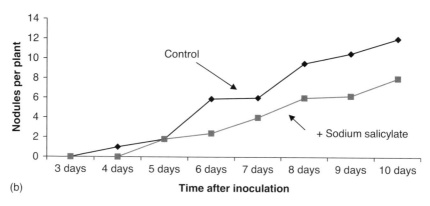

(b)

Figure 4.5   (a) Accumulation of salicylic acid in roots of alfalfa plants, noninoculated (0 hour), or inoculated with *R. meliloti* wild type (compatible), a mutant blocked in nod factor synthesis (NodC⁻), and *R. leguminosarum* bv. *trifolii* (incompatible). (b) Effect on nodulation of addition of 25 μM sodium salicylate to the plant growth solution 24 hours before inoculation of alfalfa plants with *R. meliloti* GR4. (Adapted from Martínez-Abarca *et al.* (1998), with permission of APS.)

Rhizobium strain, resulted in significantly reduced nodule formation (Figure 4.5). These results led the authors to suggest that, in addition to their well-established role in nodulation in legumes, nod factors suppress SA accumulation, thereby allowing nodule formation in a compatible interaction.

In terms of effects of SA-mediated defenses on mycorrhizal infection and colonization of plant roots, published studies are even scarcer. Treatment of barley with the SA functional analog ASM was found to have no effect on mycorrhizal infection of barley roots (Sonnemann *et al.*, 2002).

4.2.1.5   Effects of SA- and JA-mediated defenses on bacterial communities associated with plants

Induced resistance generates broad-spectrum defense against microbial pathogens, and it seems reasonable to assume that, as such, it is likely to affect a much wider range of microbes, such as bacterial communities associated with plants. Although any effects of induced resistance are likely to be greatest for endophytic bacterial communities, it might also affect epiphytic communities and rhizosphere bacterial communities. Using two mutants of *Arabidopsis thaliana* deficient in SA and JA signaling pathways, Kniskern *et al.* (2007) compared communities of endophytic and epiphytic bacteria. This showed that induction of SA-mediated defenses reduced endophytic bacterial community diversity, whereas epiphytic bacterial diversity was greater in plants deficient in JA-mediated defenses. As a result, it is possible that natural variation in the induction of systemic defenses among plants within a population or between populations might generate spatial and temporal variation in the community dynamics of microbes that depend on plants for survival.

Rhizosphere bacterial communities can also be affected by induced resistance. This was demonstrated by work that found visible differences in the rhizosphere community fingerprints of different *Arabidopsis* SAR mutants, although there was no clear decrease of rhizosphere diversity because of constitutive SAR expression (Hein *et al.*, 2008).

4.2.2   Triggering SA-dependent defenses does not always compromise defense against insect herbivores

One might be forgiven for assuming from the previous sections that eliciting SA-dependent defenses always leads to suppression of JA-dependent defenses. Nothing could be further from the truth. In fact, some workers could find no effect (Ajlan & Potter, 1992; Inbar *et al.*, 1998), while others reported positive effects. For example, inoculation of tomato leaves with the biotrophic pathogen *P. syringae* pv. *tomato* induced resistance against not just *P. syringae* pv. *tomato*, but also against the corn earworm, *Helicoverpa zea* (Figure 4.6). Antiherbivore defenses were activated in protected plants since both PI mRNA levels and activity of polyphenol oxidase were increased systemically (Stout *et al.*, 1999). Other work on tomato showed that, although induction of SAR had a positive effect on performance of the cabbage looper, there was no effect on thrips, spider mites, or hornworm caterpillars (Thaler *et al.*, 2002).

More recent experiments by Van Oosten *et al.* (2008) showed that eliciting SAR produced different effects on specialist and generalist herbivores. Thus, SAR had no significant effect on the level of resistance in *Arabidopsis* against the specialist insect herbivore, *Pieris rapae*, although it led to significant growth reductions in the generalist *S. exigua*. These findings were thought to be in line with the observation that *P. rapae* larvae are generally insensitive to JA-dependent defenses and, therefore, are also likely to be insensitive to the effect that SA exerts on JA signaling. The effects of SAR on *S. exigua* were correlated with increased expression of the JA-dependent genes *PDF1.2* and *HEL*. At first sight, the increased expression of JA-dependent defenses in plants expressing SA-mediated SAR might seem incongruous. However, SAR induced by prior inoculation with *P. syringae* pv. *tomato* has been shown to lead, not just to elevated levels of SA, but also to increases in JA and ET (De Vos *et al.*, 2005), and it is possible therefore that the increased expression of *PDF1.2* and *HEL* observed by Van Oosten *et al.* (2008) was the result of pathogen-induced increases in JA and ET.

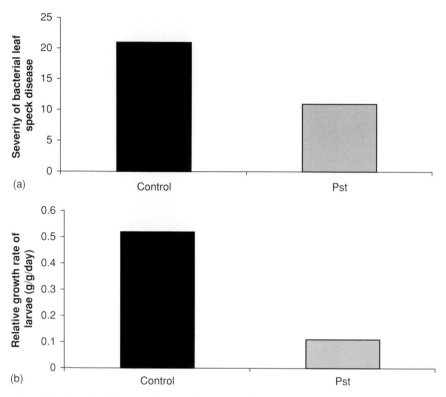

Figure 4.6 (a) Severity of bacterial speck disease (number of lesions per leaflet) caused by *P. syringae pv. tomato* (*Pst*) on leaflets from undamaged leaves, and from leaves previously inoculated with *Pst*. (b) Relative growth rates of *H. zea* larvae reared for 24 hours on leaflets from undamaged leaves, and from leaves previously inoculated with *Pst*. (Data reproduced from Stout *et al.* (1999), with permission of Elsevier.)

4.2.3 Trade-offs and positive outcomes associated with triggering JA-dependent defenses

Although much is known about the effects of SA-mediated signaling on JA-mediated defenses, much less is known about the effects of JA-mediated signaling on SA-mediated defenses. JA treatment was shown to suppress SA-dependent expression of *PR* genes, while SA-mediated gene expression and defenses were found to be enhanced in JA- and coronatine-insensitive mutants (Niki *et al.*, 1998; Kloek *et al.*, 2001; Li *et al.*, 2004). In another study, an *Arabidopsis* mutant in which JA responses are constitutively expressed (*cev1*) was used to study resistance to *Erysiphe cichoracearum*, *P. syringae*, and the aphid, *Myzus persicae*. Although this constitutive activation of JA defenses provided resistance to all three attackers, this occurred despite the suppression of SA-dependent defenses (Ellis *et al.*, 2002).

In contrast to the paucity of available data on effects of activation of JA-dependent defenses on SA-dependent responses, there are more reports of positive outcomes. For example, defoliation of soybean by larvae of the moth *Pseudoplusia includens* led to substantial reductions in natural infection by the pathogen *Diaporthe phaseolorum*, while laboratory studies showed that herbivory of *Rumex obtusifolius*, by the beetle *Gastrophysa viridula*, reduced infection by the rust fungus *Uromyces rumicis* (Hatcher *et al.*, 1994; Padgett *et al.*, 1994). In fact, field studies using the latter system showed that herbivory by *G. viridula* reduced infection not just by the biotroph *U. rumicis*, but also to the hemibiotroph *Venturia rumicis*, and the necrotroph *Ramularia rubella* (Hatcher & Paul, 2000). These reports are supported by work on broad beans, *Vicia faba*, and the rust fungus, *Uromyces fabae*. Mechanical wounding of broad bean leaves led to local and systemic increases in JA concentrations and also to substantial reductions in rust infection in both the wounded first leaf and unwounded second leaves (Figure 4.7) (Walters *et al.*, 2006).

An example of induced resistance being effective simultaneously against pathogens and insects comes from work on cucumber. Here, induction of ISR against bacterial wilt disease, caused by *Erwinia tracheiphila*, was associated with reduced feeding by the cucumber beetle, the vector for the bacterial pathogen. Apparently, induction of ISR was associated with reduced concentrations of the secondary plant metabolite cucurbitacin, a powerful feeding stimulant for cucumber beetles (Zehnder *et al.*, 2001). However, because ISR against *E. tracheiphila* was also effective in the absence of the beetle vector, it seems that disease protection was provided by a combination of reduced beetle feeding and induction of defenses. Perhaps this dual effectiveness against a pathogen and a pest should not come as a surprise, since ISR is JA/ET-dependent, as is defense against chewing insects.

Experiments reported by De Vos *et al.* (2006) highlight the importance of not making generalizations and assumptions based on existing knowledge. They examined the effect of feeding by caterpillars of *P. rapae*, which leads to enhanced production of JA and ET, but unaltered levels of SA, on resistance against various pathogens on *Arabidopsis*. To their surprise, they found no protection afforded against the necrotroph *A. brassicicola*, despite the fact that it is resisted via JA-dependent defenses, while local and systemic protection was observed against the biotrophic turnip crinkle virus, which is normally resisted via SA-activated defenses. Clearly, the spectrum of effectiveness in protection they observed was very different from that predicted on the basis of existing knowledge. This study demonstrates something that should be abundantly clear by this stage in reading this book: that defenses triggered upon attack are amazingly complex and much research is required to unravel what has taken evolution many millions of years to produce. The next section (Section 4.2.4) looks at what is currently known about how plants manage to integrate the various signaling events to come up with a specific defense response.

Before we leave this section, let us turn our attention briefly to plant responses to herbivory and competition with neighboring plants. Plant responses to herbivory and competition both require resources, so it is pertinent to ask if there is a trade-off between the two. In fact, it is known that resource allocation to defense against herbivory can reduce plant competitive ability, and competition with neighboring plants can compromise defense responses (Cipollini, 2004; Zavala *et al.*, 2004; Kurashige & Agrawal, 2005). But what are the sensing mechanisms underlying these trade-offs between herbivory and competition? Plants use specific photoreceptors to sense light spectral changes that signal the onset of periods of increased competition from other plants. When these spectral signals are perceived, the plant will activate rapid morphological responses (e.g., elongation of stems and petioles, increased leaf angles) that increase the plant's ability to find and exploit light

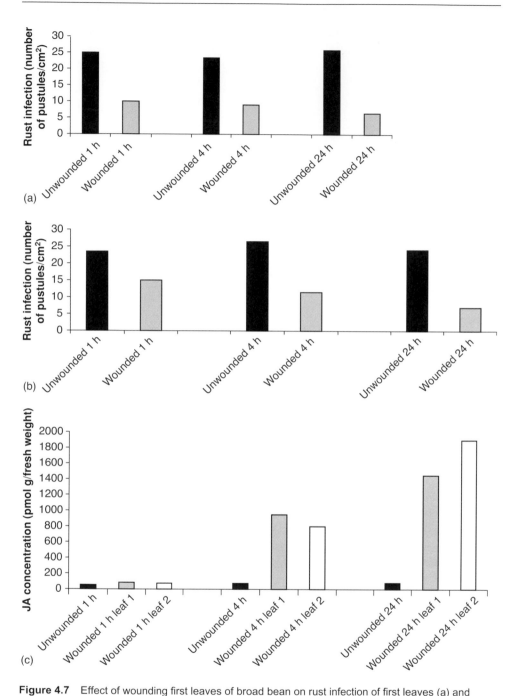

Figure 4.7   Effect of wounding first leaves of broad bean on rust infection of first leaves (a) and second leaves (b). Leaves were inoculated with rust 1, 4, and 24 hours after wounding. (c) Effect of wounding first leaves on JA concentrations in first and second leaves of broad bean. (Reproduced from Walters *et al.* (2006), with permission of Oxford University Press.)

opportunities (Smith, 1995). One of the signals perceived is far-red radiation (FR). Plants are able to remotely detect the proximity of neighboring plants by perception of an increase in the ratio of FR to red (R) light, using the photoreceptor phytochrome B (Ballaré *et al.*, 1990). Work on wild tobacco (*Nicotiana longiflora*) showed that reflected FR induced a dramatic down-regulation of chemical defenses and increased the performance of the specialist herbivore, *M. sexta* (Izaguirre *et al.*, 2006). This effect of FR on plant defense was found to be mediated by phytochrome B. Subsequent work showed that the plant solves the dilemma of whether to invest resources in competition or defense against herbivory by modulating its sensitivity to JA using information on the risk of competition, sensed by phytochrome (Moreno *et al.*, 2009). This selective desensitization to JA would save plant resources by reducing investment in defense, while avoiding the inhibitory effects of JA on plant growth.

4.2.4   Putting it all together: orchestrating the appropriate defense response

The extensive cross-talk between the SA and JA signaling pathways highlights the complex nature of signaling for disease and pest resistance. But can plants use this cross-talk to deal with multiple attackers, and if so, how is it achieved? A glimpse of how plants integrate insect- and pathogen-induced signals into specific defense responses was provided by work on *Arabidopsis*. Researchers tracked the dynamics of SA, JA, and ET signaling following attack by pathogens and insect pests. When they compared global gene expression profiles, considerable overlap was found in the changes induced by pathogens and insects. Thus, all the different pathogen and insect attackers stimulated JA biosynthesis, although most of the changes in JA-responsive gene expression were attacker specific (De Vos *et al.*, 2005). The workers suggested that although SA, JA, and ET play a primary role in orchestrating plant defense, the final defense response is shaped by other regulatory mechanisms, such as cross-talk between different pathways (Figure 4.8). For example, ET produced by *Arabidopsis* following damage by *P. rapae* was found to prime the plant for enhanced SA-mediated defenses activated following infection by turnip crinkle virus (De Vos *et al.*, 2006).

As mentioned above, cross-talk between signaling pathways is thought to help plants to decide on the most appropriate defensive strategy to employ, depending on the lifestyle of the attacker. In an interesting twist, attackers appear to have evolved the ability to manipulate plants for their own benefit, by suppressing induced defenses or modulating the defense signaling network (Pieterse & Dicke, 2007). For example, it seems that herbivorous nymphs of the silverleaf whitefly (*Bemisia tabaci*) may activate the SA signaling pathway as a decoy strategy to sabotage JA-dependent defenses and so enhance insect performance (Zarate *et al.*, 2007). However, plants might fight back and suppress specific direct defenses in the favor of more general indirect defense strategies when attacked by specialist herbivores, which are resistant to their host plant's direct, chemical defenses (Kahl *et al.*, 2000). Perhaps unsurprisingly, pathogens can trick plants into mounting inappropriate defenses by producing hormones or their functional mimics to manipulate the signaling network of the plant (Robert-Seilaniantz *et al.*, 2007). A good example is the production of coronatine, a functional mimic of JA, by virulent *P. syringae* (Nomura *et al.*, 2005).

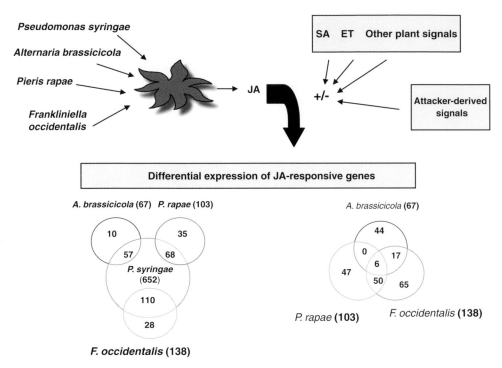

Figure 4.8   Differential expression of JA-responsive genes upon attack by JA-inducing pathogens and insects. Attack on *Arabidopsis* by *P. syringae*, *A. brassicicola*, *P. rapae*, or *F. occidentalis* resulted in a strong increase in the production of JA, and a concomitant change in the expression of a large number of JA-responsive genes (numbers are given within parenthesis). Nevertheless, the overlap among the JA-responsive genes between the different *Arabidopsis*-attacker combinations was relatively low (number of overlapping genes between the indicated *Arabidopsis*–attacker combinations is given in the Venn diagrams). SA and ET have been demonstrated to cross-communicate with the JA pathway. Hence, depending on the amount and timing of their production, SA and ET may have positive or negative effects on the expression of specific sets of JA-responsive genes. In addition, so-far-unidentified plant- or attacker-derived signals, or physiological conditions that are inflicted by the attacker, may be involved in modulating JA-responsive gene expression. (Redrawn from De Vos *et al.* (2005), with permission.)

The intricate complexity of signaling cross-talk in interactions between a plant and its attackers comes from some very interesting work examining the responses of tobacco, *Nicotiana attenuata*, to the two most common lepidopteran herbivores in the plant's native habitat: the specialist tobacco hornworm, *M. sexta*, and the generalist beet armyworm, *S. exigua* (Diezel *et al.*, 2009). Feeding by tobacco hornworm larvae elicits marked bursts of JA and ET, while attack by larvae of the beet armyworm elicits smaller bursts of JA and ET, but a much greater SA burst. Oral secretion (OS) from the insects was found to mimic these JA, ET, and SA responses. There was, however, an important difference in the OS from *S. exigua* compared to that from *M. sexta*. The *S. exigua* OS contained the enzyme glucose oxidase (GOX), the activity of which leads to the production of H_2O_2 and SA. This burst

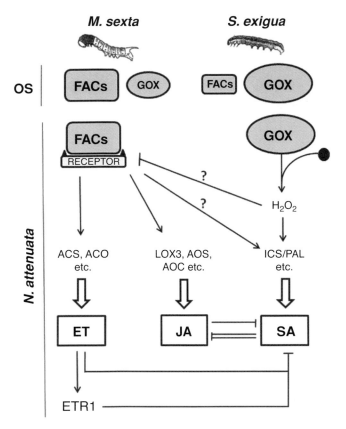

Figure 4.9   A model of OS-elicited JA/ET/SA cross-talk in tobacco hornworm and beet armyworm-attacked *N. attenuata* leaves. When tobacco hornworm larvae feed on *N. attenuata* plants, FACs and GOX proteins from the larval OS contaminate wounds, which results in the amplification of levels of JA and ET, but not SA. OS from beet armyworm larvae, compared to those from tobacco hornworm larvae, contain less of the FACs known to elicit JA and ET production, but more GOX activity, which, together with Glc (•), results in higher H_2O_2 and SA accumulations. Increased SA levels are known to antagonize the JA burst and the deployment of JA-dependent defenses. During tobacco hornworm feeding, perception of the FAC-induced ET burst through ETR1 suppresses SA production. The herbivore-elicited ET burst, therefore, is an important regulator of herbivory-elicited SA/JA cross-talk. ACS, 1-amino-cyclopropane-1-carboxylic acid synthase; ACO, 1-amino-cyclopropane-1-carboxylic acid oxidase; LOX3, lipoxygenase 3; AOS, allene oxide synthase; AOC, allene oxide cyclase; ICS, isochorismate synthase; PAL, phenylalanine ammonia lyase. (Reproduced from Diezel *et al.* (2009), with permission.)

of SA antagonizes the JA burst and the deployment of JA-dependent defenses. In contrast, OS produced by *M. sexta* during feeding induces an ET burst, which is perceived by the plant, leading to the suppression of SA generation. Therefore, by suppressing the SA burst, the ET burst produced, following *M. sexta* feeding, allows JA-mediated defense activation to occur (Figure 4.9) (Diezel *et al.*, 2009). This work suggests that, in addition to cross-talk between SA and JA, cross-talk between SA and ET is responsible for tuning the responses of *N. attenuata* plants to the elicitors present in the OS of the two herbivores.

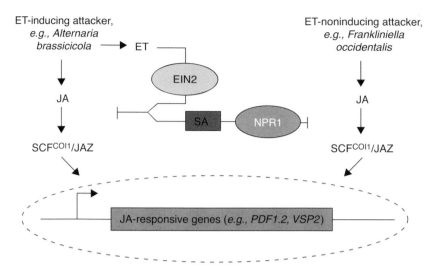

Figure 4.10 A working model illustrating the role of ET in modulating the NPR1 dependency of SA/JA cross-talk. Attack of *Arabidopsis* by the necrotrophic fungus *A. brassicicola* and the herbivorous insect *F. occidentalis* results in the biosynthesis of JA and the activation of the JA signaling pathway in which the E3 ubiquitin ligase SCF^{COI1} and jasmonate ZIM-domain (JAZ) proteins that repress the transcription of JA-responsive genes are central components (see Box 3.3; Chini et al., 2007; Thines et al., 2007). Activation of the JA signaling cascade leads to the activation of JA-responsive genes such as *PDF1.2* and *VSP2*. SA suppresses JA-responsive gene expression in an NPR1-dependent manner. However, when ET signaling is stimulated, such as upon infection by the ET-inducer *A. brassicicola*, the NPR1 dependency of SA/JA cross-talk is bypassed, resulting in wild-type levels of suppression of JA signaling in the *npr1-1* mutant background. (Reproduced from Leon-Reyes *et al.* (2009), with permission.)

In Section 4.2.1.2, we saw that the regulatory protein NPR1 is an important regulator of SA/JA cross-talk. Work on *Arabidopsis* demonstrated that ET strongly affects the requirement for NPR1 in the antagonistic effect of SA on JA-dependent defenses. Exogenous application of gaseous ET or production of ET during pathogen attack bypassed the dependency of SA/JA cross-talk on NPR1 (Leon-Reyes *et al.*, 2009). This research shows that the final outcome of the interaction between the SA and JA signaling pathways during the interaction between a plant and its attackers can be shaped by ET. This modulating role of ET in SA/JA cross-talk was highlighted by the observation that the antagonistic effect of SA on MeJA-induced resistance against thrips that do not induce ET production was controlled by NPR1, while SA-mediated suppression of JA-dependent resistance against the necrotroph *A. brassicicola*, which induces both JA and ET, functioned independently of ET (Figure 4.10) (Leon-Reyes *et al.*, 2009) (Box 4.1).

Box 4.1   Out of sight but not out of mind: effects of root herbivory on shoot defenses

In terms of studying plant defense, and in particular induced resistance, the vast majority of effort has concentrated on shoots. This is perhaps unsurprising, given the perceived difficulties associated with working on roots. As a PhD student in the late 1970s, I studied the effects of a foliar infection on root physiology and biochemistry. I was often asked, including by those who should have known better, why I was bothering to work on what happens in roots when it was the shoot, and not the root, that was infected. The fact is that what happens in any one part of a plant can exert an effect, sometimes profound, on other, distal parts of the plant. The phenomenon of induced resistance is a good case in point—attack on one leaf can affect defense responses in other leaves. It stands to reason therefore that should roots be attacked, consequences are likely to be detected above ground. This has indeed be shown to be the case, with increasing evidence suggesting that root herbivory modulates shoot defenses (Wäckers & Bezemer, 2003; van Dam et al., 2005; Rasmann & Turlings, 2007).

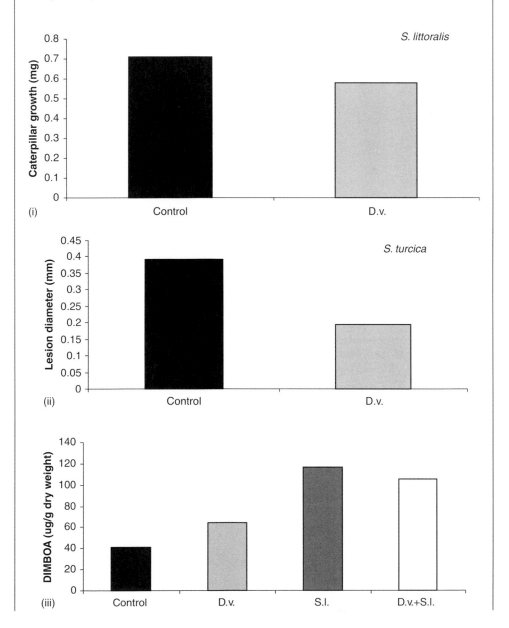

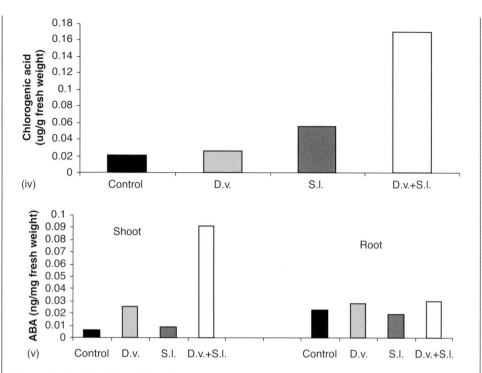

Figure A   Root herbivore-induced resistance in maize against *S. littoralis* and *S. turcica*. Leaves were challenged with *S. littoralis* caterpillars and *S. turcica* spores 4 days after application of *D. virgifera virgifera* larvae to the roots. (i) Average growth of *S. littoralis* caterpillars over a feeding period of 11 hours on control plants and on *D. v. virgifera*-infested plants (D.v.). (ii) Average lesion diameter on leaves of plants infected with *S. turcica* 3 days after inoculation. (iii and iv) Average concentrations of DIMBOA and chlorogenic acid in leaves of herbivore-infested plants. Leaves were harvested after 4 days of below-ground infestation by *D. v. virgifera* (*D.v.*), 2 days of above-ground infestation by *S. littoralis* (*S.l.*), or simultaneous infestation by *D. v. virgifera* and *S. littoralis* (*D.v. + S.l.*). (v) Average ABA concentrations in shoots and roots of herbivore-infested maize plants. Leaves were harvested after 4 days of below-ground infestation by *D. v. virgifera* (*D.v.*), 2 days of above-ground infestation by *S. littoralis* (*S.l.*), or simultaneous infestation by *D. v. virgifera* and *S. littoralis* (*D.v. + S.l.*). (Redrawn from Erb *et al.* (2009), with permission of Blackwell Publishing Ltd.)

Recent work on maize showed that root herbivory by the western root cornworm, *Diabrotica virgifera virgifera*, induced above-ground resistance against the generalist herbivore *S. littoralis* and the necrotrophic pathogen *Setosphaeria turcica* (Figure Ai and Aii) (Erb *et al.*, 2009). This was associated with increased levels of 2,4-dihydroxy-7-methoxy-1,4-benzoxazin-3-one (DIMBOA) in shoots of roots damaged by *D. v. virgifera*, and in addition, the induction of chlorogenic acid was primed for subsequent infestation by *S. littoralis* (Figure Aiii and Aiv).

Interestingly, herbivory by *D. v. virgifera* led to increased abscisic acid (ABA) levels in roots and shoots, and also primed plants for augmented ABA production after subsequent attack by *S. littoralis* caterpillars (Figure Av). Infestation of roots by *D. v. virgifera* led to JA accumulation locally, but not systemically, in leaves. In fact, ABA was the only hormone to accumulate systemically following root herbivory by *D. v. virgifera*. So, why was ABA accumulating in these plants? The researchers found that root infestation by *D. v. virgifera* caused significant reductions in leaf water content. They suggested that induction of ABA represented the response of the plant to water stress induced by *D. v. virgifera* herbivory on the roots. Since they also found that exogenously applied ABA induced resistance against *S. turcica*, and ABA is known to be involved as a regulatory signal in defense against pathogens (Ton *et al.*, 2005), the authors further suggested that *D. v. virgifera*-induced ABA was sufficient to induce resistance to the pathogen. However, ABA alone did not appear to be responsible for the induced resistance against *S. littoralis*. Nevertheless, ABA is a strong candidate to act as a systemic signal in this particular system, although, as the authors point out, the above-ground resistance observed is likely to involve additional layers of regulation (Erb *et al.*, 2009).

As indicated in Section 4.2.3 above, there are examples where attack by chewing insects, or mechanical damage to leaves, leads to enhanced resistance to both biotrophic and necrotrophic pathogens. The work of Erb *et al.* (2009) is in line with such observations, although it goes much further, taking a leap in the right direction, and considers the whole plant. Future work considering how plants defend themselves against multiple attackers needs to take into account both shoots and roots. Such work is challenging, but necessary, if we are to truly understand what plants in the real world have to deal with.

4.3   Can beneficial plant–microbe interactions induce resistance in plants?

4.3.1   Introduction

Interactions between plant roots and beneficial microbes include mycorrhizal symbioses, nodulation of legume roots by bacteria of the genera *Rhizobium* and *Bradyrhizobium*, and interactions with plant growth-promoting rhizobacteria (PGPR). There are many reports of disease control provided by mycorrhizal infection and colonization, and there is also increasing evidence that mycorrhizas can induce resistance in their host plants. As we saw in Chapter 2, certain strains of PGPR can also induce resistance (known as induced systemic resistance; ISR). In addition, certain fungal endophytes and various plant growth-promoting fungi are also known to provide protection against pathogens and pests. Since we have already dealt with ISR in some detail, let us turn our attention to the protection provided by mycorrhizas and beneficial fungi.

4.3.2   Induction of resistance by mycorrhizas

Arbuscular mycorrhizas (AM) are found in most terrestrial plant species. The organisms responsible, AM fungi (AMF), are obligate biotrophs since they are dependent on the host plant for reproduction and survival (Smith & Read, 1997). The AM symbiosis provides benefits to the host plant in terms of fitness, for example, through enhanced mineral nutrition, as well as increased ability to cope with biotic and abiotic stresses (Pozo & Azcón-Aguilar, 2007). AM can provide growth benefits to plants, particularly under nutrient-limiting conditions, although it is well established that plants vary in their responsiveness to AMF and many plant and environmental factors can influence the response (Smith & Read, 1997; Jakobsen *et al.*, 2002).

Various reports have indicated an effect of mycorrhizal infection on plant defense against pathogens (Pozo & Azcón-Aguilar, 2007). Thus, colonization of tomato roots by the AMF *Glomus mosseae* induced cell defense responses, new isoforms of the hydrolytic enzymes chitinase and β-1,3-glucanase, and local and systemic protection against *Phytophthora parasitica* (Cordier *et al.*, 1998; Pozo *et al.*, 2002). Similarly, infection of barley roots

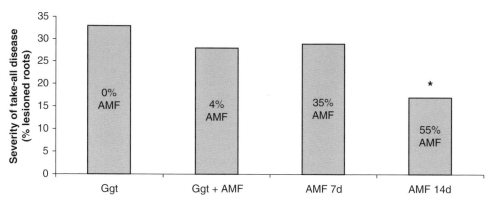

Figure 4.11   Effect of prior inoculation with an arbuscular mycorrhizal fungus (AMF) on the severity of take-all disease on roots of barley. Split root systems of barley were used, with one side of the split root system sequentially inoculated with the AMF *G. mosseae*, and the other side inoculated with the take-all fungus *G. graminis* (Ggt). Treatments were inoculation with Ggt only (Ggt only); simultaneous inoculation with AMF and Ggt (SI); inoculation with AMF 7 days prior to inoculation with Ggt (AMF 7d); inoculation with AMF 14 days prior to inoculation with Ggt. Percentages within the histogram bars show the amount of the root colonized by AMF. The asterisk (*) indicates significantly different from other treatments. (Reproduced from Khaosaad *et al.* 2007, with permission of Elsevier.)

with *G. mosseae* induced systemic protection against the take-all fungus *Gaeumannomyces graminis* f.sp. *tritici* (Figure 4.11) (Khaosaad *et al.*, 2007). But infection by AM fungi does not only affect resistance responses in the root. Mycorrhizal symbiosis in *Medicago truncatula* was associated with a complex pattern of changes in gene expression in both roots and shoots, and was also associated with increased resistance to the shoot pathogen *Xanthomonas campestris* (Liu *et al.*, 2007). Importantly, these effects on disease were not due to altered phosphorus nutrition in the plant as a result of the AM symbiosis.

Mycorrhizal infection and colonization have also been shown to modify the effectiveness of induced resistance. Thus, at low and medium levels of colonization of barley roots by *Glomus etunicatum*, ASM had either no effect or decreased foliar infection by powdery mildew, while high levels of mycorrhizal colonization increased mildew infection (Sonnemann *et al.*, 2005). It seems that although infection by AMF usually leads to reduced infection by soilborne pathogens, effects on shoot pathogens are dependent on the lifestyle of the attacker (Pozo & Azcón-Aguilar, 2007). Thus, resistance induced by AMF in shoots appears to be effective against necrotrophic pathogens and generalist chewing insects, but less so against biotrophic pathogens. Interestingly, it is proposed that the establishment of a functional mycorrhizal symbiosis requires a partial suppression of SA-dependent responses in the plant, compensated for by an enhancement of JA-regulated responses (Pozo & Azcón-Aguilar, 2007). Therefore, the resistance induced by AMF correlates with a potentiation of JA-dependent defenses and the priming of tissues for effective defense activation following challenge (Figure 4.12).

It is important to remember that AMF positively affects the nutritional status of the host plant, and that host resistance can be limited by resource supply. Therefore, mutualistic microbes might enhance host resistance by direct resistance induction or by reducing resource limitations and, thus, allocation costs of resistance (see Chapter 5). Disentangling these two possible mechanisms will require future research.

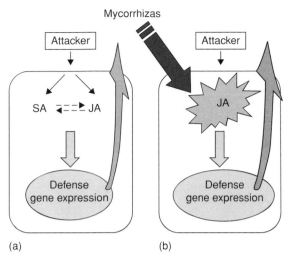

Figure 4.12   A model illustrating priming of JA-dependent responses in mycorrhizal plants. (a) Upon attacker recognition, the plant produces the defense-related signals JA, ET, and SA in different proportions. Cross-talk among the pathways coordinated by these signals fine-tunes the appropriate response. (b) Mycorrhiza formation primes the tissues for a quicker and more effective activation of JA-dependent defense responses upon attack, resulting in enhanced resistance. (Reproduced from Pozo & Azcón-Aguilar (2007), with permission of Elsevier.)

4.3.3   Resistance induced by endophytic and other beneficial fungi

Fungi of the family Clavicipitaceae have a long history of associations with plants, ranging from mutualism to antagonism, and there are many reports of protection against pathogens and pests provided by Ascomycete endophytes (Schardl *et al.*, 2004). These particular fungal endophytes are confined to the upper parts of plants, grow intercellularly, and have a narrow host range. It seems that the beneficial effects of these fungal endophytes are due to the antimicrobial and insecticidal effects of the alkaloids they produce. However, there are many reports that fungal endophytes, both shoot- and root-inhabiting, can induce resistance (Walters, 2009).

A root-colonizing basidiomycete, *Piriformospora indica*, was shown to provide a strong growth-promoting effect, but was also found to induce resistance to foliar pathogen infection (Waller *et al.*, 2005). Thus, powdery mildew infection of barley leaves was reduced significantly in plants where the roots were colonized by *P. indica*. The pathogen appeared to be arrested by a combination of the HR and cell wall-associated defenses (Figure 4.13) (Waller *et al.*, 2005).

There are also many reports of induction of systemic resistance in a variety of plants against parasitic nematodes using fungal endophytes (Sikora *et al.*, 2008). For example, four endophytic isolates of the fungus *Fusarium oxysporum* were shown to induce systemic resistance in banana to the burrowing nematode *Radopholus similis*, while the endophytic *F. oxysporum* strain 162 induced resistance in tomato to the nematode *Meloidogyne incognita* (Vu *et al.*, 2006; Dababat & Sikora, 2007).

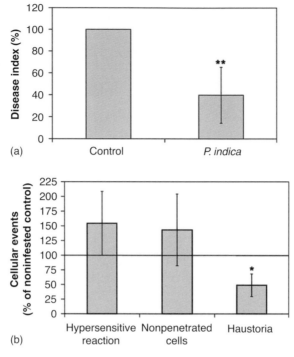

Figure 4.13 Systemic disease resistance conferred by *P. indica*. (a) Severity of powdery mildew infection (disease index) was calculated as colonies produced by the powdery mildew fungus *Blumeria graminis* f.sp. *hordei* on the youngest leaf of 3-week-old barley plants (cultivar Ingrid), with roots either not infested (control) or infested (*P. indica*). (b) Cellular responses to powdery mildew attack were evaluated by counting cells showing an active defense response, a hypersensitive response of the whole cell (hypersensitive reaction), a local defense stopping a penetration attempt (nonpenetrated cell), or a successful penetration (haustoria), visible as the successful formation of a fungal haustorium in the cell. Single asterisk (*) and double asterisks (**) indicate statistically significant differences between leaves of infested and noninfested plants. (From Waller *et al.* (2005), with permission of the National Academy of Sciences, USA.)

In addition to fungal endophytes, certain plant growth-promoting fungi have also been reported to induce resistance in plants. For example, preinoculation of cucumber plants with the beneficial fungus *Trichoderma asperellum* T203 led to a JA/ET-dependent systemic resistance that was associated with primed *PR* gene expression in response to pathogen attack (Shoresh *et al.*, 2005). A similar response was observed in *Arabidopsis*, where a nonpathogenic strain of *Penicillium* induced ISR against the bacterial pathogen *P. syringae*. The induced ISR was shown to be dependent on JA, ET, and NPR1, and was associated with priming for JA/ET-inducible gene expression (Hossain *et al.*, 2008).

4.4 Conclusions

Plants are faced with the difficulty of fending off multiple attackers while trying to en-sure that interactions with beneficial microorganisms are not compromised. In general, it appears that defense against biotrophic pathogens and certain insect pests (e.g., aphids) is accomplished via the SA signaling pathway, while warding off necrotrophs involves activating JA/ET-inducible defenses. As we have seen, there are cases where activation of SA-mediated defenses compromises JA-mediated defenses, although examples in the opposite direction are rarer. As discussed above, cross-talk seems to be used to fine-tune the plant's response to particular attackers.

However, it is interesting to note that simultaneous activation of ISR and SAR resulted in enhanced induced protection against the pathogen *P. syringae* pv. *tomato* DC3000. In this case, the SA-dependent SAR pathway and the JA/ET-dependent ISR pathway operated independently and additively to provide increased protection against this particular pathogen. It would seem that negative cross-talk between SA- and JA-dependent defenses is confined to specific combinations of inducers, plants, and attackers. Further, it would appear that the ability to defend against attackers that are resisted through JA-mediated defenses is only compromised when there is very strong activation of SA-dependent defenses (Pieterse & Van Loon, 2007). As we have seen above, in other cases, there is little, if any, antagonism between the SA and JA pathways.

Systemic resistance induced by beneficial microbes such as mycorrhizal fungi and PGPR appears to be regulated predominantly via the JA/ET pathway, and is based on priming for enhanced defense. There is, therefore, overlap between signaling pathways in response to parasitic and beneficial organisms, highlighting the fine balance between defense against attackers and acquisition of benefits via mutualistic associations. It is noteworthy that the resistance induced by beneficial microbes appears to be based on priming, since activating defenses only when the attacker is present will use fewer resources than triggering defenses in the absence of an aggressor. This issue of the costs associated with plant defenses is important and is dealt with in the next chapter.

Recommended reading

Bostock RM, 2005. Signal crosstalk and induced resistance: straddling the line between cost and benefit. *Annual Review of Phytopathology* **43**, 545–580.

Koornneef A, Pieterse CMJ, 2008. Cross talk in defense signaling. *Plant Physiology* **146**, 839–844.

Pieterse CMJ, Schaller A, Mauch-Mani B, Conrath U, 2006. Signaling in plant resistance responses: divergence and cross-talk of defense pathways. In: Tuzun S, Bent E, eds. *Multigenic and Induced Systemic Resistance in Plants*. New York: Springer, pp. 166–196.

Pozo MJ, Azcón-Aguilar C, 2007. Unraveling mycorrhiza-induced resistance. *Current Opinion in Plant Biology* **10**, 393–398.

Robert-Seilaniantz A, Navarro L, Bari R, Jones JD, 2007. Pathological hormone imbalances. *Current Opinion in Plant Biology* **10**, 372–379.

References

Ajlan AM, Potter DA, 1992. Lack of effect of tobacco mosaic virus-induced systemic acquired resistance on arthropod herbivores in tobacco. *Phytopathology* **82**, 647–651.

Ballaré CL, Scopel AL, Sánchez RA, 1990. Far-red radiation reflected from adjacent leaves: an early signal of competition in plant canopies. *Science* **247**, 329–332.

Bostock RM, 2005. Signal crosstalk and induced resistance: straddling the line between cost and benefit. *Annual Review of Phytopathology* **43**, 545–580.

Chini A, Fonseca S, Fernandez G, Adie B, Chico JM, Lorenzo O, Garcia-Casado G, Lopez-Vidriero I, Lozano FM, Ponce MR, Micol JL, Solano R, 2007. The JAZ family of repressors is the missing link in jasmonate signalling. *Nature* **448**, 666–672.

Cipollini D, 2004. Stretching the limits of plasticity: can a plant defend against both competitors and herbivores? *Ecology* **85**, 28–37.

Cipollini D, Enright S, Traw B, Bergelson J, 2004. Salicylic acid inhibits jasmonic acid-induced resistance of *Arabidopsis thaliana* to *Spodoptera exigua*. *Molecular Ecology* **13**, 1643–1653.

Cordier C, Pozo MJ, Barea JM, Gianinazzi S, Gianinazzi-Pearson V, 1998. Cell defense responses associated with localized and systemic resistance to *Phytophthora parasitica* induced in tomato by an arbuscular mycorrhizal fungus. *Molecular Plant-Microbe Interactions* **11**, 1017–1028.

Cui J, Jander G, Racki LR, Kim PD, Pierce NE, Ausubel FM, 2002. Signals involved in *Arabidopsis* resistance to *Trichoplusia ni* caterpillars induced by virulent and avirulent strains of the phytopathogen *Pseudomonas syringae*. *Plant Physiology* **129**, 1–14.

Dababat AEA, Sikora RA, 2007. Induced resistance by the mutualistic endophyte, *Fusarium oxysporum* strain 162, toward *Meloidogyne incognita* on tomato. *Biocontrol Science and Technology* **17**, 969–975.

De Vos M, Van Oosten VR, Van Poecke RMP, Van Pelt JA, Pozo MJ, Mueller MJ, Buchala AJ, Métraux J-P, Van Loon LC, Dicke M, Pieterse CMJ, 2005. Signal signature and transcriptome changes of *Arabidopsis* during pathogen and insect attack. *Molecular Plant-Microbe Interactions* **18**, 923–937.

De Vos M, Van Zaanen W, Koornneef A, Korzelius JP, Dicke M, Van Loon LC, Pieterse CMJ, 2006. Herbivore-induced resistance against microbial pathogens in *Arabidopsis*. *Plant Physiology* **142**, 352–363.

Diezel C, von Dahl CC, Gaquerel E, Baldwin IT, 2009. Different lepidopteran elicitors account for cross-talk in herbivory-induced phytohormone signaling. *Plant Physiology* **150**, 1576–1586.

Ellis C, Karafyllidis I, Turner JG, 2002. Constitutive activation of jasmonate signaling in an *Arabidopsis* mutant correlates with enhanced resistance to *Erysiphe cichoracearum*, *Pseudomonas syringae*, and *Myzus persicae*. *Molecular Plant-Microbe Interactions* **15**, 1025–1030.

Erb M, Flors V, Karlen D, de Lange E, Planchamp C, D'Alessandro M, Turlings TCJ, Ton J, 2009. Signal signature of aboveground-induced resistance upon belowground herbivory in maize. *Plant Journal* **59**, 292–302.

Glazebrook J, 2005. Contrasting mechanisms of defense against biotrophic and necrotrophic pathogens. *Annual Review of Phytopathology* **43**, 205–227.

Govrin EM, Levine A, 2000. The hypersensitive response facilitates plant infection by the necrotrophic pathogen. *Botrytis cinerea. Current Biology* **10**, 751–757.

Hatcher PE, Paul ND, 2000. Beetle grazing reduces natural infection of *Rumex obtusifolius* by fungal pathogens. *New Phytologist* **146**, 325–333.

Hatcher PE, Paul ND, Ayres PG, Whittaker JR, 1994. Interactions between *Rumex* spp., herbivores and a rust fungus: *Gastrophysa viridula* grazing reduces subsequent infection by *Uromyces rumicis*. *Functional Ecology* **8**, 265–272.

Hein JW, Wolfe GV, Blee KA, 2008. Comparison of rhizosphere bacterial communities in *Arabidopsis thaliana* mutants for systemic acquired resistance. *Microbial Ecology* **55**, 333–343.

Hossain MM, Sultana F, Kubota M, Koyama H, Hyakumachi M, 2008. Systemic resistance to bacterial leaf speck pathogen in *Arabidopsis thaliana* induced by the culture filtrate of a plant growth promoting fungus (PGPF) *Phoma* sp. GS8-1. *Journal of General Plant Pathology* **74**, 213–221.

Inbar M, Doostdar H, Sonoda RM, Leibee GL, Mayer RT, 1998. Elicitors of plant defensive systems reduce insect densities and disease incidence. *Journal of Chemical Ecology* **24**, 135–149.

Izaguirre MM, Mazza CA, Biodini M, Baldwin IT, Ballaré CL, 2006. Remote sensing of future competitors: impacts on plant defenses. *Proceedings of the National Academy of Sciences, USA* **103**, 7170–7174.

Jakobsen I, Smith SE, Smith FA, 2002. Function and diversity of arbuscular mycorrhizae in carbon and mineral nutrition. In: Sanders IR, Van der Heijden MGA, eds. *Mycorrhizal Ecology*, Vol. 157. Berlin, Germany: Springer-Verlag, pp. 75–92.

Jones JDG, Dangl JL, 2006. The plant immune system. *Nature* **444**, 323–329.

Kahl J, Siemens DH, Aerts RJ, Gäbler R, Kühnemann F, Preston CA, Baldwin IT, 2000. Herbivore-induced ethylene suppresses a direct defense but not a putative indirect defense against an adapted herbivore. *Planta* **210**, 336–342.

Khaosaad T, Garcia-Garrido JM, Steinkellner S, Vierheilig H, 2007. Take-all disease is systemically reduced in roots of mycorrhizal barley plants. *Soil Biology and Biochemistry* **39**, 727–734.

Kliebenstein DJ, Rowe HC, 2008. Ecological costs of biotrophic versus necrotrophic pathogen resistance, the hypersensitive response and signal transduction. *Plant Science* **174**, 551–556.

Kloek AP, Verbsky ML, Sharma SB, Schoelz JE, Vogel J, Klessig DF, Kunkel BN, 2001. Resistance to *Pseudomonas syringae* conferred by an *Arabidopsis thaliana* coronatine-insensitive (coi1) mutation occurs through two distinct mechanisms. *Plant Journal* **26**, 509–522.

Kniskern JM, Traw MB, Bergelson J, 2007. Salicylic acid and jasmonic acid signaling defense pathways reduce natural bacterial diversity on *Arabidopsis thaliana*. *Molecular Plant-Microbe Interactions* **20**, 1512–1522.

Koornneef A, Pieterse CMJ, 2008. Cross talk in defense signaling. *Plant Physiology* **146**, 839–844.

Kurashige NS, Agrawal AA, 2005. Phenotypic plasticity to light competition and herbivory in *Chenopodium album* (Chenopodiaceae). *American Journal of Botany* **92**, 21–26.

Lemaire S, 2004. The glutaredoxin family in oxygenic photosynthetic organisms. *Photosynthesis Research* **79**, 305–318.

Leon-Reyes A, Spoel SH, De Lange E, Abe H, Kobayashi M, Tsuda S, Millenaar FF, Welschen RAM, Ritsema T, Pieterse CMJ, 2009. Ethylene modulates the role of NONEXPRESSOR OF PATHOGENESIS-RELATED GENES1 in cross-talk between salicylate and jasmonate signaling. *Plant Physiology* **149**, 1797–1809.

Li J, Brader G, Palva ET, 2004. The WRKY70 transcription factor: a node of convergence for jasmonate-mediated and salicylate-mediated signals in plant defense. *The Plant Cell* **16**, 319–331.

Lian B, Zhou X, Miransari M, Smith DL, 2000. Effects of salicylic acid on the development and root nodulation of soybean seedlings. *Journal of Agronomy and Crop Sciences* **185**, 187–192.

Liu JY, Maldonado-Mendoza I, Lopez-Meyer M, Cheung F, Town CD, Harrison MJ, 2007. Arbuscular mycorrhizal symbiosis is accompanied by local and systemic alterations in gene expression and an increase in disease resistance in the shoots. *Plant Journal* **50**, 529–544.

Martínez-Abarca F, Herrera-Cervera JA, Bueno P, Sanjuan J, Bisseling T, Olivares J, 1998. Involvement of salicylic acid in the establishment of the *Rhizobium meliloti*-alfalfa symbiosis. *Molecular Plant-Microbe Interactions* **11**, 153–155.

Moreno JE, Tao Y, Chory J, Ballaré CL, 2009. Ecological modulation of plant defense via phytochrome control of jasmonate sensitivity. *Proceedings of the National Academy of Sciences, USA* **106**, 4935–4940.

Ndamukong I, Abdallat AA, Thurow C, Fode B, Zander M, Weigel R, Gatz, C, 2007. SA-inducible *Arabidopsis* glutaredoxin interacts with TGA factors and suppresses JA-responsive *PDF1.2* transcription. *Plant Journal* 128–139.

Niki T, Mitsuhara I, Seo S, Ohtsubo N, Ohashi Y, 1998. Antagonistic effect of salicylic acid and jasmonic acid on the expression of pathogenesis-related (PR) protein genes in wounded mature tobacco leaves. *Plant and Cell Physiology* **39**, 500–507.

Nomura K, Melotto M, He SY, 2005. Suppression of host defense in compatible plant-*Pseudomonas syringae* interactions. *Current Opinion in Plant Biology* **8**, 361–368.

Padgett GB, Russin BM, Snow JP, Boethel DJ, Berggren GT, 1994. Interactions among the soybean looper (Lepidoptera: Noctuidae), three cornered alfalfa hopper (Homoptera: Membracidae), stem canker, and red crown rot in soybean. *Journal of Entomological Science* **29**, 110–119.

Pieterse CMJ, Dicke M, 2007. Plant interactions with microbes and insects: from molecular mechanisms to ecology. *Trends in Plant Science* **12**, 564–569.

Pieterse CMJ, Van Loon LC, 2007. Signalling cascades involved in induced resistance. In: Walters D, Newton A, Lyon G, eds. *Induced Resistance for Plant Defence: A Sustainable Approach to Crop Protection.* Oxford: Blackwell Publishing Ltd., pp. 65–88.

Pozo MJ, Azcón-Aguilar C, 2007. Unraveling mycorrhiza-induced resistance. *Current Opinion in Plant Biology* **10**, 393–398.

Pozo MJ, Cordier C, Dumas-Gaudot E, Gianinazzi S, Barea JM, Azcón-Aguilar C, 2002. Localized versus systemic effect of arbuscular mycorrhizal fungi on defence responses to *Phytophthora* infection in tomato plants. *Journal of Experimental Botany* **53**, 525–534.

Preston CA, Lewandowski C, Enyedi AJ, Baldwin IT, 1999. Tobacco mosaic virus inoculation inhibits wound-induced jasmonic acid-mediated responses within but not between plants. *Planta* **209**, 87–95.

Ramanujam MP, Abdul Jaleel V, Kumaravelu G, 1998. Effect of salicylic acid on nodulation, nitrogenous compounds and related enzymes of *Vigna mungo. Biologia Plantarum* **41**, 307–311.

Rasmann S, Turlings TCJ, 2007. Simultaneous feeding by aboveground and belowground herbivores attenuates plant-mediated attraction of their respective natural enemies. *Ecology Letters* **10**, 926–936.

Rayapuram C, Baldwin IT, 2007. Increased SA in NPR1-silenced plants antagonizes JA and JA-dependent direct and indirect defenses in herbivore-attacked *Nicotiana attenuata* in nature. *Plant Journal* **52**, 700–715.

Robert-Seilaniantz A, Navarro L, Bari R, Jones JD, 2007. Pathological hormone imbalances. *Current Opinion in Plant Biology* **10**, 372–379.

Runyon JB, Mescher MC, De Moraes CM, 2008. Parasitism by *Cuscuta pentagona* attenuates host plant defenses against insect herbivores. *Plant Physiology* **146**, 987–995.

Schardl CL, Leutchmann A, Spiering MJ, 2004. Symbioses of grasses with seedborne fungal endophytes. *Annual Review of Plant Biology* **55**, 315–340.

Shoresh M, Yedidia I, Chet I, 2005. Involvement of jasmonic acid/ethylene signalling pathway in the systemic resistance induced in cucumber by *Trichoderma asperellum* T203. *Phytopathology* **95**, 76–84.

Sikora RA, Pocasangre L, zum Felde A, Niere B, Vu TT, Dababat AA, 2008. Mutualistic endophytic fungi and *in planta* suppressiveness to plant parasitic nematodes. *Biological Control* **46**, 15–23.

Smith H, 1995. Physiological and ecological function within the phytochrome family. *Annual Review of Plant Physiology and Plant Molecular Biology* **46**, 289–315.

Smith SE, Read DJ, 1997. *Mycorrhizal Symbiosis.* London: Academic Press.

Sonnemann I, Finkhaeuser K, Wolters V, 2002. Does induced resistance in plants affect the belowground community? *Applied Soil Ecology* **21**, 179–185.

Sonnemann I, Streicher NM, Wolters V, 2005. Root associated organisms modify the effectiveness of chemically induced resistance in barley. *Soil Biology and Biochemistry* **37**, 1837–1842.

Spoel SH, Koornneef A, Claessens SMC, Korzelius JP, Van Pelt JA, Mueller MJ, Buchala AJ, Métraux J-P, Brown R, Kazan K, Van Loon LC, Dong XN, Pieterse CMJ, 2003. NPR1 modulates cross-talk between salicylate- and jasmonate-dependent defense pathways through a novel function in the cytosol. *The Plant Cell* **15**, 760–770

Spoel SH, Johnson JS, Dong X, 2007. Regulation of tradeoffs between plant defenses against pathogens with different lifestyles. *Proceedings of the National Academy of Sciences, USA* **104**, 18842–18847.

Stout MJ, Fidantsef AL, Duffey SS, Bostock RM, 1999. Signal interactions in pathogen and insect attack: systemic plant-mediated interactions between pathogens and herbivores of the tomato, *Lycopersicon esculentum*. *Physiological and Molecular Plant Pathology* **54**, 115–130.

Stotz HU, Koch T, Biedermann A, Weniger K, Boland W, Mitchell-Olds T, 2002. Evidence for regulation of resistance in *Arabidopsis* to Egyptian cotton worm by salicylic and jasmonic acid signaling pathways. *Planta* **214**, 648–652.

Thaler JS, Fidantsef AL, Duffey SS, Bostock RM, 1999. Trade-offs in plant defense against pathogens and herbivores: a field demonstration of chemical elicitors of induced resistance. *Journal of Chemical Ecology* **25**, 1597–1609.

Thaler JS, Fidantsef AL, Bostock RM, 2002. Antagonism between jasmonate- and salicylate-mediated induced plant resistance: effects of concentration and timing of elicitation on defense-related proteins, herbivore, and pathogen performance in tomato. *Journal of Chemical Ecology* **28**, 1131–1159.

Thines B, Katsir L, Melotto M, Niu Y, Mandaokar A, Liu GH, Nomura K, He SY, Howe GA, Browse J, 2007. JAZ repressor proteins are targets of the SCF[COI1] complex during jasmonate signalling. *Nature* **448**, 661–666.

Ton J, Jakab G, Toquin V, Flors V, Iavicoli A, Maeder MN, Métraux J-P, Mauch-Mani B, 2005. Dissecting the beta-aminobutyric acid-induced priming phenomenon in *Arabidopsis*. *The Plant Cell* **17**, 987–999.

Van Baarlen P, Woltering EJ, Staats M, Van Kan JAL, 2007. Histochemical and genetic analysis of host and non-host interactions of *Arabidopsis* with three *Botrytis* species: an important role for cell death control. *Molecular Plant Pathology* **8**, 41–54.

Van Dam NM, Raaijmakers CE, Van Der Putten WH, 2005. Root herbivory reduces growth and survival of the shoot feeding specialist *Pieris rapae* on *Brassica nigra*. *Entomologia Experimentalis et Applicata* **115**, 161–170.

Van Oosten VR, Bodenhausen N, Reymond P, Van Pelt JA, Van Loon LC, Dicke M, Pieterse CMJ, 2008. Differential effectiveness of microbially induced resistance against herbivorous insects in *Arabidopsis*. *Molecular Plant-Microbe Interactions* **21**, 919–930.

Vu T, Hauschild R, Sikora RA, 2006. *Fusarium oxysporum* endophytes induced systemic resistance against *Radopholus similes* on banana. *Nematology* **8**, 847–852.

Wäckers FL, Bezemer TM, 2003. Root herbivory induces an above-ground indirect defense. *Ecology Letters* **6**, 9–12.

Waller F, Achatz B, Baltruschat H, Fodor J, Becker K, Fischer M, Heier T, Huckelhoven R, Neumann C, von Wettstein D, Franken P, Kogel KH, 2005. The endophytic fungus *Piriformospora indica* reprograms barley to salt-stress tolerance, disease resistance, and higher yield. *Proceedings of the National Academy of Sciences, USA* **102**, 13386–13391.

Walters DR, 2009. Are plants in the field already induced? Implications for practical disease control. *Crop Protection* **28**, 459–465.

Walters DR, Cowley T, Weber H, 2006. Rapid accumulation of trihydroxy oxylipins and resistance to the bean rust pathogen *Uromyces fabae* following wounding in *Vicia faba*. *Annals of Botany* **97**, 779–784.

Yuan Y, Zhong S, Li Q, Zhu Z, Lou Y, Wang L, Wang M, Li Q, Yang D, He Z, 2007. Functional analysis of rice *NPR1*-like genes reveals that *OsNPR1/NH1* is the rice orthologue conferring disease resistance with enhanced herbivore susceptibility. *Plant Biotechnology Journal* **5**, 313–324.

Zarate SI, Kempema LA, Walling LL, 2007. Silverleaf whitefly induces salicylic acid defenses and suppresses effectual jasmonic acid defenses. *Plant Physiology* **143**, 866–875.

Zavala JA, Patankar AG, Gase K, Baldwin IT, 2004. Constitutive and inducible trypsin proteinase inhibitor incurs large fitness costs in *Nicotiana attenuata*. *Proceedings of the National Academy of Sciences, USA* **101**, 1607–1612.

Zehnder GW, Murphy JF, Sikora EJ, Kloepper JW, 2001. Application of rhizobacteria for induced resistance. *European Journal of Plant Pathology* **107**, 39–50.

Chapter 5
The Evolution of Plant Defense

5.1 Introduction

All living organisms are a potential source of food. It seems plausible therefore that defense mechanisms in autotrophic organisms would have arisen soon after the origin of heterotrophy. As a result, the appearance of defense mechanisms against parasitism and predation probably pre-dates the move of green plants to the land (Heath, 1987, 1991). Although the defenses that arose initially are not necessarily those existing today, as we have seen in Chapter 2, the array of defenses available to plants is staggering. Some of these defenses are constitutive, that is, always expressed in the plant, while others are induced following attack.

Many of the defenses present in plants are active against more than one attacker. Indeed, many of the individual defense components play other roles in the plant that might originally have been, and perhaps still are, more important than their role in defense against attackers. Thus, defenses might be effective against pathogens and insect herbivores, but might also be important in plant responses to abiotic stress. For example, a waxy plant surface, capable of deterring fungal invasion or insect attack, might have originally evolved in response to environmental pressures such as drought. Other defenses exploit processes that occur during normal plant development. Deposition of hydroxyproline-rich glycoproteins might strengthen cell walls against internal mechanical stresses during morphogenesis (Ye & Varner, 1991), but their involvement in plant defense might be the result of secondary linking of their induction to stress signals arising from damage caused by fungal pathogens (Heath, 1991). It seems plausible, therefore, that if any plant defense component is used against more than one stressor, and also has a role in plant growth and development, that component is likely to be subject to multiple selective pressures, of which a particular attacker might only be a minor part.

5.2 Hypotheses of plant defense

For many decades, plant biologists have been intrigued by the interactions between plants and herbivores, and in particular by the fact that plants seem very well defended against

Plant Defense, First Edition, by Dale Walters   © 2011 by Blackwell Publishing Ltd.

attacks from herbivores. Indeed, a central goal in the study of plant–herbivore interactions has been to explain and predict phenotypic, genetic, and geographic variation in plant defense. This quest has been guided by a series of hypotheses that, individually, have served as frameworks for studying patterns of defense against herbivores, and in particular, the pattern of constitutive defense.

An assumption of most of the hypotheses discussed in the following pages is that plant defense is costly. It is important, therefore, to ask whether such costs exist. Defenses are thought to be costly to the plant because they divert energy and resources away from other plant processes (e.g., growth and reproduction). Therefore, in situations where no attackers are present, individual plants that are less well defended should be fitter than better defended plants. The costs associated with diverting resources away from growth and other plant processes toward defense are known as allocation costs (see Section 5.5.1). There are other costs associated with defense, such as ecological costs, including effects of defense on mutualistic associations, but these are dealt with in Chapter 4.

Defense is an expensive business and the biosynthesis of secondary metabolites is certainly costly to the plant, requiring precursors from primary metabolism, enzymes, and cofactors such as adenosine triphosphate (ATP) and nicotinamide adenine dinucleotide phosphate (reduced) (NADPH), to drive the biochemical reactions (Gershenzon, 1994a, 1994b). Rates of photosynthesis are usually sufficient to provide an adequate supply of carbon substrates for biosynthetic reactions such as terpenoid formation, while, because nitrogen uptake by plants is limited, synthesis of nitrogen-containing compounds such as alkaloids can compete with protein synthesis for precursors (Harborne, 1993). It has been estimated that terpenoids are less costly to synthesize than alkaloids (2.6 g of photosynthetically produced carbon per gram of secondary metabolite for terpenoids, compared to 5 g for alkaloids) (Gulmon & Mooney, 1986). With defense being so expensive, plants are faced with the dilemma of concentrating valuable resources on growth or on defense (Herms & Mattson, 1992). It is possible that the allocation costs associated with defense might maintain genetic variation within plant populations by preventing alleles that code for high levels of defense from becoming fixed. Costs of resistance have been demonstrated, as with brassica genotypes varying in myrosinase activity. Genotypes exhibiting high myrosinase activity were more resistant to herbivory than low myrosinase genotypes, but at a price, since seed production was lower in these plants (Mitchell-Olds *et al.*, 1996). However, work examining costs of trichome production in 17 potato cultivars could find no evidence for allocation costs (Kaplan *et al.*, 2009).

A potential complicating factor in attempting to detect costs of defense is tolerance (Stamp, 2003). Tolerance refers to traits that reduce the impact of damage on plant fitness, and includes flexible rates of photosynthesis and nutrient uptake, allocation patterns, and developmental plasticity (resistance refers to traits that reduce the amount of damage). Apparently, plants have one of three strategies: (1) well-developed tolerance and poor defense, (2) well-developed defense and poor tolerance, or (3) something between these two extremes (van der Meijden *et al.*, 1988). Tolerance could incur allocation costs and ecological costs, and is known to affect plant fitness. For example, grazing-tolerant plants are competitively inferior to grazing-intolerant plants in the absence of herbivores (Painter, 1987). It has been suggested that the more a plant allocates to tolerance, the more difficult it will be to detect any trade-off between growth and tolerance or between defense and tolerance (Mole, 1994; Stamp, 2003).

5.2.1   The growth–differentiation balance hypothesis

The growth–differentiation balance (GDB) hypothesis is based on the premise that there is a physiological trade-off between growth (cell division and enlargement) and differentiation processes (chemical and morphological changes leading to cell maturation and specialization) in plants (Loomis, 1953). This trade-off exists because secondary metabolism and structural reinforcement are physiologically constrained in growing cells, and because they divert resources away from the generation of new leaf area. Thus, according to Herms & Mattson (1992), plants face a dilemma: to grow fast enough to compete with other plants, or to maintain the defenses required to ward off attackers. The GDB hypothesis integrates the trade-off between growth and secondary metabolism with responses of net assimilation rate (NAR) and relative growth rate (RGR) to the availability of resources, and predicts that resource availability (nutrients and water) will exert a parabolic effect on the concentration of secondary metabolites (Figure 5.1) (Herms & Mattson, 1992). NAR reflects the balance between carbon gain in photosynthesis and carbon losses via respiration and other processes, per unit leaf area per unit time. Therefore, NAR integrates environmental effects on net carbon acquisition at the whole plant level over a specified growth period. RGR is the product of NAR and leaf area ratio (the ratio of total leaf area to total plant mass) (Lambers & Poorter, 1992).

The GDB hypothesis is based mainly on three physiological assumptions (Herms & Mattson, 1992): (1) the major determinant of phenotypic variation in RGR is the differential investment of photosynthate into new leaf area (Körner, 1991), (2) NAR (and therefore photosynthesis also) is less sensitive to the availability of nutrients than is RGR (McDonald, 1990), and (3) synthesis of secondary metabolites diverts resources away from production of new plant tissue, and vice versa (Chapin, 1989).

The GDB hypothesis predicts that rapidly growing plants will have low levels of secondary metabolites, because production of new leaves is supported by export of

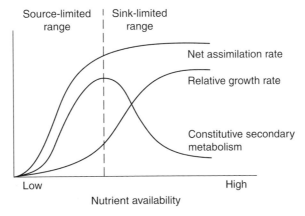

Figure 5.1 Responses of relative growth rate, net assimilation rate, and constitutive secondary metabolism across a gradient of nutrient availability as predicted by the growth–differentiation balance hypothesis. In source-limited plants, a positive correlation is predicted between growth and secondary metabolism, while in sink-limited plants the correlation is predicted to be negative. Reproduced from Glynn *et al.* (2007) and Herms & Mattson (1992), with permission.

photosynthate from source leaves, leaving little for synthesis of secondary metabolites (Figure 5.1). Exposure of plants to moderate levels of nutrient deficiency limits the growth of sink tissues such as new leaves, with little effect on NAR. Thus, photosynthate that would otherwise be exported to young, rapidly growing leaves, accumulates instead in source leaves, where they could be used to produce secondary metabolites (Figure 5.1). Under conditions of severe nutrient deficiency, where NAR is limited, both growth and secondary metabolism are predicted to decrease, since both processes are carbon limited (Herms & Mattson, 1992).

Herms & Mattson (1992) summarized the GDB hypothesis with a mechanistic conceptual model of the evolution of plant resource allocation patterns (Figure 5.2). The model incorporates the effects of natural selection by both plant competition and herbivory, with the evolutionary outcome mediated by resource availability. In the model, competition selects for growth, while herbivory selects for allocation to secondary metabolism, and as a result, gives rise to life history strategies. Plant species that are growth-dominated have adaptations that optimize the benefits of minimal investment in defense, such as inducible resistance and highly bioactive secondary metabolites. In contrast,

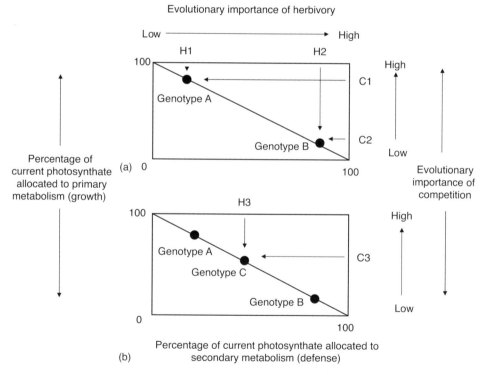

Figure 5.2 A conceptual model of the effects of competition (C) and herbivory (H) on the evolution of plant resource allocation patterns in varying environments. (a) Stable polymorphism (genotypes A and B) is maintained by disruptive selection in environments in which the evolutionary importance of herbivory relative to competition is low and high, respectively. (b) Directional selection exerted on populations (a) and (b) in an environment in which herbivory and competition are of equal importance, results in the evolution of genotype C. (Reproduced from Herms & Mattson (1992), with permission.)

differentiated-dominated plant species possess adaptations that optimize the benefits of maximal retention and economy of acquired resources.

Unfortunately, there have been few rigorous tests of the GDB hypothesis, although nutrient availability was shown to exert a quadratic effect on terpene concentrations in camphorweed, with the highest terpene concentrations occurring at a moderate level of nitrate (Mihaliak & Lincoln, 1985). A quadratic effect of nitrate on two phenolics was also observed in tomato, again with the greatest concentration of the phenolics in plants grown at moderate nitrate levels (Wilkens *et al.*, 1996). More recently, Glynn *et al.* (2007) tested the GDB hypothesis by quantifying temporal variation in RGR, NAR, and phenylpropanoid concentrations of two species of willow across five fertility levels. They found that initially, as fertility increased, RGR increased and total phenylpropanoids declined, although NAR was not affected. Thereafter, NAR and phenylpropanoids declined under low fertility, giving rise to a quadratic response of secondary metabolism across the nutrient gradient. However, the parabolic responses observed in this work were transient, leading the authors to suggest that the GDB hypothesis could be integrated with models of optimal phenotypic plasticity to predict that the parabolic response represents a temporary state imposed by carbon stress in environments where nutrient levels are extremely low (Glynn *et al.*, 2007).

5.2.2   *Optimal defense hypotheses*

It is clear that there is considerable variation among species in the distribution and abundance of defenses. Using alkaloids as a case in point, McKey (1974) argued that synthesis of these nitrogen-containing compounds is costly to the plant and, as a result, their distribution would be governed by two factors: (1) the vulnerability of a tissue to herbivores and (2) its fitness value to the plant. In other words, selection will favor defense when the benefit of that defense exceeds its cost. The optimal defense hypothesis (ODH) was formulated to address the question—"in what regions of the plant should the limited quantity of defensive compounds be concentrated?" In this hypothesis, herbivore pressure and the fitness consequences of herbivory are assumed to constitute important evolutionary forces that vary among different parts of the plant.

As indicated above, according to the ODH, the value of a plant tissue would be determined by the reduction in plant fitness resulting from the loss of that tissue (McKey, 1974). For the most part, experimental data support this prediction. For example, in wild parsnip, the reproductive parts of the plant were considered to be at most risk of attack and to be of high value to the plant. Interestingly, these parts exhibited high constitutive levels of the furanocoumarin, xanthotoxin, which was not inducible. In contrast, roots of wild parsnip were much less at risk of attack and had low constitutive levels of furanocoumarin, although it was highly inducible (Zangerl & Rutledge, 1996). Similarly, in *Brassica juncea*, allocation of the glucosinolate–myrosinase defense system to cotyledons was found to be consistent with the allocation of defenses to tissues in proportion to their value to the plant (Wallace & Eigenbrode, 2002). Thus, during periods when the cotyledons were critical for plant fitness, myrosinase activity and glucosinolate concentrations were highest, while the level of these defenses declined when the importance of the cotyledons for plant fitness also declined (Figure 5.3).

The two studies described briefly in the previous paragraph dealt with effects on direct defenses. But what about tests of this subhypothesis using indirect defenses? Heil *et al.*

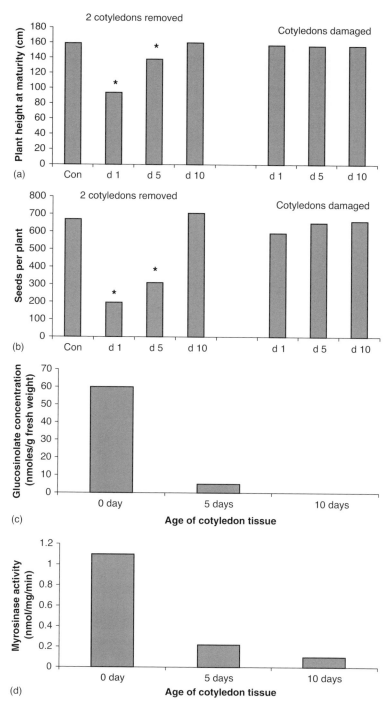

Figure 5.3   Effects of cotyledon removal and damage at different seedling ages on height at maturity
(a) and seed production per plant (b), of *B. juncea* plants. The asterisk (*) indicates significantly
different from other treatments. Glucosinolate concentration (c) and myrosinase activity (d) in
cotyledons of *B. juncea* seedlings 1, 5, and 10 days after emergence. (Reproduced from Wallace &
Eigenbrode (2002), with permission of Springer.)

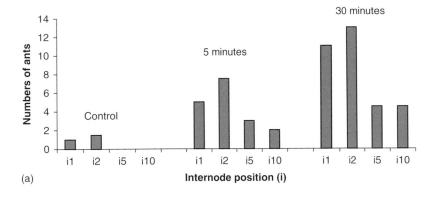

(a)

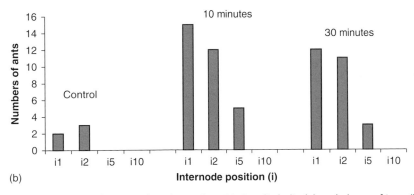

(b)

Figure 5.4 Spatiotemporal patterns in ant recruitment to termite baits (a) and pieces of tape (b) on leaves of *M. bancana* saplings. Termites represent a transient herbivore and tape mimics a long-term stress such as a climber or plant parasite. Mean ant numbers are presented for different times after placing the enemy mimic on various internodes, or leaves (1 is the youngest internode or leaf and 10 the oldest studied). Control, ant activity on untreated plants. (Adapted from Heil *et al.* (2004a), with permission of Cambridge University Press.)

(2004a) carried out such a test using the plant *Macaranga bancana*, which houses and provides nutrition for ants of the genus *Crematogaster*. In turn, the ants protect their host plant from herbivores, shoot borers, and pathogens. When the ten uppermost leaves were subjected to simulated herbivory, some 75% of all ants were concentrated on the youngest leaves (Figure 5.4). The high intensity of defense by ants on the youngest leaves can be explained by their vulnerability and their value in terms of expected future contributions to carbon assimilation (Heil *et al.*, 2004a). The spatiotemporal patterns of this indirect defense are therefore in line with the prediction of this optimal defense (OD) subhypothesis. Subsequent work by Radhika *et al.* (2008) using *Phaseolus lunatus* and *Ricinus communis* showed that extrafloral nectar (EFN) and volatile organic compounds (VOCs) are produced preferentially by young leaves and, moreover, that the EFN secretion patterns result from the reallocation of carbon to the young leaves.

  Further support was provided by recent work on fungal endophytes of meadow fescue, *Lolium pratense* (Zhang *et al.*, 2009). In this plant, the fungal endophytes *Neotyphodium uncinatum* and *Neotyphodium siegelii* produce loline alkaloids, which protect the host against insects. Levels of these alkaloids increase greatly following wounding or herbivore damage. When *L. pratense* was wounded, much greater loline levels were found in younger, compared to older, leaf blades. These data support the prediction of this subhypothesis that younger tissues lacking in physical defenses are provided with greater chemical defenses than older tissues (Zhang *et al.*, 2009).

The evolution of increased competitive ability hypothesis is an extension of the OD hypothesis and proposes that when plants are introduced to a new geographical area, relaxation of selective pressure from natural enemies will favor evolution of reduced defenses. This, in turn, will lead to greater allocation of resources to competitively advantageous traits such as growth and reproduction (Blossey & Nötzold, 1995). However, despite the potential of this hypothesis to explain the success of invasive exotic plants, tests to date have given mixed results (e.g., Willis *et al.*, 2000; Siemann & Rogers, 2003; Vila *et al.*, 2003; Meyer *et al.*, 2005).

5.2.3   *Plant apparency hypothesis*

In terms of the vulnerability of a plant to insect herbivores, one aspect that is likely to be important is its conspicuousness or apparency. This concept was put forward as a factor likely to influence the evolution of defenses (Feeny, 1976). Thus, plants that are not easily detected by herbivores are less likely to suffer damage and loss from herbivory and so have less need for defenses. Such plants would invest in what are known as qualitative defenses, small molecules such as glucosinolates, which are relatively inexpensive to synthesize. In contrast, apparent plants would possess a range of dosage-dependent or quantitative defenses, which would interfere with the ability of herbivores to acquire nutrients (Feeny, 1976). Plants would need heavy investment in such defenses because (1) the compounds, such as tannins, are large and biosynthetically expensive to make, and (2) a lot of the particular defense would be required to be effective. On the other hand, evolved tolerances to certain defense compounds by specialized herbivores that function as counteradaptations are common in the case of qualitative defenses but much less so in the case of quantitative defenses. The defenses would reduce herbivore growth rates, thereby subjecting the herbivores to increased predation. However, although there is some support for the hypothesis, not all the evidence supports plant apparency, likely because the true apparency of a plant to its herbivores is extremely difficult to quantify and changes with changing biotic contexts (such as season and surrounding vegetation).

5.2.4   *The carbon–nutrient balance hypothesis*

The carbon–nutrient balance (CNB) hypothesis provides a framework for the influence of carbon and nutrient supply on defense expression in plants (Bryant *et al.*, 1983; Tuomi *et al.*, 1988, 1991). It postulates that if the carbon–nutrient ratio of a plant controls the allocation of resources to the various functions of the plant, this will affect the ability of the plant to express its genetic potential for defense. The hypothesis assumes (1) that a plant will preferentially invest in carbon-based defenses when nutrients limit growth more than photosynthesis, whereas in situations in which photosynthesis is limited by factors other than nutrients, the "free" nutrients will be allocated to defense, and (2) that herbivory is a primary selective force for constitutive secondary metabolites, and that herbivory is reduced by defenses. However, it does not assume that either the total amount of defense or the type of defense, for example, nitrogen-containing versus non-nitrogen containing compounds, is selected for by herbivory.

A number of predictions are made concerning the way in which stressful environments affect the amount and general type of defense (nitrogen-containing as opposed to non-nitrogen-containing defenses) (Bryant *et al.*, 1983). For plant genotypes with phenotypic plasticity in defense, it is predicted that any effects of resource availability on the carbon–nutrient ratio can result in a change in the total level of defense. Thus, fertilizer application or shade would decrease the carbon–nutrient ratio, thereby reducing excess carbon production. This would lead to a reduction in non-nitrogen-containing defenses and an increase in the availability of nitrogen for defenses. For genotypes with little or no phenotypic plasticity in defense, it is predicted that effects of resource availability on the carbon–nutrient ratio in the plant do not lead to altered levels of defense. So, plants adapted to low-resource environments would have low intrinsic growth rates and little capacity for regrowth following herbivory. As a result, this would favor selection for maintaining high levels of defense.

Loss of leaf area following herbivory can result in reduced rates of photosynthesis, and this, in turn, might affect the plant's stores of carbon and nutrients. According to the CNB hypothesis, the resulting alteration in carbon–nutrient balance might then lead to a decline in leaf protein and a nonspecific accumulation of non-nitrogen-containing defenses (Bryant *et al.*, 1983; Tuomi *et al.*, 1984). However, accumulation of such defenses is dependent on the availability of soil nutrients and where carbon is stored in the plant (Tuomi *et al.*, 1988). Thus, because deciduous trees store their reserves of carbon in roots and stems, leaf damage should lead to marked and long-term increases in non-nitrogen-containing defenses if nutrients are limiting. Any increase in these defenses would be less marked if nutrient supply was not limiting and the carbon–nutrient balance can be restored quickly to preherbivory levels (Stamp, 2003).

It is important to remember that the CNB predicts that, compared to constitutive defense, plant species can have some combination of fixed and flexible allocation to defense, and this can vary between a complete fixed allocation and a complete flexible allocation (Stamp, 2003). For example, it appears that in some plant species, the level of secondary metabolite production is fixed (Holopainen *et al.*, 1995). Thus, some species or plant populations exhibit much less variation in terpene production in response to changes in the environment than others (Muzika *et al.*, 1989). As a result, any test of the CNB hypothesis should first establish the baseline genetic defense, that is, the level and range of defenses in plants grown at optimal nutrition and growth rate (Stamp, 2003).

Many experimental studies have examined the CNB hypothesis and the results are equivocal (see Koricheva *et al.*, 1998; Stamp, 2003). However, because it focused attention on the influence of resources on constitutive and inducible defenses, it contributed to the growth rate hypothesis (see next).

5.2.5   *The growth rate hypothesis*

The growth rate (GR) hypothesis, also known as the resource availability hypothesis, suggests that plants that have evolved in low-resource or stressful environments exhibit inherently slower growth rates than plants that have evolved in more productive environments (Grime, 1977, 2001; Coley *et al.*, 1985). Consequently, it is predicted that stress-adapted species would evolve particular suites of resistance traits, including higher levels of constitutive resistance and lower levels of induced resistance than faster growing species. The rationale for slow-growing species possessing high levels of constitutive defense is that such species are less able to replace tissue lost to herbivory than faster growing species, and they might also lack the metabolic capacity for effective induced resistance (Karban & Baldwin, 1997; Grime, 2001). Moreover, due to more intense competition experienced in productive habitats compared to stressful habitats (Grime, 1977, 2001; Herms & Mattson, 1992), it is predicted that plant species from productive environments should allocate more of their resources to growth than constitutive defense, and instead, should use inducible defenses, thereby minimizing allocation costs (Coley, 1987; Karban & Baldwin, 1997).

Some studies provide evidence in support of the GR hypothesis. For example, experiments on saplings of 47 tree species found a negative correlation between growth rate and tannins, as predicted by the hypothesis (Coley, 1988). Long-lived leaves of slow-growing species had higher concentrations of tannins than the short-lived leaves of fast-growing species. A negative correlation was not observed, however, between the rate of herbivore damage and tannin concentration. More recent work using species from resource-rich habitats where competition is likely to be intense, and habitats likely to provide a stressful environment for plants, found that, with the exception of one plant genus, species from stress-inducing habitats grew more slowly than their counterparts from resource-rich habitats (Van Zandt, 2007). Stress-adapted species also tended to have greater levels of constitutive resistance and lower inducible defenses than species from the resource-rich environments. Work on 18 grass species found a strong negative relationship between the overall investment in defense and growth rate, thus supporting predictions of the GR hypothesis (Figure 5.5) (Massey *et al.*, 2007). Thus, grasses with faster growth rates had lower

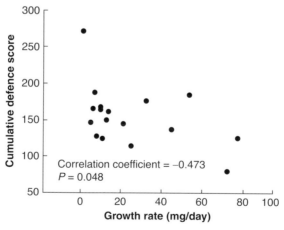

Figure 5.5 Relationship between cumulative defense score (CDS) and plant growth rates of 18 grass species. In order to assess the overall defense strategy for a particular species, data on all three primary defenses (silica concentration, phenolic concentration, and toughness) were combined to generate a CDS for each species. The correlation coefficient and *P* value for the relationship are shown. (Adapted from Massey *et al.* (2007), with permission of Blackwell Publishing Ltd.)

levels of defenses and were more palatable to mammalian herbivores (voles) than slower growing grass species. Interestingly, most support for the GR hypothesis has come from studies conducted in tropical habitats (e.g., Coley, 1987; Bryant *et al.*, 1989; Fine *et al.*, 2006). In contrast, evidence from studies in temperate habitats has been mixed, with some work supporting the GR hypothesis (e.g., Shure & Wilson, 1993; Fraser & Grime, 1999), but many studies failing to provide support (e.g., Almeida-Cortez *et al.*, 1999; Hendriks *et al.*, 1999; Messina *et al.*, 2002). It has been suggested that these differences between tropical and temperate studies might lie in the greater variation in tropical regions for traits of relevance to the hypothesis, such as growth rates, leaf toughness, and leaf lifespan (Van Zandt, 2007). Overall, it would appear that, from an evolutionary perspective, the effects of resource supply on growth are considerably greater than effects of defense cost (Coley, 1987). It also suggests that because intrinsic plant growth rates and related characteristics, for example, leaf lifespan, are molded so strongly by the abiotic environment, they determine the basic defensive profiles of plants (Stamp, 2003).

5.2.6   *Hypotheses of plant defense—where next?*

As pointed out by Stamp (2003), the various hypotheses outlined above are not mutually exclusive and it is appropriate to attempt to integrate them (Price, 1991; Herms & Mattson, 1992; Tuomi, 1992; Mole, 1994). However, individually, none of the hypotheses has been firmly rejected and each has contributed to our current understanding of plant defense. They could still be useful, but only if there is recognition that each of the hypotheses has its limitation and each will make different contributions to understanding. Our knowledge of plant defense could be increased further by conducting well-designed experiments based on a clear understanding of the hypotheses and thoughtful dissection of the various studies that have used them in the past.

5.3   Evolution of plant defense strategies

So far in this book, we have focused largely on plant defense at the species level. In terms of the evolution of plant defenses, there is renewed interest in patterns that occur across plant species, that is, the macroevolution of plant defense strategies. Agrawal (2006) argues that studying macroevolutionary patterns of plant defense is necessary if we are to answer outstanding questions on the evolution of plant defense. In the sections below, we will examine, briefly, the various phylogenetic hypotheses that have been put forward to account for the evolution of plant defense.

5.3.1   *The univariate trade-off hypothesis*

According to this hypothesis, if one defense is sufficient, then in plant species expressing high levels of that defense, selection should not occur for other, redundant, defenses. Further, there should be a trade-off between defense strategies across species, generated by the allocation costs associated with producing a given defense. Thus, scarce resources could be used for plant reproduction rather than a redundant defense. In this model, therefore, there would be a negative association between two traits. However, despite early support for this hypothesis, subsequent studies provided limited evidence of a trade-off between defense traits, with some studies failing to provide any supporting evidence (Rehr *et al.*, 1973; Steward & Keeler, 1988; Heil *et al.*, 2002; Agrawal & Fishbein, 2006). For example, in a phylogenetically controlled analysis of 31 species of cotton (*Gossypium*) species, no evidence was found for a trade-off between extrafloral nectaries and toxic leaf glands or trichomes, although there was a trade-off between toxic glands and trichomes (Rudgers *et al.*, 2004). However, a recent study on lima bean has provided evidence for a trade-off between defense traits (Ballhorn *et al.*, 2008). In this study, a direct defense (release of hydrogen cyanide, HCN) and an indirect defense (emission of VOCs) were quantified in cultivated and wild-type accessions of lima bean. Clear evidence of a trade-off between HCN release and emission of VOCs was found (Figure 5.6). In terms of the mechanism responsible for this trade-off, there was no evidence for the involvement of pleiotropy (a single gene affecting multiple phenotypic traits), suggesting that resource limitation and/or ontogeny were responsible. In lima beans, direct defense by HCN and indirect defense via VOCs appear to be targeted against different types of herbivores, and the relative dominance

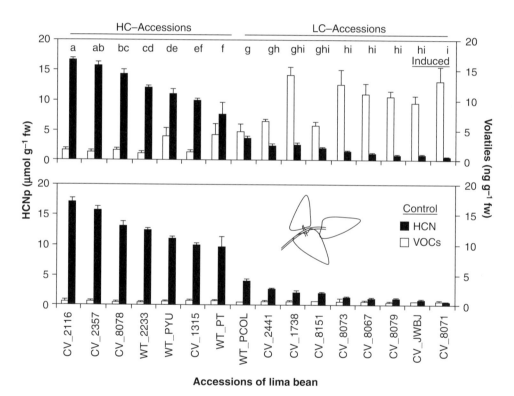

Figure 5.6 Variability of HCNp (cyanogenic potential—concentration of cyanogenic precursors) and VOC emission among lima bean accessions. Secondary leaves of cultivated (CV) and wild-type (WT) lima beans were analyzed for HCNp and total emission of volatiles. Data are presented separately for JA-induced (upper panel) and control plants (lower panel). Lima bean accessions marked with different letters in the upper panel differ significantly in their HCNp. HC accessions, high cyanogenic accessions; LC, low cyanogenic accessions. (Reproduced from Ballhorn *et al.* (2008), with permission of Blackwell Publishing Ltd.)

of these two defenses changed during plant ontogeny. This suggests that adaptations to varying enemy pressures are the most likely explanations for the trade-off (Ballhorn *et al.*, 2008).

There are three possible explanations for the lack of trade-offs in plant defense traits observed in various studies: (1) Plant defense is rarely effective as a single trait and can be most effective when acting synergistically (Berenbaum & Zangerl, 1996). Because of this, there should be no expectation that two defensive traits should be negatively correlated as long as they can interact with each other. (2) Most traits have more than one function. For example, in addition to a role in defense, trichomes also reflect UV light and serve as a barrier against transpiration. Therefore, it cannot be assumed that two traits involved in defense are redundant, when both abiotic and biotic selective forces are taken into consideration. (3) Several defensive traits are probably needed to deal with the range of possible attackers.

5.3.2 The resistance–regrowth trade-off hypothesis

Because various plant species are able to successfully overcome high levels of defoliation due to their regrowth capacity, an hypothesis was formulated that defense mechanisms and regrowth capacity are alternative strategies (van der Meijden *et al.*, 1988). In other words, given limited resources, there should be a trade-off between resistance traits and tolerance traits (traits that improve regrowth capacity following herbivory). Thus, plant species with a resistance strategy would be subject to little selection for tolerance, since they are not exposed to much damage from herbivory, while species with a tolerance strategy would not experience strong selection for resistance, because tissue loss via herbivory does not reduce their fitness. Implicit in this hypothesis is that both resistance and tolerance are strategies, each comprising several traits.

This hypothesis was tested using five species of biennials, which were compared for herbivory, regrowth, and energy and nutrient saving, in their natural environment (coastal sand dunes) (van der Meijden *et al.*, 1988). It was predicted that (1) herbivory and re-growth should be positively correlated and (2) root–shoot ratio and regrowth should be positively correlated. Observations agreed with these predictions. Thus, *Verbascum thapsus* was heavily attacked but exhibited high regrowth ability, whereas *Senecio jacobaea* was largely resistant, but exhibited very poor ability to regrow following damage (Figure 5.7). In contrast, studies using a range of potato cultivars could find no evidence of a trade-off between resistance (trichome density) to leafhoppers and tolerance (Kaplan *et al.*, 2009).

This hypothesis gave rise to the development of hypotheses for the evolution of convergent plant defense syndromes (see below).

5.3.3 The plant apparency hypothesis

As we saw in Section 5.2.3 above, Feeny (1976) proposed a hypothesis suggesting that plants that are apparent to herbivores (i.e., easy to see) will convergently evolve a suite of defenses that are effective against most herbivores. These defenses would quantitatively reduce the nutritional quality and edibility of the plant via, for example, increased leaf toughness, high levels of tannins, and low levels of water in the tissue. On the other hand, plants that are not so apparent or conspicuous would, despite being highly nutritious, operate

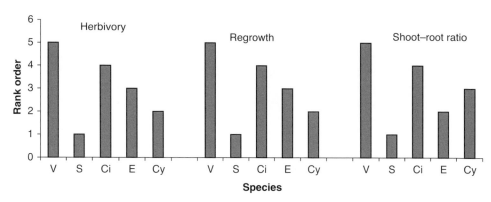

Figure 5.7   Rank orders of herbivory, regrowth, and shoot–root ratio in rosette plants of the following species: *V. thapsus* (V), *S. jacobaea* (S), *Cirsium vulgare* (Ci), *Echium vulgare* (E), and *Cynoglossum officinale* (Cy). Note that some species, for example, *V. thapsus*, are subject to heavy attack by herbivores, but possess very high regrowth ability, while other species, for example, *S. jacobaea*, are largely resistant to herbivore attack, but if damaged, possess little ability to regrow. (Adapted from van der Meijden *et al.* (1988), with permission of Blackwell Publishing Ltd.)

qualitative barriers to herbivore feeding such as alkaloids and cardenolides. This hypothesis has been tested on various occasions with positive results. For example, Bustamante *et al.* (2006) conducted a study of a range of species from the flora of Chile. These ranged from species differing in longevity (annual or perennial), leaf-shedding manner (evergreen or deciduous), latitudinal range, and level of drought experienced in the native habitat. They constructed an index of alkaloids/tannins (an A/T index) and found that unapparent plants such as annual herbs had a higher mean A/T index than apparent plants such as trees and perennial shrubs (Figure 5.8) (Bustamante *et al.*, 2006). In an interesting study on inducibility of defense-related enzymes in wild plants against pathogen attack, Heil & Ploss (2006) found that plants characterized by different life histories appear to have evolved different strategies to defend themselves against pathogens. They found that large, summer

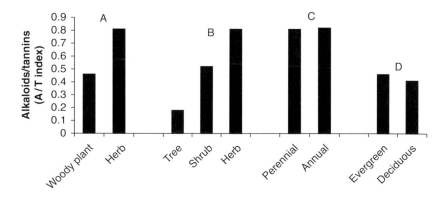

Figure 5.8   Mean A/T index (alkaloids/tannins) for species differing in life form (A and B), herbs differing in longevity (C), and woody plants differing in leaf-shedding manner (D). (Reproduced from Bustamante *et al.* (2006), with permission of Blackwell Publishing Ltd.)

flowering, perennial plants were significantly more inducible than plants that flowered in the spring. The latter grew in an environment characterized by low pathogen pressure, and as a result, probably escaped infection. For these plants, the growing season would also be short, allowing little time for induction of defenses. Here, constitutive defenses would be more appropriate. In contrast, the summer flowering perennials had a much longer growing season, and also experienced greater pathogen pressure. Under these conditions, inducible defenses would be useful.

5.3.4   *The resource availability hypothesis*

The resource availability hypothesis, also referred to as the GR hypothesis (see Section 5.2.3), proposes that resource-limited species will have slower growth rates and higher optimal levels of defense, reflecting the decreased ability of a resource-limited plant to compensate for tissue lost to herbivory (Janzen, 1974; Coley *et al.*, 1985). As we saw in Section 5.2.3, this hypothesis has received experimental support from a number of studies. The core of this hypothesis is that the abiotic environment is the driving force behind plant evolutionary strategies and, as a result, sets the template for the types of defenses that can evolve. Here, trade-offs are invoked at the multivariate level. Research by Coley and colleagues has shown that, in the tropics, trees fall along an escape-defense continuum. At one end of this continuum, escape species (those growing under high light conditions) are predicted to possess little in the way of chemical defenses, but exhibit rapid synchronous leaf expansion, and are of poor nutritional quality, while at other end of the continuum, defense species (those growing under low light conditions) have high levels of chemical defenses and asynchronous leaf expansion (Kursar & Coley, 2003; Coley *et al.*, 2005). Rather than "escape," it is probably more appropriate to think of these species as being "highly competitive"; that is, they cannot invest much in defense (1) because they have to grow sufficiently fast to outcompete neighboring plants, and (2) because they can easily replace tissue lost to herbivores.

The plant apparency hypothesis and the resource availability hypothesis show that unrelated plant species have converged evolutionarily on suites of similar defense strategies.

5.3.5   *Plant defense syndromes*

As we have seen in this and previous chapters, when defending themselves against herbivores, or indeed most attackers, plants utilize a broad arsenal of defensive traits. Agrawal & Fishbein (2006) argue that it is, therefore, more useful to think about plant defense as a suite of traits, which might include aspects of its direct and indirect defenses. This is important because synergistic interactions between multiple traits could provide much greater levels of defense than would be achieved if traits were present independently (Gunasena *et al.*, 1988; Berenbaum *et al.*, 1991; Stapley, 1998). However, the prediction of possible trade-offs among different defense strategies, due to cost and/or redundancy, has led them to usually being considered as single traits. According to Agrawal & Fishbein (2006), this is not strictly accurate, because plants do simultaneously employ multiple defense traits, in which case, as with most adaptations, they might be organized into coadapted complexes (Dobzhansky, 1970). In fact, a suite of multiple covarying traits associated with a particular ecological interaction is known as a syndrome. Thus, plants inhabiting particular abiotic or biotic environments might convergently evolve a set of particular defense traits (Fine *et al.*, 2006). Within a syndrome, any two traits might be positively or negatively correlated across taxa, or not correlated at all. However, in the plant defense syndrome model, no two plant defenses would be redundant, but rather should be negatively associated across species (e.g., Twigg & Socha, 1996; Rudgers *et al.*, 2004). It might be possible for plant defense syndromes, as a whole, to trade-off, providing they are really alternative adaptive strategies (Agrawal & Fishbein, 2006).

Defense syndromes can be examined within communities of diverse plant species, as well as within clades of closely related species. Either way, it is predicted that plant defense traits can consistently covary across species, because of shared evolutionary ancestry or due to adaptive convergence. Agrawal & Fishbein (2006) examined potential defense syndromes in 24 species of milkweeds (*Asclepias* spp.). Their analysis revealed three distinct clusters of species. The defense syndromes of these species clusters were associated with either low nutritional quality or a balance of higher nutritional quality coupled with physical or chemical defenses (Figure 5.9). Interestingly, this phenogram based on defense traits was not congruent with a molecular phylogeny of the milkweeds, and suggested to the authors

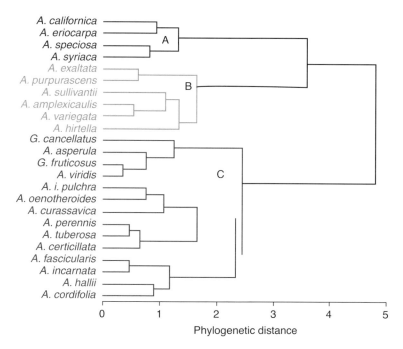

Figure 5.9 A defense phenogram that depicts similarity among 24 species of *Asclepias* generated by hierarchical cluster analysis of 7 defense-related traits. Tightly clustered species are defensively similar and can be considered to form defense syndromes. A–C: A, species with combinations of high nitrogen (i.e., low C/N), coupled with high physical defense traits (trichomes, latex); B, species with very high C/N ratio, coupled with tough leaves and low water content (difficult to eat); C, species with low C/N and specific leaf area, coupled with high cardenolides. (From Agrawal & Fishbein (2006), with permission.)

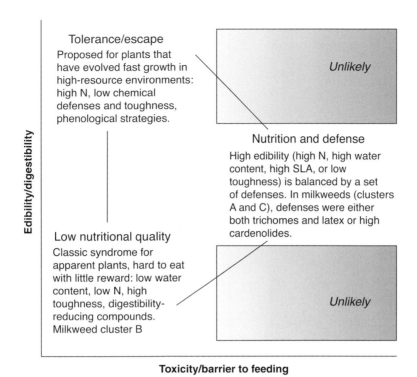

Figure 5.10   The plant defense syndrome triangle. Low nutritional defense syndrome is consistent with that outlined for apparent plants by Feeny (1976); a similar group was found in the study by Agrawal & Fishbein (2006). Tolerance follows the fast-growth and high-edibility pattern outlined by Coley *et al.* (1985). Nutrition and defense are a strategy that couples a toxic defense or barrier to feeding with relatively high edibility and digestibility. SLA, specific leaf area. (From Agrawal & Fishbein (2006), with permission.)

a convergence on defense syndromes. Based on this work, Agrawal & Fishbein (2006) proposed a "defense syndrome triangle" hypothesis, which includes all defense categories (Figure 5.10).

Agrawal & Fishbein (2006) envisage the identification of syndromes as a starting point to test alternative hypotheses for why plant defenses have converged. Thus, the abiotic and biotic environment could drive the evolution of particular syndromes. For example, plants that share guilds of herbivores (those that attack the plant in a similar way) might use similar defenses. The hope is that by identifying convergent plant defense syndromes, it might be possible to understand the evolutionary association between communities of herbivores and adaptive variation in plant species.

5.4   Patterns of plant defense evolution

It has been suggested that broad patterns of plant defense will emerge in plant phylogenies that might not be obvious from studies of a few co-occurring species within a community.

Agrawal (2006) suggests that consideration of these patterns of plant defense will illustrate how understanding broader phylogenetic relationships can help us to better understand the evolution of plant defenses.

5.4.1 Adaptive radiation

In their classic work, Ehrlich & Raven (1964) hypothesized that plant lineages diversify at a greater rate when the lineages are temporarily freed from herbivores via the origin of a novel defense. Eventual circumvention of this defense by herbivores then allows some herbivore lineages to radiate onto an underused resource. In this way, the major radiations of plants and herbivorous insects might reflect the historical sequence of adaptive zones that each has presented to the other.

As we saw in Chapter 2, secretory canals are an effective defense against herbivorous insects and pathogens. They have repeatedly evolved and are highly convergent in flowering plants, occurring in 10% of all species. Farrell *et al.* (1991) figured that secretory canals might be expected to allow plant radiation in an adaptive zone of reduced herbivory and disease. They compared the diversities of lineages that possess independently evolved secretory canals with their sister groups. The study showed that plant clades with latex and resin canals were significantly more species-rich than sister clades that lacked such secretory canals.

In a recent study, Agrawal *et al.* (2009) studied seven milkweed traits, ranging from seed size to defense traits such as cardenolides and latex, all known to exert an impact on herbivores. They found early bursts of trait evolution for two traits, and stabilizing selection for several others. They also modeled the relationship between trait change and species diversification, while allowing rates of trait evolution to vary during the radiation. This revealed that species-rich lineages underwent a proportionately greater decline in latex and cardenolides compared to species-poor lineages. Further, the rate of trait change was most rapid early in the radiation. The authors suggest that this result might mean that reduced investment in defensive traits accelerated diversification early in the adaptive radiation of milkweeds.

5.4.2 Escalation of defense potency

If the novel plant defense is eventually overcome by counteradaptations of insect herbivores, Ehrlich & Raven (1964) suggested that only the evolution of an additional novel, more powerful defense would allow that plant lineage to continue to diverge. Therefore, because it was the initial escape from insect herbivory that allowed the subsequent radiation of each clade, in evolutionary terms, any escalation of defense complexity and potency should only occur among broad phylogenetic classes of plants (Agrawal, 2006).

Several studies have provided evidence in support of increasing defense potency (Nelson *et al.*, 1981; Berenbaum, 1983; Farrell & Mitter, 1998). In one such study, Armbruster (1997) studied the ecology and evolution of relationships among a group of vines belonging to the genus *Dalechampia*. This work showed that multiple lines of defense appear to have evolved in sequence in this genus. Deployment of triterpene resins to defend staminate flowers was the first defense system to appear. In fact, this particular defense was a preadaptation, which allowed the evolution of a resin-based, pollinator reward system. It seems therefore that pollination by resin-collecting bees originated as a transfer exaptation; that is, a new function replaced the old function. Once this resin defense system of flowers was lost by conversion to a reward system, a sequence of defensive innovations followed over time, including (1) nocturnal closure of large, involucral bracts to protect staminate and pistillate flowers—the bracts probably originated as a floral advertisement system, and assumed a defensive function through additional exaptation; that is, a new function was added to the old function; (2) deployment of resin to protect developing ovaries and seeds; (3) deployment of sharp, detaching trichomes on enveloping sepals to defend developing seeds; and (4) deployment of resin to protect leaves and developing shoot tips. Thus, at least one pollinator-reward system originated by modification of a defense system, and several defense systems originated by modification of pollinator and advertisement systems (Armbruster, 1997). Evolutionary lability in ecological traits and relationships, as exhibited by *Dalechampia*, appears to be common, since several studies have indicated similar degrees of lability in the evolution of pollinator relationships and breeding systems (e.g., Bogler *et al.*, 1995; Kohn *et al.*, 1996). In most of these studies, exaptation probably plays important roles in evolutionary shifts in ecology.

5.4.3   *Phylogenetic conservatism*

Phylogenetic conservatism in plant defense postulates that because the biochemical pathways involved in the synthesis of defensive chemicals are so complex, it is likely that they evolved only once or a few times. Thereafter, the pathways were probably modified within clades, although they might not have become any more complex. A good example of phylogenetic conservatism is the domination of particular classes of defense chemicals in certain plant families, such as the presence of glucosinolates in the Brassicaceae.

Some impressive examples of phylogenetic conservatism have been found. For example, Wink (2003) reconstructed molecular phylogenies of three plant families, the Fabaceae, Solanaceae, and Lamiaceae, and employed them as a framework to map and interpret the distribution of major defense chemicals that are typical of these families: quinolizidine alkaloids and nonprotein amino acids for legumes, tropane, and steroidal alkaloids for Solanaceae, and iridoids and essential oils for labiates. Wink (2003) found that the distribution of the respective compounds was almost mutually exclusive in the families, implying a strong phylogenetic and ecological component. Interestingly, some exceptions were observed, in that certain compounds are absent (or present) in a given taxon, although all neighboring and ancestral taxa express (or do not express) the particular trait. According to Agrawal (2006), such exceptions can reveal remarkable examples of convergence. Thus, in addition to the Solanaceae, tropane alkaloids have been found in a few species in distantly related plant families such as the Brassicaceae and Proteaceae (Wink, 2003).

5.4.4   *Phylogenetic escalation and decline of plant defense strategies*

Agrawal & Fishbein (2008) noted that although macroevolutionary patterns are an explicit component of plant defense theories, phylogenetic analyses had not been attempted to test predictions concerning investment in resistance traits, recovery following damage, or plant growth rate. They constructed a molecular phylogeny of 38 species of milkweed and tested four major predictions of plant defense theory: (1) Do individual resistance traits trade-off due to redundancy, or do they evolve together repeatedly as a suite of covarying traits? (2) As predicted by the resource availability hypothesis, does plant growth rate covary with investment in resistance traits? (3) Do regrowth ability and resistance trade-off as alternative defense strategies to cope with herbivores? (4) Is there evidence for phenotypic escalation in defense trait expression as plant lineages diversify? The study found that, contrary to the first prediction, there were no trade-offs between the three most potent resistance traits (cardenolides, latex, and trichomes). They could also find no support for the plant GR hypothesis, and did not find a significant trade-off between regrowth and resistance. However, they did find evidence for an evolutionary escalation, although not in resistance traits, but in regrowth after damage. Their key finding therefore was a pattern of phyletic decline in resistance and an escalation of regrowth ability. As a result, they proposed that these countervailing evolutionary trends could result from the dominance of the milkweed herbivore fauna by specialist insects (Agrawal & Fishbein, 2008). Each of the specialist herbivores employs various mechanisms to circumvent the negative effects of the defenses, and as a result, herbivory on these plants is common and frequently intense. They predict that for plants with a herbivore fauna dominated by specialists, there will be macroevolutionary relaxation of existing and poorly functioning defenses.

5.5   Why do plants have induced defenses?

Induced responses to attack (not just to herbivory) are common and well documented in plants (see Chapters 2 and 3), and yet there has been little attention paid to the constraints and selection pressures that could favor an inducible over a constitutive defense strategy. A frequently asked question is, if induced responses are effective at protecting plants against attackers, why aren't they on all the time? After all, it seems a bit risky to depend on defenses that only get switched on once the plant is attacked. There are many reasons why induced defenses might be favored over constitutive defenses. These have been considered in some detail by Karban & Baldwin (1997), Agrawal & Karban (1999), and Zangerl (2003). Expression of a trait, for example inducibility of defenses, might enhance plant fitness (beneficial), reduce plant fitness (costly), or have no effect on fitness (neutral), and it stands to reason, therefore, that to understand the circumstances that favor inducibility of defenses, it is necessary to consider the possible costs and benefits that contribute to overall fitness. The various explanations for the evolution of induced defense responses cover both costs and benefits, and in the sections below, we will examine them briefly.

5.5.1   *Costs*

Constructing defenses is an expensive business, requiring energy and metabolites. As we have seen already in this chapter, the need for plants to defend themselves against attackers places them in a dilemma—do they use the energy and resources for growth and reproduction, or do they divert some of it toward producing defenses. In fact, the prevalence of inducible defenses had been taken as evidence that plant defense traits are costly to produce (Karban & Baldwin, 1997). Inducible defenses are produced upon attack, thereby ensuring that energy and resources are only used in defense when needed. Once the attack is over, the expectation is that defenses would return to baseline levels reasonably quickly, over perhaps a few days or weeks. The rapid induction of defenses was demonstrated clearly by Green & Ryan (1972), when they found the induction of proteinase inhibitor (PI) in young leaflets of wounded tomato plants within 12 hours. Their data suggested the rapid passage of a signal out of wounded leaves to unwounded leaves on the same plant. The movement of this signal out of wounded leaves had a half-life of 3.5 hours; if leaves were detached immediately after wounding, there was no PI accumulation in unwounded leaves, whereas, if leaves were detached 10 hours or more following wounding, there was no effect on PI accumulation in unwounded leaves (Green & Ryan, 1972). Wounding of first (youngest) leaves of *Vicia faba* was shown to induce resistance to rust infection in unwounded second leaves (Walters *et al.*, 2006). Not only was this effect rapid, so too was the accumulation of antifungal trihydroxy oxylipins in unwounded second leaves (Figure 5.11).

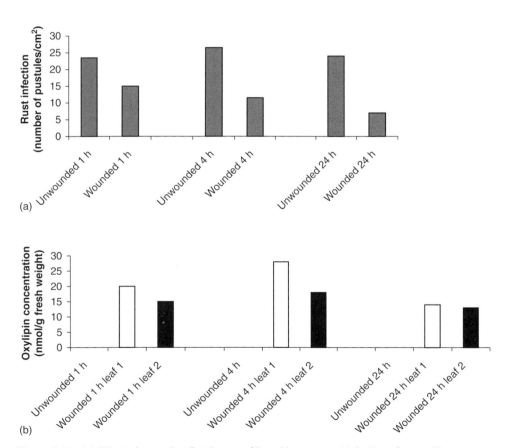

Figure 5.11   (a) Effect of wounding first leaves of broad bean on rust infection of second leaves. Second leaves were inoculated with rust 1, 4, and 24 hours after wounding. (b) Effect of wounding first leaves of broad bean on concentrations of 9,12,13-trihydroxy-10(*E*)-octadecanoic acid (TriHOE) in first (white bar) and second leaves (black bar). TriHOE was not detected in unwounded plants. (Reproduced from Walters *et al.* (2006), with permission of Oxford University Press.)

It is predicted that among plant species, there should be a negative correlation between levels of constitutive and inducible defenses. However, although some experimental data have provided support for this prediction, in many cases, the prediction did not hold true (see Karban & Baldwin, 1997). This might reflect the evolutionary and ecological implications of possessing a system of defense that can deal with a range of different attackers, but which also incurs costs (Agrawal & Karban, 1999; Agrawal, 2000; see also Chapter 4). For example, induced resistance to pathogens might protect the plant against other pathogens, but not insect herbivores, while in other plants, induced resistance to herbivores might not offer protection against pathogens. Such costs, or trade-offs, are dealt with more fully in Chapter 4.

So, what exactly are these costs? In evolutionary terms, costs can be any trade-off between resistance and another fitness-relevant process. The costs of any particular trait are normally outweighed by its benefits, and therefore can only be quantified in an environment in which the beneficial effects cannot affect fitness. There are several different types of cost:

1. Allocation costs—These result from the allocation of limited resources (energy and metabolites) to resistance instead of to other fitness-relevant processes such as growth or reproduction. There is also a cost associated with maintaining an inducible capability, that is, in maintaining the genetic and physiological apparatus required to perceive attack and to respond to it. Indeed, Agrawal *et al.* (2002) showed that more inducible genotypes of wild radish had lower fitness in the absence of attack than less inducible genotypes, while Tian *et al.* (2003) showed that *Arabidopsis* plants transformed with the *Rpm1* gene, a resistance gene that binds and detects bacterial effectors in gene-for-gene interactions, produced 7% less seeds than empty vector controls in a field experiment.
2. Autotoxicity costs—These arise from negative effects of a resistance trait on the plant's own metabolism.
3. Ecological costs—These are the result of negative effects of resistance on any of the plant's interactions with its environment that would affect its fitness under natural conditions. Examples include plant–rhizobial associations, interactions between plants and mycorrhizal fungi (see Chapter 4), and pollinators in the case of indirect defense via ants.
4. Genetic costs—These involve heritable effects of a resistance trait that correlate negatively with plant fitness (i.e., negative pleiotropic effects).
5. Opportunity costs—These relate to short-term growth reductions that result from the production of defenses, and which might compromise the ability of the plant to compete with neighboring plants. These represent missed opportunities and can reduce plant fitness.

Costs should be detectable when defense is expressed under enemy-free conditions, but should be counterbalanced by the beneficial effects when the plant is under attack. Therefore, the following should hold true: (a) plant growth and/or fitness in the absence of attackers should be lower in plants expressing induced resistance or in plants in which such resistance is constitutively overexpressed, (b) costs should be higher for plants growing under nutrient-limiting conditions compared to those growing under nutrient-sufficient conditions, (c) investment in resistance should be constrained by resources, (d) there should be a negative correlation between the expression of different resistance traits, which depend on the same resources.

Although induced defense may be less costly than constitutive defense, inducibility will still incur costs. Costs of defense can be estimated, but measuring them is difficult because they are not always manifested as detectable reductions in plant fitness (Zangerl, 2003). Plants can achieve their fitness in different ways, and so some plants might be well-defended or tolerant of herbivory (or disease), or they might be good competitors but poor at defending themselves (Fineblum & Rausher, 1995; Cipollini & Bergelson, 2001). Therefore, defense costs are epistatically entangled with the costs and benefits of alternative strategies of enhancing fitness (Simms & Triplett, 1994), and as Zangerl (2003) points out, disentangling these costs can only be achieved using isogenic lines, which are rarely available for wild plants. Nevertheless, numerous studies have attempted to measure the costs of inducible defense, and in this section, we will examine some of these studies. Ecological costs are dealt with in Chapter 4 and are not considered further here.

5.5.1.1   *Allocation costs associated with induced responses to herbivory*

In studying the costs of induced responses to herbivory, a number of experimental approaches are possible. The most obvious approach would be to expose experimental plants to the herbivore for a prescribed period to induce a response, after which the plant's performance would be measured. However, experiments with natural herbivory can be difficult to control, and so some studies looking at costs of induced responses have used alternative approaches such as mechanical wounding and application of chemical elicitors. But differences will exist in the amount and type of damage in plants subjected to insect herbivory and mechanical wounding. There is also the issue of discriminating between the costs of damage to plant tissue and the costs of producing the defense. One way of achieving discrimination is to use a chemical elicitor, although since plants have evolved their defenses in response to natural herbivory, responses to chemical elicitation might differ from those in plants subjected to the natural situation (Heil, 2002). Whatever approach is used, it is important to recognize that each has advantages and disadvantages (see Cipollini *et al.*, 2003). In most studies of costs of induced responses, the plant's performance is measured in terms of growth and fitness, the latter usually as seed production.

A number of studies have failed to detect allocation costs associated with induced plant responses (e.g., Simms, 1992; Karban, 1993; Gianoli & Niemeyer, 1997). In a study using wild tobacco, the induction of the defensive metabolite nicotine incurred costs that might have represented an allocation trade-off (Baldwin *et al.*, 1990). However, in this work, if plants were damaged early in the season, costs could not be detected, whereas if they were damaged just prior to flowering, costs were incurred. Baldwin *et al.* (1990) measured costs in terms of female fitness (seed production), and it has been suggested that this approach might have missed effects on male fitness characters, for example, time to first flower, pollen size, and pollen number (Agrawal *et al.*, 1999). Interestingly, research examining costs of induced responses in wild radish, incorporating both female and male fitness characters, found that induced plant responses reduced plant fitness when both female and male characters were considered together (Agrawal *et al.*, 1999). If the traits were considered individually, induction costs were only detected for time to first flower and number of pollen grains produced per flower, both male fitness traits.

Therefore, these costs would not have been detected if only female fitness characters had been measured.

Some studies measuring female fitness characters have found costs associated with induced responses. For example, Zavala *et al.* (2004) examined costs associated with the induction of trypsin proteinase inhibitor (TPI) in *Nicotiana attenuata*. They found that TPI production was increased following application of methyl jasmonic acid, and this was associated with reduced production of seed capsules, especially when plants were facing competition from neighboring plants.

In a long-term, large-scale field experiment in Kenya, the potential costs of defense against herbivory by large mammalian herbivores were examined in the savanna tree *Acacia drepanolobium* (Goheen *et al.*, 2007). Grazing by native herbivores triggers the production of longer spines, an induced defensive response. The experiment showed that in the absence of native herbivores, individual trees were twice as likely to reproduce, and those that did, produced a greater biomass of seeds (Figure 5.12). Here, spine length was significantly and negatively related both to the occurrence and magnitude of reproduction.

In order to circumvent the problem of distinguishing between the cost of losing leaf tissue and the cost of producing an inducible defense, the response of willow damaged by the leaf beetle, *Phratora vulgatissima*, was compared with plants damaged mechanically to the same extent, but without eliciting the defense (production of trichomes) (Björkman *et al.*, 2008). The workers calculated that the cost of producing the defense was a 20% reduction in shoot length growth and biomass production, while the cost of leaf area removal was an 8% reduction in shoot length growth. The results suggest that the cost of leaf area removal was relatively small compared with the costs of induced trichome production.

It is predicted that the costs of defense should be increased by interplant competition, since this should increase resource limitation (Herms & Mattson, 1992). A number of studies have examined how the optimal deployment of defenses might depend on competition, with some studies demonstrating increased costs (e.g., Van Dam & Baldwin, 1998; Cipollini, 2007) and others reporting no effect on costs (e.g., Karban, 1993; Siemens *et al.*, 2002). A recent study examined the costs and benefits of production of the defense metabolite sinigrin in *Brassica nigra* in response to herbivory by the slug, *Deroceras reticulatus* (Lankau & Kliebenstein, 2009). This study used groups of plants selected for high and low sinigrin levels and found that plant fitness, measured as total seed production, was not significantly affected by herbivory in the high sinigrin plants, but was significantly reduced by herbivory in the low sinigrin plants (Figure 5.13). However, costs disappeared under interplant competition, and plants with low sinigrin tended to outperform plants with high sinigrin in the presence of both competition and herbivory (Figure 5.13). Thus, the fitness costs and benefits of sinigrin production in *B. nigra* agreed with plant defense theory in the absence, but not in the presence, of interplant competition.

As we saw in Chapter 4, plants are able to remotely detect the proximity of neighboring plants by perception of an increase in the ratio of far-red radiation (FR) to red (R) light, using the photoreceptor phytochrome B. Interestingly, Izaguirre *et al.* (2006) showed that reflected FR down-regulated chemical defenses and increased herbivore performance, and moreover, that this effect was mediated by phytochrome B. They found that the plant solves the dilemma of whether to invest resources in competition or defense against herbivory by

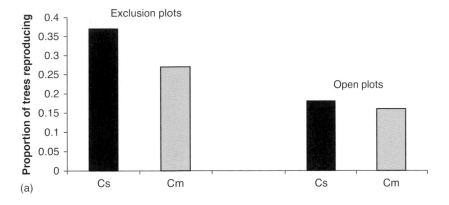

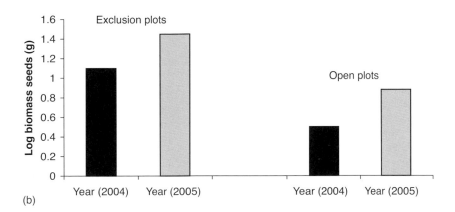

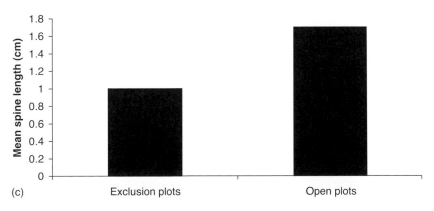

Figure 5.12 Reproduction and mean spine lengths for the savanna tree *A. drepanolobium* in plots accessible to native herbivores (open plots) and plots from which native herbivores were excluded for 10 years (exclusion plots). (a) Average proportion of trees reproducing across plots, segregated by ant occupant (Cs, *Crematogaster sjostedti*; Cm, *Crematogaster mimosae*) and pooled between 2004 and 2005, (b) average seed production for reproductive trees across plots in 2004 and 2005, (c) spine lengths for trees averaged across plots, measured in 2005. (Redrawn from Goheen *et al.* (2007), with permission of Blackwell Publishing Ltd.)

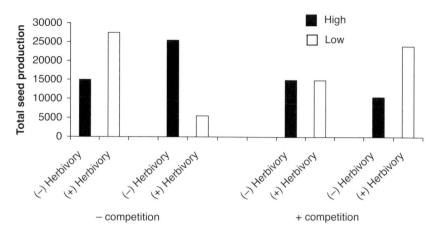

Figure 5.13   Mean fitness (total weight of seeds produced in mg) of families of *B. nigra* from high sinigrin (black bars) and low sinigrin (white bars) selection groups across herbivory and competition treatments. (Redrawn from Lankau & Kliebenstein (2009), with permission of Blackwell Publishing Ltd.)

modulating its sensitivity to jasmonic acid (JA) using information on the risk of competition, sensed by phytochrome (Moreno *et al.*, 2009).

5.5.1.2   *Allocation costs associated with induced responses to pathogens*

In contrast to induced resistance to herbivory, the situation with respect to pathogens is less clear (Walters & Heil, 2007). More than 25 years ago, Smedegaard-Petersen & Stolen (1981) observed a 7% reduction in grain yield in barley plants inoculated with the powdery mildew fungus *Erysiphe graminis* f.sp. *hordei* compared with uninoculated control plants. They suggested that the reduction in grain yield was the result of increased dark respiration, required to provide resistance to pathogen infection. However, because these studies were conducted in the presence of the pathogen, they cannot be used to quantify the costs associated with induced resistance. Later work showed that the chemical inducer acibenzolar-S-methyl (ASM) applied to wheat, in the absence of pathogen pressure, reduced plant growth and yield and provided a clear indication that use of ASM incurred allocation costs (Heil *et al.*, 2000). Since that work, similar results have been reported in other crop plants (e.g., Prats *et al.*, 2002; Buzi *et al.*, 2004; Hukkanen *et al.*, 2007), although some studies could find no evidence for costs incurred following treatment of plants with ASM (Iriti & Faoro, 2003).

However, although some studies have demonstrated costs associated with induced resistance, other studies have shown that whether costs are incurred depends on the type of induced resistance, that is, priming or direct induction of resistance. Thus, work on *Arabidopsis* has shown that priming incurs less fitness costs than direct induced resistance, and moreover, the benefits of priming-mediated resistance outweigh the costs when the plant is under pathogen pressure (Van Hulten *et al.*, 2006). Similarly, work on barley showed that not only did priming not incur costs, but it provided benefits when plants were under high disease pressure (Figure 5.14) (Walters *et al.*, 2009).

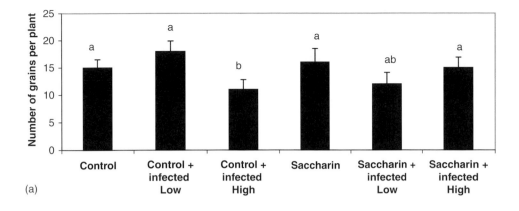

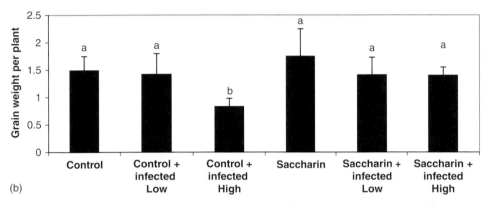

Figure 5.14 Effect of saccharin treatment of leaves 1–4 of barley on (a) numbers of grains per ear and (b) grain weight per plant. Leaves 5–7 were inoculated with the fungal pathogen *Rhynchosporium secalis* 2 days after saccharin treatment, using a low spore concentration (1×10^4 spores/mL) or a high spore concentration (5×10^5 spores/mL). Bars carrying different letters are significantly different at $P \leq 0.05$. Note that compared to control plants under high disease pressure, saccharin-treated plants under high disease pressure produced significantly more grains per plant and had significantly greater grain weights per plant. (Reproduced from Walters *et al.* (2009), with permission of Elsevier.)

What about the mechanisms by which systemic acquired resistance (SAR)-associated costs are incurred? Gene array studies have shown that genes involved in photosynthesis and growth are down-regulated during the expression of induced resistance to pathogens or herbivores, or following application of chemical elicitors (Scheideler *et al.*, 2002; Heidel & Baldwin, 2004; de Nardi *et al.*, 2006). Interestingly, plants may compensate for this effect, at least in uninfected leaves (e.g., Roberts & Walters, 1986). For instance, inoculation of the lower leaves of broad bean with rust led to increases in both photosynthesis and resistance to rust infection in upper leaves (Murray & Walters, 1992), and the authors suggested that, without the increased rates of photosynthesis, assimilates to fund defense responses would need to be found from existing resources, diverting them away from plant growth and development (Figure 5.15 and Table 5.1).

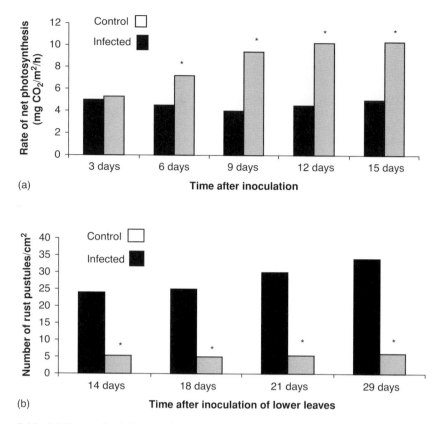

Figure 5.15   (a) Rates of net photosynthesis in upper leaves (leaf 3) of broad beans. In controls, the lower leaves (leaves 1 and 2) were uninfected, while in infected plants, the lower leaves were infected with the rust, *Uromyces viciae-fabae*. (b) Rust infection (number of pustules per square centimeter) of the upper leaves (leaf 3) of broad beans following prior inoculation of the lower leaves (leaves 1 and 2) with rust. In infected plants, the upper leaves were challenge-inoculated 1 day after inoculation of the lower leaves. Significant differences from the control are shown as *. (Redrawn from Murray & Walters (1992), with permission of Blackwell Publishing Ltd.)

Table 5.1   Effect of shading the upper leaves (leaf 3) of broad beans on the induction of systemic resistance to rust (*Uromyces viciae-fabae*) infection in those leaves

Treatment	Irradiance (μ mol m^{-2} s^{-1})	Rate of net photosynthesis (mg CO$_2$ m^{-2} h^{-1})	Rust infection (%)
Control	690	6.5	47.8
Infected (no shade)	690	11.7	3.4
Infected (partial shade)	500	7.6	16.2
Infected (heavy shade)	20	0.8	40.1

Source: From Murray & Walters (1992).
In infected plants, upper leaves were challenge-inoculated 2 days after inoculation of the lower leaves (leaves 1 and 2). In controls, the lower leaves were not inoculated with rust. Rust infection was assessed 9 days after inoculation. Note that shading not only reduced the increase in photosynthesis in the upper leaves of infected plants, but also reduced the amount of systemic protection against rust infection.

Perhaps unsurprisingly, genes involved in energy metabolism are up-regulated in plants expressing systemic resistance, highlighting the need to provide energy for resource-demanding defense responses (Schenk *et al.*, 2003; de Nardi *et al.*, 2006). It seems likely therefore that a "switch from housekeeping to pathogen defense metabolism" may be a prerequisite for the full commitment of a plant to transcriptional activation of resistance pathways (Logemann *et al.*, 1995; Scheideler *et al.*, 2002).

During SAR, whether costs are incurred and the magnitude of such costs will depend on environmental factors, both abiotic and biotic. It is also likely to be genotype-dependent, although this has received little attention to date. Whether costs were incurred in wheat and *Arabidopsis* following induction of resistance depended on nitrogen supply (Heil *et al.*, 2000; Dietrich *et al.*, 2005). Similarly, successful induction of resistance by ASM in barley was dependent on establishment of a mycorrhizal association (Sonnemann *et al.*, 2005).

5.5.2   *Targeting of inducible direct defenses*

Plants possess a broad range of defenses, and it is likely that they will be differentially effective against different types of attackers. Ensuring that the most appropriate defenses are induced when a plant is attacked by a particular herbivore or pathogen could therefore be highly beneficial. If all the defenses to a particular type of attacker, for example, herbivores, are positively correlated, the cost savings of not producing the defense until required would be the main factor favoring inducibility. If the defenses are not correlated, inducing only those defenses that would be effective against the particular attacker would minimize costs. If, however, the defenses are negatively correlated, the situation is complicated by the existence of an additional cost—the cost associated with increased susceptibility to a different herbivore. Here, inducing the defense only when the susceptible herbivore is attacking the plant minimizes both the cost of the defense and the cost of increased susceptibility to other herbivores (Zangerl, 2003).

We saw in Chapter 3 that plants use several different transduction pathways to activate defenses against different attackers. Thus, the salicylic acid (SA) pathway is generally effective against pathogens, with little effect against chewing herbivores, while the JA pathway is effective against most herbivorous insects and less so against pathogens. Targeting can be more specific and, for example, the SA pathway tends to be effective against biotrophic pathogens, while the JA/ET pathway tends to be more effective against necrotrophs. Moreover, plants can use cross-talk between these different signaling pathways to fine-tune their defense responses to a particular attacker (Korneef & Pieterse, 2008). Proving that plants can target defenses among different species of attackers, for example, herbivores, requires demonstration that different herbivores elicit qualitatively different defenses, that these defenses are effective against the particular herbivore and not all other herbivores, and that induction of all the defenses simultaneously, in the absence of damage, is more costly than induction of a specific defense.

There is evidence that some defense chemicals, while effective at deterring most generalist herbivores, can attract specialist herbivores (e.g., Karban & Niho, 1995). There are also many examples of herbivores being attracted to the VOCs induced by herbivory (e.g., Frati *et al.*, 2009; Garcia-Robledo & Horvitz, 2009). Thus, a potential benefit of employing induced defenses is that they are only deployed when needed, without attracting specialist herbivores when the plant is not under attack.

5.5.3   *Targeting of inducible indirect defenses*

Although several positive and even synergistic interactions between plant defenses and natural enemies of insect herbivores have been found (e.g., Hare, 1992; Duffey *et al.*, 1995), other studies have demonstrated that plant defense can adversely affect parasitoids or predators of herbivorous insects (e.g., Reitz & Trumble, 1997; Krips *et al.*, 1999; Havill & Raffa, 2000). For example, parasitoids suffered high mortality due to plant trichomes on wild tomato (Kauffman & Kennedy, 1989), and consumption of herbivores feeding on resistant plants resulted in decreased survival and fecundity, and increased development time of predators and parasites (Barbosa *et al.*, 1991; Stamp *et al.*, 1997). Inducibility of defenses might be favored as a strategy to reduce such negative impacts of plant defense on predators and parasitoids of herbivorous insects.

   Plants can emit volatile signals to attract predators and parasitoids of herbivorous insects. Thus, feeding by spider mites on lima bean induces the emission of several terpenoids and MeSA, and these volatiles attract the predatory mite *Phytoseiulus persimilis* (Van den Boom *et al.*, 2004). Arthropod predators and parasitoids are known to be adept at learning to associate nonhost cues with the presence of hosts (e.g., Turlings *et al.*, 1993; Hu & Mitchell, 2001). The cost of constitutive expression of these volatile signals is the likelihood that predators and parasitoids would learn to ignore signals that provide no useful information, because herbivorous insects will not be there all the time. In contrast, emission of the signals only when the plant is under attack provides the predators with useful information, alerting them to the presence of herbivorous insects.

5.5.4   *Dispersal of damage*

Localized, as opposed to systemic, induction of defenses might be beneficial if it results in herbivorous insects moving around the plant, thereby dispersing their damage, or moving to neighboring plants (Edwards & Wratten, 1983; Van Dam *et al.*, 2001). Indeed, there is some evidence that dispersed damage has less effect on plant fitness than concentrated damage (Mauricio *et al.*, 1993; Meyer, 1998), although it is not clear if this is a more widely applicable result. Moreover, dispersed damage might not always equate to less impact. For example, insects moving around the plant might also be transmitting bacterial, fungal, and viral pathogens (Garnier *et al.*, 2001). But what about evidence that localized damage influences the insect's feeding patterns? Although this is more difficult to determine, some workers did find that herbivores were likely to move away from damaged sites on leaves (Bergelson *et al.*, 1986).

5.5.5   *Possible role of pathogenic bacteria in the*
evolution of SAR

Because SA- and JA-mediated defenses can incur substantial costs, there should be corresponding benefits that prevent these traits from being lost in natural plant populations. Traw *et al.* (2007) tested the potential benefits of SAR and JA-mediated defense by using chemical elicitors to induce the two pathways in *Arabidopsis* plants transplanted into a large natural population of *Arabidopsis* known to contain pathogenic bacteria. They generated differences in natural bacterial abundance by treating plants with an antibiotic. Plants treated with SA to induce SAR exhibited reduced bacterial levels in their leaves and increased fitness compared to controls, while no such effects were observed in plants treated with JA. Based on these results, the authors suggest that induction of SAR might be important in regulating the growth of microbial communities and further suggest a role for bacterial pathogens in the evolution of SAR.

5.5.6   *Conclusion*

Although the evolution of induced defenses might have been favored for a number of reasons, information is patchy, and to date, most studies have concentrated on the costs of defense. Few studies have attempted to examine the phylogenetics of induced plant defenses. In one such study, Thaler & Karban (1997) examined constitutive and induced resistance of 21 species of *Gossypium* to herbivory by spider mites. They found a positive correlation between constitutive and induced resistance across the 21 cotton species and suggested that constitutive resistance was the ancestral state. This contrasts with the prediction that, because defenses are costly, there should be a trade-off between constitutive and induced defenses (Koricheva *et al.*, 2004). A later study by Heil *et al.* (2004b) on Mesoamerican *Acacia* showed that EFN production moved from an inducible strategy to a constitutive one during the evolution of obligate myrmecophytism. In this work, constitutively low species showed strong induction and constitutively high species showed no induction, suggesting a trade-off between the two modes of defense. Recently, a study of defense (cardenolides) in 12 species of tropical and temperate milkweeds found no evidence for a trade-off between constitutive and induced cardenolides in the shoots and roots (Figures 5.16 and 5.17) (Rasmann *et al.*, 2009).

Given the paucity of data, it is difficult to assess the relative importance of the factors discussed above in promoting the evolutionary origin and maintenance of induced defenses. Clearly, more effort needs to be concentrated on studying, not just the role of allocation costs, but also other potential benefits and constraints to the evolution of inducible defenses.

5.6   The coevolutionary arms race

Ehrlich & Raven (1964) proposed that insect herbivores and their host plants are locked in an arms race through reciprocal evolution, or coevolution. They observed that phylogenetically related butterflies tended to have as their hosts, phylogenetically related plants. They argued that such patterns of plant use arose as a result of coevolution between plants and insects, a process involving alternating adaptive radiations. They suggested that, initially, a genetic event (e.g., a mutation) occurs in a plant species that confers on it resistance to most or all of its insect herbivores. This releases the plant from herbivory, enabling it to occupy new adaptive zones and radiate. Subsequently, a genetic event occurs in an insect species conferring on it the ability to overcome the novel defense. This releases the insect from its competitors and it then undergoes adaptive radiation, eventually being able to utilize many, if not all, of the plant species.

The extent to which a plant and a herbivore coevolve will be determined by three factors. First, the plant and the herbivore must be able to influence each other's fitness. Second, there must be genetic variation for characteristics of both the plant and the insect that influence the outcome of the interaction. Third, there must be a response by each organism to selection from the other. In fact, different types of coevolution have been recognized: (1) classical coevolution, where there is reciprocal evolution between pairs of species (Ehrlich & Raven,

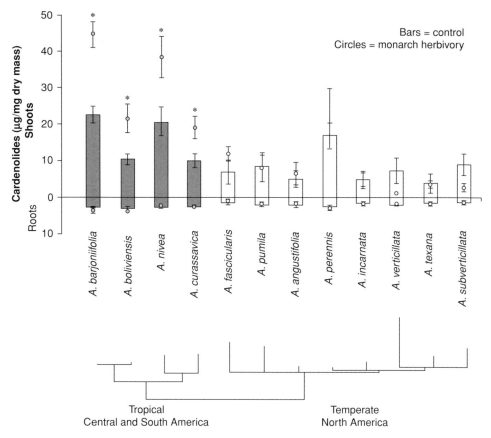

Figure 5.16   Root and shoot cardenolide concentrations of 12 *Asclepias* species (milkweeds, series Incarnatae) in undamaged control plants (bars) and plants damaged by monarch caterpillars (circles). Asterisks above circles indicate significant increases in shoot cardenolide concentrations after induction ($P < 0.05$). Below the data is the corresponding phylogram for the 12 species pruned from a comprehensive phylogeny of *Asclepias* (Agrawal & Fishbein, 2008); the left-hand vertical scale indicates base pair substitutions. Tropical species are represented with solid bars and circles, and temperate species with open bars and circles. (From Rasmann *et al.* (2009), with permission.)

1964), (2) diffuse coevolution, which is considered in a community context rather than as an interaction between two species (Fox, 1988), and (3) the geographical mosaic theory of coevolution. This considers the spatial variation that occurs within populations so that there is a continually shifting geographical pattern of coevolution between two or more species (Thompson, 1994).

Phenotypic plasticity is the ability of an organism to express different phenotypes depending on the environment: single genotypes can alter their physiology, biochemistry, development, morphology, or behavior, in response to environmental cues (Agrawal, 2001). Therefore, individuals that interact may adjust their phenotype in response to their respective partner, and can reflect an evolutionary response to variation encountered by individuals. Reciprocal phenotypic change between individuals of interacting species represents an

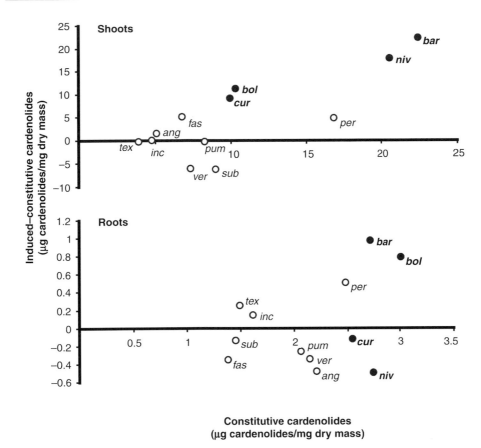

Figure 5.17   Raw data for mean investment of constitutive versus induced (control minus herbivory) cardenolides in shoots and roots for the 12 species of *Asclepias* in the Incarnatae series. Black circles represent species of the tropical clade; open circles represent species of the temperate zone. Three-letter codes refer to the first three letters of species names (see Figure 5.19). Both relationships (roots and shoots) had statistically significant regression slopes. (From Rasmann *et al.* (2009), with permission.)

interaction norm where the response of one species to the other creates the environment to which the other species may then respond (Figure 5.18) (Agrawal, 2001). In plant–herbivore interactions, plants might induce defenses that are dependent on the density of attackers, and herbivores might induce counterdefenses that are dependent on the concentrations of plant defenses consumed (Figure 5.19). According to Agrawal (2001), the continuous range of phenotypes induced by each partner exemplifies a major requirement of the ecological arms race hypothesis because it allows for escalating phenotypic change.

Testing the hypothesis of coevolution has proved to be extremely difficult. The paragraphs below focus on evidence in support of, and against, a coevolutionary arms race between plants and herbivorous insects. However, much is now known about plant–pathogen coevolution, especially in relation to the evolution of resistance (*R*) and avirulence (*Avr*) genes. This is dealt with in Box 5.1.

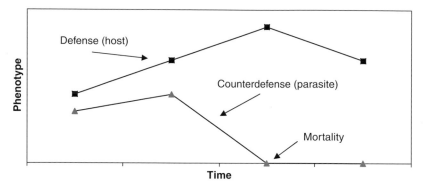

Figure 5.18 One possible course of reciprocal phenotypic change between individuals in an antagonistic species interaction. The dynamic nature of such increase in defense following attack and decrease following removal of parasites has been demonstrated for spines on *A. drepanolobium* that are induced by vertebrate herbivores (Young & Okello, 1998). Ecological reciprocity may take place in all interactions, irrespective of the sign of effect on an individual species. However, phenotypic change in response to a species interaction need not be directional (Adler & Karban, 1994). (From Agrawal (2001), with permission of The American Association for the Advancement of Science.)

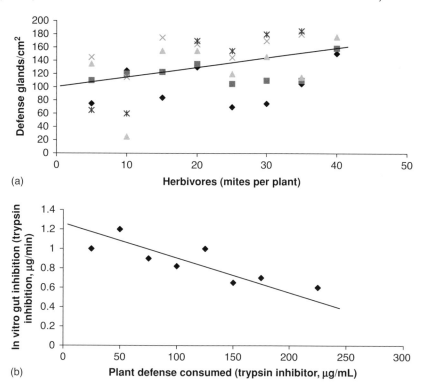

Figure 5.19 Potential for an arms race between plants and herbivores. (a) Phenotypic escalation in a plant defense (pigmented glands containing hemigossypolone in cotton) that is dependent on the number of attacking mite herbivores. (b) Phenotypic escalation in a herbivore's (*Trichoplusia ni*) counterdefense (production of inhibitor-insensitive proteases) dependent on the concentration of plant defense (proteinase inhibitors) consumed. (Redrawn from Agrawal (2001), with permission of The American Association for the Advancement of Science.)

Box 5.1   Coevolution between plants and pathogens

As we saw in Chapter 2, once a pathogen has broken through the outer layers of the plant, it is subject to molecular recognition by individual plant cells. Plants have evolved two classes of immune receptors to detect non-self-molecules: (1) pattern–recognition receptors that are located on the cell surface. These sense microbes by perception of pathogen-associated molecular patterns (PAMPs). This first level of immunity represents a basal resistance and is known as PAMP-triggered immunity. Virulent pathogens can successfully infect host plants by evading recognition or suppressing PAMP-triggered signaling, the latter probably mediated by secretion of virulence effectors. (2) In turn, some plants have evolved resistance (R) proteins capable of recognizing these effector proteins, resulting in a second line of defense known as effector-triggered immunity (ETI). In turn, pathogens have evolved effectors capable of suppressing ETI. Plant *R* genes encode the second layer of the plant immune system, and pathogen *Avr* genes encode the effector proteins whose normal function is to interfere with host plant cells to promote successful infection (Jones & Dangl, 2006). The presence of corresponding *R* and *Avr* genes in host and pathogen populations implies the possibility of coevolution driven by selection pressure on the pathogen to escape recognition by host *R* gene products, and concomitant pressure on the host to respond to new virulent strains of the pathogen. Indeed, there is often considerable intraspecific polymorphism at *R* gene loci, suggesting that plant recognition and pathogen evasion are a major battleground in the struggle between plants and pathogens (Bergelson *et al.*, 2001). Because of this, a great deal of research has been devoted toward understanding the evolution of *R* genes (for excellent reviews of *R* gene evolution, see McDowell & Simon, 2006, and Dodds & Thrall, 2009).

Although plant–pathogen coevolution is often described as rapidly evolving, implying rapid turnover of alleles at both *R* and *Avr* loci, studies at the population level have demonstrated the existence of remarkably long-lived polymorphism at *R* gene loci. For example, in *Arabidopsis thaliana*, *RPM1* is a single-copy gene in resistant ecotypes, which is absent in susceptible (*rpm1*) ecotypes. In *Brassica oleracea* and *Arabidopsis lyrata*, the *RPM1* loci are syntenic with functional *RPM1* in *A. thaliana*, indicating that the *rpm1* null allele was created by deletion of a functional *RPM1* gene (Grant *et al.*, 1998; Stahl *et al.*, 1999). Examination of nucleotide divergence in the flanking regions of this null allele shows that the allele has been maintained for nearly 10 million years, which puts its origin close to the period of *Arabidopsis* speciation (Stahl *et al.*, 1999). Such an ancient origin for the allele is not consistent with the arms race model, since this predicts that overcome or defeated *r* alleles are replaced by new *R* alleles, with obsolete alleles being removed from the population. Instead, a "trench warfare" model was proposed for the evolution of *RPM1*, in which both functional and null alleles are long-lived, although their frequencies fluctuate over time, due to recurrent cycles of negative frequency-dependent selection (Stahl *et al.*, 1999). But why should a null allele of a resistance gene be maintained for such a long period? A possible explanation is that the functional allele is selectively disadvantageous in the absence of pathogen pressure (McDowell & Simon, 2006). Indeed, analysis of transgenic plants under field conditions with no obvious pathogen pressure has shown that the *RPM1* transgene conferred a substantial decrease in seed production. These data are consistent with a model in which balancing selection has driven long-term maintenance of the null *rpm1* allele, in order to counterbalance an effective, but costly functional *RPM1* allele (Tian *et al.*, 2003). There is evidence that, in *Arabidopsis*, duplicated *R* genes have frequently been deleted during the evolution of its genome, suggesting that superfluous *R* genes might carry a cost (Nobuta *et al.*, 2005). However, work on two other *Arabidopsis R* genes (*RPS2* for resistance to *Pseudomonas syringae* and *RPP5* for resistance to *Hyaloperonospora parasitica*) found that neither was costly in the absence of pathogen pressure (Korves & Bergelson, 2004).

In *Arabidopsis*, the *RPP13* gene and the corresponding downy mildew *Avr* gene are highly polymorphic and subject to diversifying selection (Rose *et al.*, 2004). Similarly, in the interaction between flax and the rust *Melampsora lini*, there is considerable polymorphism at the *R* gene and *Avr* gene loci and these are also subject to diversifying selection (Dodds & Thrall, 2009). In the flax–rust interaction, it seems that R proteins interact directly with the corresponding Avr protein. This provides a molecular basis for explaining the arms race between plants and pathogens, where coevolution is driven by selection of new *Avr* alleles to avoid this interaction and the subsequent selection of new *R* alleles that have reestablished contact. This would have given rise to the extensive diversification in *R* and *Avr* genes present in the flax–rust system (Dodds & Thrall, 2009).

The studies outlined above highlight the apparent diversity in evolutionary trajectories of different *R* genes and underpin the difficulty in formulating universally applicable models for *R* gene evolution.

Because there is no interaction between plants and herbivores during the period of plant diversification, coevolution as reciprocal adaptations between interacting populations cannot be causing plant diversification. Nevertheless, population coevolution might lead to increased numbers of plant and insect species, providing the interacting species are isolated. Here, reduced gene flow among populations of the plants and herbivores could lead to each plant–insect interaction following a different evolutionary trajectory, resulting in reproductive isolation and speciation (Strauss & Zangerl, 2002). This could result in plant and insect phylogenies exhibiting similar or congruent branching patterns—parallel cladogenesis. Farrell & Mitter (1998) used a phylogenetic approach in a study of beetles (*Tetraopes* spp.) feeding on milkweeds (*Asclepias* spp.). They found strikingly similar patterns of insect and plant diversification, and interestingly, also found that the most recently derived beetle species are not only associated with the most recently evolved *Asclepias*, but the most recent plant groups also possessed the most toxic cardenolides. Because no information was available on the ability of the beetles to cope with the cardenolides, the study could not distinguish between similar radiation and coevolution. Although this study showed congruence between insect and plant lineages, there are numerous examples that could find no such congruence (e.g., Becerra, 1997; Garin *et al.*, 1999).

In studies of coevolution based on comparison of phylogenies, there is no evidence that the insect exerts selection on its host and vice versa. An alternative approach is to examine microevolutionary forces affecting contemporary plant–insect interactions at the population level. Berenbaum & Zangerl (1998) studied different populations of wild parsnip and its specialist insect herbivore, the parsnip webworm (*Depressaria pastinacella*), comparing the production of defensive furanocoumarins by the plant populations and the ability of webworm populations to detoxify the furanocoumarins. Patterns of furanocoumarin production matched patterns of furanocoumarin detoxification by the herbivores (Figure 5.20). The plant populations were found to be polymorphic, with four major phenotypes related to furanocoumarin composition occurring, while the populations of insect herbivores also comprised different phenotypes with respect to their ability to metabolize the different furanocoumarins. This indicates spatial variation in both the plant and the herbivore. Further, as can be seen in Figure 5.21, when considering plant and herbivore clusters, there was a high degree of frequency matching in three out of four of the populations. This suggests that coevolutionary processes account for the geographic mosaic evident in the differences found among populations.

Cornell & Hawkins (2003) carried out a meta-analysis of the floristic distribution and toxicity of phytochemicals to insect herbivores on the one hand, and herbivore specialization on the other. They found that secondary metabolites with a narrow distribution, representing newly evolved compounds, are more toxic than metabolites with a wider distribution, representing compounds that evolved longer ago. This provides strong support for the escape and radiation prediction of the theory of coevolution.

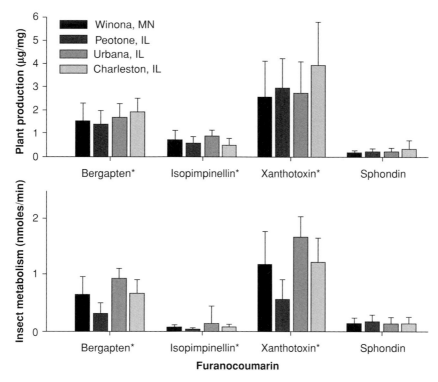

Figure 5.20   Mean furanocoumarin content (μg/mg) of wild parsnip seeds and furanocoumarin metabolism (nmol/min) by parsnip webworm larvae from four midwestern US populations. Furanocoumarins for which significant differences were found between populations are indicated by the asterisk (*). (From Berenbaum & Zangerl (1998), with permission of the National Academy of Sciences, USA.)

The prerequisites for coevolution, that a herbivore should influence plant characteristics directly, that it should discriminate between different plant morphs, and that plant characteristics influence the animal's fitness (Freeland, 1991), highlight an obvious difficulty in identifying coevolution between plants and mammalian herbivores—they are not specialist grazers, but usually feed on many plant species. Perhaps unsurprisingly, therefore, few studies support coevolution in plant–mammalian herbivore interactions, one such being a study of grasses and mammalian grazers in the Serengeti (McNaughton, 1984).

The main argument against coevolution is that the partners in the interaction, for example, plants and insect herbivores, are not equal and are involved in asymmetrical interactions in which the plants exert selection on the insects, but the insects do not exert sufficient selection on the plants for reciprocal selection to occur (Strong *et al.*, 1984; Jermy, 1988). Because of this, it is considered that insects do not influence the evolution of plants. Instead, it is proposed that sequential evolution occurs, where evolution of herbivorous insects follows the evolution of plants, but the reverse does not occur (Jermy, 1988). This assumes that insects take advantage of the available niches that plants provide in abundance. In this theory, any adaptation to the nutritional quality of a new host plant is a secondary process, host plant selection being governed mainly by the chemosensory system of the insect.

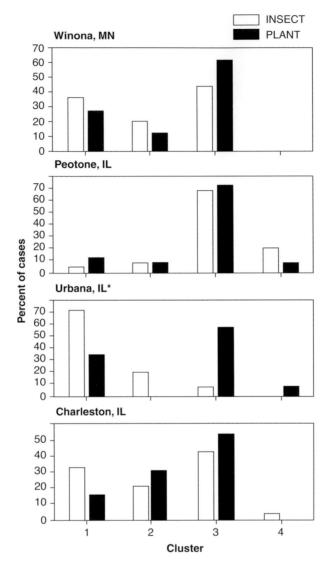

Figure 5.21 Phenotype frequency distributions of insects (parsnip webworm) and plants (wild parsnip) for each of four populations (Winona, Peotone, Urbana, and Charleston). Each of the four groups of phenotype (clusters) represents phenotypes that are similar with respect to production of furanocoumarins (wild parsnip) or detoxification of the furanocoumarins (webworms). The asterisk (*) indicates significant mismatch between plant and insect patterns. (From Berenbaum & Zangerl (1998), with permission of the National Academy of Sciences, USA.)

However, this theory does not take account of the fact that insect behavior can be modified through associative learning, where insects may learn to use certain cues to avoid a host plant that is toxic or otherwise nutritionally unsuitable. Therefore, toxins and nutrients in plant tissue have a feedback mechanism that can influence host selection behavior. As a result, nutritional quality is a major component in the evolution of plant preference.

The theory of coevolution and the arms race analogy has given rise to considerable debate on the basic evolutionary processes that influence plant–insect (and plant–pathogen) interactions. It has provided a useful framework for study of the evolutionary relationship between groups of host plants and their specialized herbivores. However, parallel diversification due to pairwise coevolution is considered to be the exception rather than the rule (Farrell & Mitter, 1994). Nevertheless, the debate on coevolution continues and will stimulate further work on the evolution of interactions between plants and their attackers.

5.7 Conclusions

The evolution of plant defense against attackers has received much attention over the past 30 or more years and has generated a range of hypotheses aimed at guiding research on predicting the phenotypic, genetic, and geographic variation in plant defense. Over the years, a great many studies have tested these hypotheses, greatly increasing our knowledge of the evolution of plant defenses, but yielding apparently contradictory results. This, in turn, has led to the suggestion that an all-encompassing theory of plant defense might be unrealistic (Berenbaum, 1995). However, as pointed out by Stamp (2003), none of the various hypotheses have been firmly rejected and future, careful research will surely lead to increased understanding of plant defense and its evolution.

It is clear from this chapter that in terms of the evolution of plant defense, there has been a concentration of effort on interactions between plants and herbivorous insects. However, this does not mean that the hypotheses of plant defense described earlier bear no relevance to the evolution of plant defenses to pathogens, or indeed any other attackers. Previous work on plant–pathogen interactions can be considered in the light of these hypotheses (see Heil & Walters, 2009), and more research in future should test the hypotheses using systems other than interactions between plants and herbivorous insects.

It should be apparent by now that most of the work on plant defense has concentrated on shoots. Perhaps this should not be surprising, since shoots are more accessible than roots. However, shoots do not exist separately from roots. Both suffer from attack by pathogens and herbivores, and, as a result, defenses can be altered systemically. Types of plant defenses can also differ between shoots and roots, even within a single species. As urged by a number of workers, if we are to gain a fuller understanding of plant defense and its allocation within the plant, future research should include both roots and shoots (Bezemer & van Dam, 2005; Rasmann & Agrawal, 2008).

One area involving plant–pathogen interactions where much effort has been directed, and much is now known, is the evolution of *R* genes. This has been covered only briefly in this chapter, but work in this area will hopefully reveal new principles that might be useful for breeding durable resistance to pathogens, either by conventional breeding or by transgenic means (McDowell & Simon, 2006; see Chapter 6).

Recommended reading

Agrawal AA, 2001. Phenotypic plasticity in the interactions and evolution of species. *Science* **294**, 321–326.

Agrawal AA, 2006. Macroevolution of plant defense strategies. *Trends in Ecology and Evolution* **22**, 103–109.

Berenbaum MR, 1995. The chemistry of defense: theory and practice. *Proceedings of the National Academy of Sciences, USA* **92**, 2–8.

Heil M, 2002. Ecological costs of induced resistance. *Current Opinion in Plant Biology* **5**, 345–350.

Herms DA, Mattson WJ, 1992. The dilemma of plants: to grow or defend. *The Quarterly Review of Biology* **67**, 283–335.

Karban R, Baldwin IT, 1997. *Induced Responses to Herbivory.* Chicago, IL: University of Chicago Press.

Rasmann S, Agrawal AA, 2009. Plant defense against herbivory: progress in identifying synergism, redundancy, and antagonism between resistance traits. *Current Opinion in Plant Biology* **12**, 473–478.

Stamp N, 2003. Out of the quagmire of plant defense hypotheses. *The Quarterly Review of Biology* **78**, 23–55.

Zangerl AR, 2003. Evolution of induced plant responses to herbivores. *Basic and Applied Ecology* **4**, 91–103.

References

Adler FR, Karban R, 1994. Defended fortresses or moving targets? Another model of inducible defenses inspired by military metaphors. *American Naturalist* **144**, 813–832.

Agrawal AA, 2000. Specificity of induced resistance in wild radish: causes and consequences for two specialist and two generalist caterpillars. *Oikos* **89**, 493–500.

Agrawal AA, 2001. Phenotypic plasticity in the interactions and evolution of species. *Science* **294**, 321–326.

Agrawal AA, 2006. Macroevolution of plant defense strategies. *Trends in Ecology and Evolution* **22**, 103–109.

Agrawal AA, Fishbein M, 2006. Plant defense syndromes. *Ecology* **87**, 132–149.

Agrawal AA, Fishbein M, 2008. Phyogenetic escalation and decline of plant defense strategies. *Proceedings of the National Academy of Sciences, USA* **105**, 10057–10060.

Agrawal AA, Karban R, 1999. Why induced defenses may be favored over constitutive strategies in plants. In: Tollrian R, Harvell CD, eds. *The Ecology and Evolution of Inducible Defenses.* Princeton, NJ: Princeton University Press, pp. 45–61.

Agrawal AA, Strauss SY, Stout MJ, 1999. Costs of induced responses and tolerance to herbivory in male and female fitness components of wild radish. *Evolution* **53**, 1093–1104.

Agrawal AA, Conner JK, Johnson MTJ, Wallsgrove R, 2002. Ecological genetics of an induced plant defense against herbivores: additive genetic variance and costs of phenotypic plasticity. *Evolution* **56**, 2206–2213.

Agrawal AA, Fishbein M, Halitschke R, Hastings AP, Robosky DL, Rasmann S, 2009. Evidence for adaptive radiation from a phylogenetic study of plant defenses. *Proceedings of the National Academy of Sciences, USA* **106**, 18067–18072.

Almeida-Cortez JS, Shipley B, Arnason JT, 1999. Do plant species with high relative growth rates have poorer chemical defences? *Functional Ecology* **13**, 819–827.

Armbruster WS, 1997. Exadaptations link evolution of plant-herbivore and plant-pollinator interactions: a phylogenetic inquiry. *Ecology* **78**, 1661–1672.

Baldwin IT, Sims CL, Kean SE, 1990. The reproductive consequences associated with inducible alkaloidal responses in wild tobacco. *Ecology* **71**, 252–262.

Ballhorn DJ, Kautz S, Lion U, Heil M, 2008. Trade-offs between direct and indirect defences of lima bean (*Phaseolus lunatus*). *Journal of Ecology* **96**, 971–980.

Barbosa P, Gross P, Kemper J, 1991. Influence of plant allelochemicals on the tobacco hornworm and its parasitoid. *Cotesia congregata. Ecology* **72**, 1567–1575.

Becerra JX, 1997. Insects on plants: macroevolutionary chemical trends in host use. *Science* **276**, 253–256.

Berenbaum M, 1983. Coumarins and caterpillars—a case for coevolution. *Evolution* **37**, 163–179.

Berenbaum MR, 1995. The chemistry of defense: theory and practice. *Proceedings of the National Academy of Sciences, USA* **92**, 2–8.

Berenbaum MR, Nitao JK, Zangerl AR, 1991. Adaptive significance of furanocoumarin diversity in *Pastinaca sativa* (Apiaceae). *Journal of Chemical Ecology* **17**, 207–215.

Berenbaum MR, Zangerl AR, 1996. Phytochemical diversity: adaptation or random variation? In: Romeo JT, Saunders JA, Barbosa P, eds. *Phytochemical Diversity and Redundancy in Ecological Interactions.* New York: Plenum Press, pp. 1–24.

Berenbaum MR, Zangerl AR, 1998. Chemical phenotype matching between a plant and its insect herbivore. *Proceedings of the National Academy of Sciences, USA* **95**, 13743–13748.

Bergelson J, Dwyer G, Emerson JJ, 2001. Models and data on plant-enemy coevolution. *Annual Review of Genetics* **35**, 469–499.

Bergelson J, Fowler S, Hartley S, 1986. The effects of foliage damage on casebearing moth larvae, *Coleophora serratella*, feeding on birch. *Ecological Entomology* **11**, 241–250.

Bezemer TM, van Dam NM, 2005. Linking aboveground and belowground interactions via induced plant defenses. *Trends in Ecology and Evolution* **20**, 617–624.

Björkman C, Dalin P, Ahrné K, 2008. Leaf trichome responses to herbivory in willows: induction, relaxation and costs. *New Phytologist* **179**, 176–184.

Blossey B, Nötzold R, 1995. Evolution of increased competitive ability in invasive nonindigenous plants—a hypothesis. *Journal of Ecology* **83**, 887–889.

Bogler DJ, Neff JL, Simpson BB, 1995. Multiple origins of the yucca-yucca moth association. *Proceedings of the National Academy of Sciences, USA* **92**, 6864–6867.

Bryant JP, Chapin FS III, Klein DR, 1983. Carbon nutrient balance of boreal plants in relation to vertebrate herbivory. *Oikos* **40**, 357–368.

Bryant JP, Kuropat PJ, Cooper SM, Frisby K, Owen-Smith N, 1989. Resource availability hypothesis of plant antiherbivore defense tested in a South African savanna ecosystem. *Nature* **340**, 227–229.

Bustamante RO, Chacón P, Niemeyer HM, 2006. Patterns of chemical defences in plants: an analysis of the vascular flora of Chile. *Chemoecology* **16**, 145–151.

Buzi A, Chilosi G, De Sillo D, Magro P, 2004. Induction of resistance in melon to *Didymella bryoniae* and *Sclerotinia sclerotiorum* by seed treatments with acibenzolar-*S*-methyl and methyl jasmonate but not with salicylic acid. *Journal of Phytopathology* **152**, 34–42.

Chapin FS III, 1989. The cost of tundra plant structures: evaluation of concepts and currencies. *American Naturalist* **133**, 1–19.

Cipollini D, 2007. Consequences of the overproduction of methyl jasmonate on seed production, tolerance to defoliation and competitive effect and response of *Arabidopsis thaliana. New Phytologist* **173**, 146–153.

Cipollini D, Purrington CB, Bergelson J, 2003. Costs of induced responses in plants. *Basic and Applied Ecology* **4**, 79–89.

Cipollini DF, Bergelson J, 2001. Plant density and nutrient availability constrain constitutive and wound-induced expression of trypsin inhibitors in *Brassica napus. Journal of Chemical Ecology* **27**, 593–610.

Coley PD, 1987. Interspecific variation in plant anti-herbivore properties: the role of habitat quality and rate of disturbance. *New Phytologist* **106** (suppl.), 251–263.

Coley PD, 1988. Effects of plant growth rate and leaf lifetime on the amount and type of anti-herbivore defense. *Oecologia* **74**, 531–536.

Coley PD, Bryant JP, Chapin FS III, 1985. Resource availability and plant antiherbivore defense. *Science* **230**, 895–899.

Coley PD, Lovkam J, Rudolph K, Bromberg K, Sackett TE, Wright L, Brenes-Arguedas T, Dvorett D, Ring S, Clark A, Baptiste C, Pennington RT, Kursar TA, 2005. Divergent strategies of young leaves in two species of *Inga*. *Ecology* **86**, 2633–2643.

Cornell HV, Hawkins BA, 2003. Herbivore responses to plant secondary compounds: a test of phytochemical coevolution theory. *American Naturalist* **161**, 507–522.

de Nardi B, Dreos R, Del Terra L, Martellossi C, Asquini E, Tornincasa P, Gasperini D, Pacchioni B, Raihinavelu R, Pallavicini A, Graziosi G, 2006. Differential responses of *Coffea arabica* L. leaves and roots to chemically induced systemic acquired resistance. *Genome* **49**, 1594–1605.

Dietrich R, Ploss K, Heil M, 2005. Growth responses and fitness costs after induction of pathogen resistance depend on environmental conditions. *Plant, Cell and Environment* **28**, 211–222.

Dobzhansky TG, 1970. *Genetics of the Evolutionary Process.* New York: Columbia University Press.

Dodds P, Thrall P, 2009. Recognition events and host-pathogen co-evolution in gene-for-gene resistance to flax rust. *Functional Plant Biology* **36**, 395–408.

Duffey SS, Hoover K, Bonning B, Hammock BD, 1995. The impact of host plant on the efficacy of baculoviruses. In: Roe RM, Kuhr RJ, eds. *Review of Pesticides and Toxicology*, Vol. 3. Raleigh, NC: Toxicology Communications, pp. 137–275.

Edwards PJ, Wratten SD, 1983. Wound induced defences in plants and their consequences for patterns of insect grazing. *Oecologia* **59**, 88–93.

Ehrlich PR, Raven PH, 1964. Butterflies and plants: a study in coevolution. *Evolution* **18**, 586–608.

Farrell BD, Dussourd DE, Mitter C, 1991. Escalation of plant defense: do latex and resin canals spur plant diversification? *American Naturalist* **138**, 881–900.

Farrell BD, Mitter C, 1994. Adaptive radiation in insects and plants—time and opportunity. *American Zoologist* **34**, 57–69.

Farrell BD, Mitter C, 1998. The timing of insect-plant diversification: might *Tetraopes* (Coleoptera: Cerambycidae) and *Asclepias* (Asclepiadaceae) have co-evolved? *Biological Journal of the Linnean Society* **63**, 553–577.

Feeny P, 1976. Plant apparency and chemical defense. In: Wallace JW, Mansell RL, eds. *Recent Advances in Phytochemistry*. New York: Plenum Press, pp. 1–40.

Fine PVA, Miller ZJ, Mesones I, Irazuzta S, Appel HM, Stevens MHH, Saaksjarvi I, Schultz JC, Coley PD, 2006. The growth-defense trade-off and habitat specialization by plants in Amazonian forests. *Ecology* **87** (suppl.), 150–162.

Fineblum WL, Rausher MD, 1995. Tradeoff between resistance and tolerance to herbivore damage in a morning glory. *Nature* **377**, 517–520.

Fox LR, 1988. Diffuse coevolution within complex communities. *Ecology* **69**, 906–907.

Fraser LH, Grime JP, 1999. Aphid fitness on 13 grass species: a test of plant defence theory. *Canadian Journal of Botany* **77**, 1783–1789.

Frati F, Chamberlain K, Birkett M, Dufour S, Mayon P, Woodcock C, Wadhams L, Pickett J, Salerno G, Conti E, Bin F, 2009. *Vicia faba-Lygus rugulipennis* interactions: induced plant volatiles and sex pheromone enhancement. *Journal of Chemical Ecology* **35**, 201–208.

Freeland WJ, 1991. Plant secondary metabolites: biochemical co-evolution with herbivores. In: Palo RT, Robbins CT, eds. *Plant Defenses against Mammalian Herbivory*. Boca Raton, FL: CRC Press, pp. 61–81.

Garcia-Robledo C, Horvitz C, 2009. Host plant scents attract rolled-leaf beetles to Neotropical gingers in a Central American tropical rain forest. *Entomologia Experimentalis et Applicata* **131**, 115–120.

Garin CF, Juan C, Petitpierre E, 1999. Mitochondrial DNA phylogeny and the evolution of host plant use in paleartic *Chrysolina* (Coleoptera, Chrysomelidae) leaf beetles. *Journal of Molecular Evolution* **48**, 435–444.

Garnier M, Foissac X, Gaurivaud P, Laigret F, Renaudin J, Saillard C, Bove JM, 2001. Mycoplasmas, plants, insect vectors: a matrimonial triangle. *Comptes Rendus de L'Academie des Sciences Serie III—Sciences de la Vie* **324**, 923–928.

Gershenzon J, 1994a. Metabolic costs of terpenoid accumulation in higher plants. *Journal of Chemical Ecology* **20**, 1281–1328.

Gershenzon J, 1994b. The cost of plant chemical defense against herbivory: a biochemical perspective. In: Bernays EA, ed. *Insect-Plant Interactions*. Boca Raton, FL: CRC Press, pp. 105–173.

Gianoli E, Niemeyer HM, 1997. Lack of costs of herbivory-induced defenses in a wild wheat: integration of physiological and ecological approaches. *Oikos* **80**, 269–275.

Glynn C, Herms DA, Orians CM, Hansen RC, Larsson S, 2007. Testing the growth-differentiation balance hypothesis: dynamic responses of willows to nutrient availability. *New Phytologist* **176**, 623–634.

Goheen JR, Young TP, Keesing F, Palmer TM, 2007. Consequences of herbivory by native ungulates for the reproduction of a savanna tree. *Journal of Ecology* **95**, 129–138.

Grant MR, McDowell JM, Sharpe AG, de Torres Zabala M, Lydiate DJ, Dangl JL, 1998. Independent deletions of a pathogen-resistance gene in *Brassica* and *Arabidopsis*. *Proceedings of the National Academy of Sciences, USA* **95**, 15843–15848.

Green TR, Ryan CA, 1972. Wound-induced proteinase inhibitor in plant leaves: a possible defense mechanism against insects. *Science* **175**, 776–777.

Grime JP, 1977. Evidence for the existence of three primary strategies in plants and its relevance to ecological and evolutionary theory. *American Naturalist* **111**, 1169–1194.

Grime JP, 2001. *Plant Strategies, Vegetation Processes, and Ecosystem Properties*. New York: John Wiley & Sons, Ltd.

Gulmon SL, Mooney HA, 1986. Costs of defense on plant productivity. In: Givnish TJ, ed. *On the Economy of Plant Form and Function*. Cambridge: Cambridge University Press, pp. 681–698.

Gunasena GH, Vinson SB, Williams HJ, Stipanovic RD, 1988. Effects of caryophyllene, caryophyllene oxide, and their interaction with gossypol on the growth and development of *Heliothis virescens* (F) (Lepidoptera: Noctuidae). *Journal of Economic Entomology* **81**, 93–97.

Harborne JB, 1993. *Introduction to Ecological Biochemistry*. London: Academic Press.

Hare JD, 1992. Effects of plant variation on herbivore-natural enemy interactions. In: Fritz RS, Simms EL, eds. *Plant Resistance to Herbivores and Pathogens: Ecology, Evolution, and Genetics*. Chicago, IL: University of Chicago Press, pp. 278–300.

Havill NP, Raffa KF, 2000. Compound effects of induced plant responses on insect herbivores and parasitoids: implications for tritrophic interactions. *Ecological Entomology* **25**, 171–179.

Heath MC, 1987. Evolution of parasitism in the fungi. In: Rayner ADM, Brasier CM, Moore D, eds. *Evolutionary Biology of the Fungi*. Cambridge: Cambridge University Press, pp. 149–160.

Heath MC, 1991. Evolution of resistance to fungal parasitism in natural ecosystems. *New Phytologist* **119**, 331–343.

Heidel AJ, Baldwin IT, 2004. Microarray analysis of salicylic acid- and jasmonic acid-signalling in responses of *Nicotiana attenuata* to attack by insects from multiple feeding guilds. *Plant, Cell and Environment* **27**, 1362–1373.

Heil M, 2002. Ecological costs of induced resistance. *Current Opinion in Plant Biology* **5**, 345–350.

Heil M, Delsinne T, Hilpert A, Shürkens S, Andarry C, Linsenmair KE, Sousa M, McKey D, 2002. Reduced chemical defence in ant-plants? A critical re-evaluation of a widely accepted hypothesis. *Oikos* **99**, 457–468.

Heil M, Feil D, Hilpert A, Linsenmair KE, 2004a. Spatiotemporal patterns in indirect defence of a South-East Asian ant-plant support the optimal defence hypothesis. *Journal of Tropical Ecology* **20**, 573–580.

Heil M, Greiner S, Meimberg H, Krüger R, Noyer J-L, Heubl G, Linsenmair KE, Boland W, 2004b. Evolutionary change from induced to constitutive expression of an indirect plant resistance. *Nature* **430**, 205–208.

Heil M, Hilpert A, Kaiser W, Linsenmair KE, 2000. Reduced growth and seed set following chemical induction of pathogen defence: does systemic acquired resistance (SAR) incur allocation costs? *Journal of Ecology* **88**, 645–654.

Heil M, Ploss K, 2006. Induced resistance enzymes in wild plants—do "early birds" escape from pathogen attack? *Naturwissenschaften* **93**, 455–460.

Heil M, Walters DR, 2009. Ecological consequences of plant defence signalling. In: Van Loon LC, ed. *Advances in Botanical Research*, Vol. 51. Burlington, VT: Academic Press, pp. 667–716.

Hendriks RJJ, de Boer NJ, van Groenendael JM, 1999. Comparing the preferences of three herbivore species with resistance traits of 15 perennial dicots: the effects of phylogenetic constraints. *Plant Ecology* **143**, 141–152.

Herms DA, Mattson WJ, 1992. The dilemma of plants: to grow or defend. *The Quarterly Review of Biology* **67**, 283–335.

Holopainen JK, Rikala R, Kainulainen P, Oksanen J, 1995. Resource partitioning to growth, storage and defence in nitrogen fertilized Scots pine and susceptibility of the seedlings to the tarnished plant bug, *Lygus rugulipennis*. *New Phytologist* **131**, 521–532.

Hu GY, Mitchell ER, 2001. Responses of *Diadegma insulare* (Hymenoptera: Ichneumonidae) to caterpillar feeding in a flight tunnel. *Journal of Entomological Science* **36**, 297–304.

Hukkanen AT, Kokko HI, Buchala AJ, McDougall GJ, Stewart D, Karenlampi SO, Karjalainen RO, 2007. Benzothiadiazole induces the accumulation of phenolics and improves resistance to powdery mildew in strawberries. *Journal of Agricultural and Food Chemistry* **55**, 1862–1870.

Iriti M, Faoro F, 2003. Does benzothiadiazole-induced resistance increase fitness cost in bean? *Journal of Plant Pathology* **85**, 265–270.

Izaguirre MM, Mazza CA, Biodini M, Baldwin IT, Ballaré CL, 2006. Remote sensing of future competitors: impacts on plant defenses. *Proceedings of the National Academy of Sciences, USA* **103**, 7170–7174.

Janzen DH, 1974. Tropical blackwater rivers, animals, and mast fruiting by the Dipterocarpaceae. *Biotropica* **6**, 69–103.

Jermy T, 1988. Can predation lead to narrow food specialization in phytophagous insects? *Ecology* **69**, 902–904.

Jones JDG, Dangl JL, 2006. The plant immune system. *Nature* **444**, 323–329.

Kaplan I, Dively GP, Denno RF, 2009. The costs of anti-herbivore defense traits in agricultural crop plants: a case study involving leafhoppers and trichomes. *Ecological Applications* **19**, 864–872.

Karban R, 1993. Costs and benefits of induced resistance and plant density for a native shrub, *Gossypium thurberi*. *Ecology* **74**, 1–8.

Karban R, Baldwin IT, 1997. *Induced Responses to Herbivory*. Chicago, IL: University of Chicago Press.

Karban R, Niho C, 1995. Induced resistance and susceptibility to herbivory—plant memory and altered plant development. *Ecology* **76**, 1220–1225.

Kauffman WC, Kennedy GG, 1989. Toxicity of allelochemicals from wild insect-resistant tomato, *Lycopersicon hirsutum* f. *glabratum* to *Compoletis sonorensis*, a parasitoid of *Heliothis zea*. *Journal of Chemical Ecology* **15**, 2051–2060.

Kohn JR, Graham SW, Morton B, Doyle JJ, Barrett CH, 1996. Reconstruction of the evolution of reproductive characters in Pontederiaceae using phylogenetic evidence from chloroplast DNA restriction-site variation. *Evolution* **50**, 1454–1469.

Koricheva J, Larsson S, Haukioja E, Keinänen M, 1998. Regulation of woody plant secondary metabolism by resource availability: hypothesis testing by means of meta-analysis. *Oikos* **83**, 212–226.

Koricheva J, Nykanen H, Gianoli E, 2004. Meta-analysis of trade-offs among plant antiherbivore defenses: Are plants jacks-of-all-trades, masters of all? *American Naturalist* **163**, E64–E75.

Korneef A, Pieterse CMJ, 2008. Cross talk in defense signaling. *Plant Physiology* **146**, 839–844.

Körner CH, 1991. Some often overlooked plant characteristics as determinants of plant growth: a re-consideration. *Functional Ecology* **5**, 162–173.

Korves TM, Bergelson J, 2004. A novel cost of R gene resistance in the presence of disease. *American Naturalist* **163**, 489–504.

Krips OE, Willems PEL, Dicke M, 1999. Compatibility of host plant resistance and biological control of the two-spotted spider mite *Tetranychus urticae* in the ornamental crop gerbera. *Biological Control* **16**, 155–163.

Kursar TA, Coley PD, 2003. Convergence in defense syndromes of young leaves in tropical rainforests. *Biochemical Systematics and Ecology* **31**, 929–949.

Lambers H, Poorter H, 1992. Inherent variation in growth rate between higher plants: a search for physiological causes and ecological consequences. *Advances in Ecological Research* **23**, 187–261.

Lankau RA, Kliebenstein DJ, 2009. Competition, herbivory and genetics interact to determine the accumulation and fitness consequences of a defence metabolite. *Journal of Ecology* **97**, 78–88.

Logemann E, Wu S-C, Schröder J, Schmelzer E, Somssich IE, Hahlbrock K, 1995. Gene activation by UV light, fungal elicitor or fungal infection in *Petroselium crispum* is correlated with repression of cell-cycle-related genes. *Plant Journal* **8**, 865–876.

Loomis WE, 1953. Growth and differentiation—an introduction and summary. In: Loomis WE, ed. *Growth and Differentiation in Plants*. Ames, IA: Iowa State College Press, pp. 1–17.

Massey FP, Ennos AR, Hartley SE, 2007. Grasses and the resource availability hypothesis: the importance of silica-based defences. *Journal of Ecology* **95**, 414–424.

Mauricio R, Bowers MD, Bazzaz FA, 1993. Pattern of leaf damage affects fitness of the annual plant *Raphanus sativus* (Brassicaceae). *Ecology* **74**, 2066–2071.

McDonald AJS, 1990. Phenotypic variation in growth rate as affected by N-supply: its effects on net assimilation rate (NAR), leaf weight ration (LWR) and specific leaf area (SLA). In: Lambers H, Konings H, Pons TL, eds. *Causes and Consequences of Variation in Growth Rate and Productivity of Higher Plants*. The Hague, the Netherlands: SPB Publishing, pp. 35–44.

McDowell JM, Simon SA, 2006. Recent insights into *R* gene evolution. *Molecular Plant Pathology* **7**, 437–448.

McKey D, 1974. Adaptive patterns in alkaloid physiology. *American Naturalist* **108**, 305–320.

McNaughton SJ, 1984. Grazing lawns: animals in herds, plant form and coevolution. *American Naturalist* **124**, 863–886.

Messina FJ, Durham SL, Richards JH, McArthur ED, 2002. Trade-off between plant growth and defense? A comparison of sagebrush populations. *Oecologia* **131**, 43–51.

Meyer GA, 1998. Pattern of defoliation and its effect on photosynthesis and growth of goldenrod. *Functional Ecology* **12**, 270–279.

Meyer GA, Clare R, Weber E, 2005. An experimental test of the evolution of increased competitive ability in goldenrod, *Solidago gigantean*. *Oecologia* **144**, 299–307.

Mihaliak CA, Lincoln DE, 1985. Growth pattern and carbon allocation to volatile leaf terpenes under nitrogen-limiting conditions in *Heterotheca subaxillaris* (Asteraceae). *Oecologia* **66**, 423–426.

Mitchell-Olds T, Siemens D, Pedersen D, 1996. Physiology and costs of resistance to herbivory and disease in *Brassica*. *Entomologia Experimentalis et Applicata* **80**, 231–237.

Mole S, 1994. Trade-offs and constraints in plant-herbivore defense theory: a life-history perspective. *Oikos* **71**, 3–12.

Moreno JE, Tao Y, Chory J, Ballaré CL, 2009. Ecological modulation of plant defense via phytochrome control of jasmonate sensitivity. *Proceedings of the National Academy of Sciences, USA* **106**, 4935–4940.

Murray DC, Walters DR, 1992. Increased photosynthesis and resistance to rust infection in upper, uninfected leaves of rusted broad bean (*Vicia faba* L.). *New Phytologist* **120**, 235–242.

Muzika RM, Pregitzer KS, Hanover JW, 1989. Changes in terpene production following nitrogen fertilization of grand fir (*Abies grandis* (Dougl.) Lindl.) seedlings. *Oecologia* **80**, 485–489.

Nelson CJ, Seiber JN, Brower LP, 1981. Seasonal and intraplant variation of cardenolide content in the California milkweed *Asclepias eriocarpa*, and implications for plant defense. *Journal of Chemical Ecology* **7**, 981–1010.

Nobuta K, Ashfield T, Kim S, Innes RW, 2005. Diversification of non-TIR class NB-LRR genes in relation to whole-genome duplication events in *Arabidopsis*. *Molecular Plant-Microbe Interactions* **18**, 103–109.

Painter EL, 1987. *Grazing and Intraspecific Variation in Four North American Grass Species*, PhD thesis. Fort Collins, CO: Colorado State University.

Prats E, Rubiales D, Jorrín J, 2002. Acibenzolar-S-methyl induced resistance to sunflower rust (*Puccinia helianthii*) is associated with an enhancement of coumarins on foliar surface. *Physiological and Molecular Plant Pathology* **60**, 155–162.

Price PW, 1991. The plant vigor hypothesis and herbivore attack. *Oikos* **62**, 244–251.

Radhika V, Kost C, Bartram S, Heil M, Boland W, 2008. Testing the optimal defence hypothesis for two indirect defences: extrafloral nectar and volatile organic compounds. *Planta* **228**, 449–457.

Rasmann S, Agrawal AA, 2008. In defense of roots: a research agenda for studying plant resistance to belowground herbivory. *Plant Physiology* **146**, 875–880.

Rasmann S, Agrawal AA, Cook SC, Erwin AC, 2009. Cardenolides, induced responses, and interactions between above- and belowground herbivores of milkweed (*Asclepias* spp.). *Ecology* **90**, 2393–2404.

Rehr SS, Feeny PP, Janzen DH, 1973. Chemical defence in Central American non-ant acacias. *Journal of Animal Ecology* **42**, 405–416.

Reitz SR, Trumble JT, 1997. Effects of linear furanocoumarins on the herbivore *Spodoptera exigua* and the parasitoid *Archytas marmoratus*: host quality and parasitoid success. *Entomologia Experimentalis et Applicata* **84**, 9–16.

Roberts AM, Walters DR, 1986. Stimulation of photosynthesis in uninfected leaves of rust-infected leeks. *Annals of Botany* **56**, 893–896.

Rose LE, Bittner-Eddy PD, Langley CH, Holub EB, Michelmore RW, Benyon JL, 2004. The maintenance of extreme amino acid diversity at the disease resistance gene, RPP13, in *Arabidopsis thaliana*. *Genetics* **166**, 1517–1527.

Rudgers JA, Strauss SY, Wendel JE, 2004. Trade-offs among anti-herbivore resistance traits: insights from Gossypieae (Malvaceae). *American Journal of Botany* **91**, 871–880.

Scheideler M, Schlaich NL, Fellenberg K, Beissbarth T, Hauser NC, Vingron M, Slusarenko AJ, Hoheisel JD, 2002. Monitoring the switch from housekeeping to pathogen defense metabolism in *Arabidopsis thaliana* using cDNA arrays. *Journal of Biological Chemistry* **277**, 10555–10561.

Schenk PM, Kazan K, Manners JM, Anderson JP, Simpson RS, Wilson IW, Somerville SC, Maclean DJ, 2003. Systemic gene expression in *Arabidopsis* during an incompatible interaction with *Alternaria brassicicola*. *Plant Physiology* **132**, 999–1010.

Shure DJ, Wilson LA, 1993. Patch-size effects on plant phenolics in successional openings of the southern Appalachians. *Ecology* **74**, 55–67.

Siemann E, Rogers WE, 2003. Reduced resistance of invasive varieties of the alien tree *Sapium sebiferum* to a generalist herbivore. *Oecologia* **135**, 451–457.

Siemens DH, Garner SH, Mitchell-Olds T, Callaway RM, 2002. Cost of defense in the context of plant competition: *Brassica rapa* may grow and defend. *Ecology* **83**, 505–517.

Simms EL, 1992. Costs of plant resistance to herbivory. In: Fritz RS, Simms EL, eds. *Plant Resistance to Herbivores and Pathogens: Ecology, Evolution, and Genetics*. Chicago, IL: University of Chicago Press, pp. 392–425.

Simms EL, Triplett J, 1994. Costs and benefits of plant responses to disease: resistance and tolerance. *Evolution* **48**, 1973–1985.

Smedegaard-Petersen V, Stolen O, 1981. Effect of energy-requiring defense reactions on yield and grain quality in a powdery mildew resistant cultivar. *Phytopathology* **71**, 396–399.

Sonnemann I, Streicher NM, Wolters V, 2005. Root associated organisms modify the effectiveness of chemically induced resistance in barley. *Soil Biology and Biochemistry* **37**, 1837–1842.

Stahl EA, Dwyer G, Mauricio R, Kreitman M, Bergelson J, 1999. Dynamics of disease resistance polymorphism at the Rpm1 locus of *Arabidopsis*. *Nature* **400**, 667–671.

Stamp N, 2003. Out of the quagmire of plant defense hypotheses. *The Quarterly Review of Biology* **78**, 23–55.

Stamp NE, Yang YL, Osier TL, 1997. Response of an insect predator to prey fed multiple allelochemicals under representative thermal regimes. *Ecology* **78**, 203–214.

Stapley L, 1998. The interaction of thorns and symbiotic ants as an effective defence mechanism of swollen-thorn acacias. *Oecologia* **115**, 401–405.

Steward JL, Keeler KH, 1988. Are there trade-offs among antiherbivore defenses in Ipomoea (Convulvulaceae). *Oikos* **53**, 79–86.

Strauss SY, Zangerl AR, 2002. Plant-insect interactions in terrestrial ecosystems. In: Herrera CM, Pellmyr O, eds. *Plant-Animal Interactions: An Evolutionary Approach*. Malden, MA: Blackwell Publishing Ltd., pp. 77–106.

Strong DR, Lawton JH, Southwood TRE, 1984. *Insects on Plants: Community Patterns and Mechanisms*. Cambridge, MA: Harvard University Press.

Thaler JS, Karban R, 1997. A phylogenetic reconstruction of constitutive and induced resistance in *Gossypium*. *The American Naturalist* **149**, 1139–1146.

Thompson JN, 1994. *The Coevolutionary Process*. Chicago, IL: University of Chicago Press.

Tian D, Traw MB, Chen JQ, Kreitman M, Bergelson J, 2003. Fitness costs of R-gene-mediated resistance in *Arabidopsis thaliana*. *Nature* **423**, 74–77.

Traw MB, Kniskern JM, Bergelson J, 2007. SAR increases fitness of *Arabidopsis thaliana* in the presence of natural bacterial pathogens. *Evolution* **61**, 2444–2449.

Tuomi J, 1992. Toward integration of plant defence theories. *Trends in Ecology and Evolution* **7**, 365–367.

Tuomi J, Fagerstrom T, Niemelä P, 1991. Carbon allocation, phenotypic plasticity, and induced defenses. In: Tallamy DW, Raupp MJ, eds. *Phytochemical Induction by Herbivores*. New York: John Wiley & Sons, Ltd., pp. 85–104.

Tuomi J, Niemelä P, Chapin FS III, Bryant JP, Sirén S, 1988. Defensive responses of trees in relation to their carbon/nutrient balance. In: Mattson WJ, Levieux J, Bernard-Dagan C, eds. *Mechanisms of Woody Plant Defenses against Insects: Search for Pattern*. New York: Springer-Verlag, pp. 57–72.

Tuomi J, Niemelä P, Haukioja E, Sirén S, Neuvonen S, 1984. Nutrient stress: an explanation for plant antiherbivore responses to defoliation. *Oecologia* **61**, 208–210.

Turlings TCJ, Wäckers FL, Vet LEM, Lewis WJ, Tumlinson JH, 1993. Learning of host-finding cues by hymenopterous parasitoids. In: Papaj DR, Lewis AC, eds. *Insect Learning: Ecological and Evolutionary Processes*. New York: Chapman and Hall, pp. 51–78.

Twigg LE, Socha LV, 1996. Physical versus chemical defence mechanisms in toxic *Gastrolobium*. *Oecologia* **108**, 21–28.

Van Dam NM, Baldwin IT, 1998. Costs of jasmonate-induced responses in plants competing for limited resources. *Ecology Letters* **1**, 30–33.

Van Dam NM, Hermenau U, Baldwin IT, 2001. Instar-specific sensitivity of specialist *Manduca sexta* larvae to induced defences in their host plant *Nicotiana attenuata*. *Ecological Entomology* **26**, 578–586.

Van den Boom CEM, Van Beek TA, Posthumus MA, De Groot A, Dicke M, 2004. Qualitative and quantitative variation among volatile profiles induced by *Tetranychus urticae* feeding on plants from different families. *Journal of Chemical Ecology* **30**, 69–89.

van der Meijden E, Wijn M, Verkaar HJ, 1988. Defense and regrowth, alternate strategies in the struggle against herbivores. *Oikos* **51**, 355–363.

Van Hulten M, Pelser M, Van Loon LC, Pieterse CMJ, Ton J, 2006. Costs and benefits of priming for plant defense in *Arabidopsis*. *Proceedings of the National Academy of Sciences, USA* **103**, 5602–5607.

Van Zandt PA, 2007. Plant defense, growth, and habitat: a comparative assessment of constitutive and induced resistance. *Ecology* **88**, 1984–1993.

Vila M, Gomez A, Maron JL, 2003. Are alien plants more competitive than their native conspecifics? A test using *Hypericum perforatum* L. *Oecologia* **137**, 211–215.

Wallace SK, Eigenbrode SD, 2002. Changes in the glucosinolate-myrosinase defense system in *Brassica juncea* cotyledons during seedling development. *Journal of Chemical Ecology* **28**, 243–256.

Walters DR, Cowley T, Weber H, 2006. Rapid accumulation of trihydroxy oxylipins and resistance to the bean rust pathogen *Uromyces fabae* following wounding in *Vicia faba*. *Annals of Botany* **97**, 779–784.

Walters DR, Heil M, 2007. Costs and trade-offs associated with induced resistance. *Physiological and Molecular Plant Pathology* **71**, 3–17.

Walters DR, Paterson L, Walsh DJ, Havis ND, 2009. Priming for plant defense in barley provides benefits only under high disease pressure. *Physiological and Molecular Plant Pathology* **73**, 95–100.

Wilkens RT, Spoerke JM, Stamp NE, 1996. Differential responses of growth and two soluble phenolics of tomato to resource availability. *Ecology* **77**, 247–258.

Willis AJ, Memmott J, Forrester RI, 2000. Is there evidence for the post-invasion evolution of increased size among invasive plant species? *Ecology Letters* **3**, 275–283.

Wink M, 2003. Evolution of secondary metabolites from an ecological and molecular phyogenetic perspective. *Phytochemistry* **64**, 3–19.

Ye Z-H, Varner JE, 1991. Tissue-specific expression of cell wall proteins in developing soybean tissues. *The Plant Cell* **3**, 23–37.

Young TP, Okello BD, 1998. Relaxation of an inducible defense after exclusion of herbivores: spines on *Acacia drepanolobium*. Oecologia **115**, 508–513.

Zangerl AR, 2003. Evolution of induced plant responses to herbivores. *Basic and Applied Ecology* **4**, 91–103.

Zangerl AR, Rutledge CE, 1996. The probability of attack and patterns of constitutive and induced defense: a test of optimal defense theory. *American Naturalist* **147**, 599–608.

Zavala JA, Patankar AG, Gase K, Baldwin IT, 2004. Constitutive and inducible trypsin proteinase inhibitor production incurs large fitness costs in *Nicotiana attenuata*. *Proceedings of the National Academy of Sciences, USA* **101**, 1607–1612.

Zhang D-X, Nagabhyru P, Schardl CL, 2009. Regulation of a chemical defense against herbivory produced by symbiotic fungi in grass plants. *Plant Physiology* **150**, 1072–1082.

Chapter 6
Exploiting Plant Defense

6.1 Introduction

As we saw in Chapter 1, attack by pathogens, herbivores, nematodes, and parasitic plants results in considerable damage to plants in both natural and crop situations. It is clear that plants possess a remarkable ability to defend themselves and to deal with damage caused by attackers. However, agriculture as practiced in many parts of the world stacks the odds against plants. Huge areas of monocropping facilitate the rapid spread of pathogens and insect herbivores that manage to overcome resistance bred into our crops, or which manage to develop resistance to the crop protection chemicals we use. The ever-present threat to crops from a myriad of attackers necessitates the use of a variety of crop protection approaches. Prominent among these approaches is the use of crop varieties with resistance to particular pests and pathogens. Here, plant defense can be harnessed to protect our crops from the worst ravages of attackers. In this chapter, we examine the uses to which plant defense are currently put in the protection of our crops, and we look at how we might better use plant defense in the future to help safeguard crop production.

6.2 Using plant resistance to protect crops—breeding

6.2.1 Introduction

The natural genetic resistance of plants has undoubtedly played a major role in protecting crops since the very beginnings of agriculture, and it seems likely that systematic cultivation of crops would have been impossible without the ability of most plants to ward off attack by pathogens and pests. Crop plants would have been selected for high yields and nutritional value, together with low mammalian toxicity and adequate resistance to pests and diseases. But in our continuous efforts to develop better yielding crop cultivars, very few cultivated species have retained the levels of pest and disease resistance present in their wild ancestors.

Genetic resistance is, in fact, one of the oldest recognized forms of plant protection. According to Allard (1960), Theophrastus recognized differences in disease susceptibility among crop cultivars in the third century BC. The earliest documentary evidence on plant resistance is a paper produced in 1782 reporting resistance to Hessian fly in the USA, in the wheat variety Underhill (Havens, 1792). However, although plant resistance to diseases and

Plant Defense, First Edition, by Dale Walters   © 2011 by Blackwell Publishing Ltd.

pests was recognized in the nineteenth century, the breeding of pest-resistant plant cultivars was undertaken only after the discovery of Mendel's law of heredity in 1900. Work by Biffen (1907) reporting that resistance to yellow rust in wheat is controlled by a single recessive gene prompted agriculturalists to search for genes for disease resistance in wheat and other crops. Since then, the development of crop cultivars with resistance to pathogens and pests has become an important part of modern agriculture. In the sections below, we look at the methods used in breeding for resistance to pathogens, insect pests, nematodes, and parasitic plants.

6.2.2   Breeding for resistance

From a plant breeding perspective, resistance is any inherited characteristic of the plant that reduces the effect of disease or pest infestation. In producing a resistant plant cultivar, it is necessary to have (a) a source of genetic resistance, (b) a means of identifying and selecting the resistance, and (c) a means of introducing this resistance into a plant to produce a new cultivar that is commercially acceptable (Lucas, 1998).

6.2.2.1   Sources of resistance

Plant breeding is dependent on genetic variability within crop species and related wild relatives. Although there might already be significant genetic variation within the crop itself, for example, if the crop is outbreeding, if such variation is not present within the crop, it is often necessary to turn to alternative sources of genetic variation. Therefore, broad-based collections of germplasm are likely to be important for the success of crop improvement programs. Germplasm collections consist of (1) wild species, (2) weed races, (3) landraces, (4) unimproved or purified cultivars no longer in cultivation, (5) improved modern cultivars under cultivation, (6) breeding stocks developed by breeders, but not released for cultivation, and (7) mutants developed by mutagenic treatments, as well as those of spontaneous origin (Panda & Khush, 1995).

Breeders searching for new sources of resistance to pathogens or pests usually look to centers of diversity, where the pathogen or pest is endemic and the host plant and microbe or insect have coevolved. For example, a good deal of the brown planthopper-resistant germplasm comes from South India and Sri Lanka, where this insect is endemic (Khush, 1977). Similarly, there is substantial diversity for late blight resistance in wild potatoes in Central America, and wild grasses have been useful sources of resistance for powdery mildew and rust diseases of barley and wheat.

Landraces are collections made from fields where the farmer is unlikely to have introduced modern varieties (Wallwork, 2009). They can be quite mixed and will have been locally selected over a long period, and as a result, might contain genetic variation, which is not present in modern breeding programs. Wild relatives, including crop ancestors, are species that in nature, by and large, remain genetically isolated from the crop species. The advantage here is that they can be hybridized with the crop to allow transfer of a desired trait, without the problems of hybrid sterility.

Other sources of novel resistance genes include mutation breeding and somaclonal variation. Mutation breeding is based on the use of mutagenic treatments, such as γ-irradiation and chemicals, usually ethyl methyl sulfonate or sodium azide, which induce alterations in DNA. Sometimes, this approach can produce a loss of function, rather than a gain of a new property such as resistance. A highly successful example of mutation breeding is the development of durable resistance to powdery mildew (*Blumeria graminis* f.sp. *hordei*) controlled by resistance at the *mlo* locus in barley. Some *mlo* mutants have been widely deployed in high-yielding varieties across Europe without any evidence of the resistance being overcome by the pathogen (Jorgensen 1992). Of interest here is that the protein encoded by the *mlo* gene is different from other resistance genes identified to date. The wild-type allele codes for a cell membrane receptor and it is the nonfunctional allele, which provides the resistance (Büschges *et al.*, 1997).

Somaclonal variation involves the use of plant tissue culture to extend the range of variation available. Although it can produce alterations in the reactions of plants to pathogens, for instance, few examples of useful resistance have been generated to date. It is also possible to transfer foreign genes into plants using genetic transformation techniques. This aspect is covered later in this chapter.

6.2.2.2   Breeding methods and selection strategies

The breeding method chosen will depend on the reproductive system of the crop species. Therefore, a distinction must be made between self-pollinating species and largely cross-pollinating species. All plants in a population of outcrossing species are highly heterozygous, and plant vigor deteriorates with forced inbreeding. Plant breeding programs for such crops aim to maintain heterozygosity or to restore it as a final step in the process. In contrast, populations of self-pollinated plants possess little or no heterozygosity, and tend to consist of many closely related homozygous lines. Here, individual plants are fully vigorous homozygotes, and breeding programs aim to produce a pure line.

Marker-assisted selection

In some cases, the gene for resistance is tightly linked with a plant morphological trait, in which case the resistance and the morphological trait segregate together in the breeding populations. Plants selected on the basis of a morphological trait are resistant to the pest, for example. But such morphological markers are not common, and they tend to negatively affect the plant's performance. It is now possible, with the use of molecular markers, to efficiently select breeding lines based on a plant's genotype rather than its phenotype. These markers can only be used where the resistance genes required for resistance expression have been linked on the chromosome to markers. There also needs to be a sufficient number of markers polymorphic across the breeder's germplasm to enable the rapid and accurate identification of specific resistant genotypes. Breeders are keen to have "perfect" markers rather than linked markers. A perfect marker shows complete association with the resistance gene and provides unambiguous information on the presence of specific resistance alleles. Many such markers are based on single-nucleotide polymorphisms that underlie variation in the resistance genes (Wallwork, 2009). Molecular markers allow a breeder to reliably pyramid two or more resistance genes into a single variety, thus increasing the chances of the variety possessing durable resistance.

Mass selection

The idea here is to grow a large number (several hundred or preferably 2000–3000) of early generation progeny under high pathogen or pest pressure, thereby ensuring that the most resistant lines survive. Although it has the advantage of requiring little technology, the method is inefficient for the production of varieties where quality or other required traits are not naturally selected for in the specific screening environment (Wallwork, 2009). Alfalfa germplasm resistant to the spotted alfalfa aphid was developed through mass selection (Hanson *et al.*, 1972).

Pedigree breeding

This method is used most widely with self-pollinated crops and involves the selection of individual plants in the F_2 and subsequent generations, enabling a precise pedigree of each line to be traced through the plant breeding program. It allows for plant selection where maximum diversity is expressed among the segregating progeny. Pedigree breeding is likely to be used where the resistant donor parent is adapted to the region of cultivation, or where both adapted parents have some useful resistance. In terms of disease, selection for resistance in early generations ensures that resistance is retained during the selection processes for other plant traits of agronomic importance. Since this is likely to be achieved using high throughput, low-cost screening systems, it is useful for foliar pathogens that are easy to observe and which can be inoculated reliably in nurseries.

A number of insect-resistant varieties of various crops have been developed using pedigree breeding, such as rice varieties resistant to the green leafhopper *Nephotettix virescens*, three biotypes of the brown planthopper *Nilaparvata lugens*, and the stem borer *Scirpophaga incertulas* (Khush, 1989).

Backcrossing

Backcrossing is particularly suited for transferring specific genes for resistance into a susceptible variety that is otherwise high-yielding and well-adapted. The resistance donor is crossed with the variety to be improved, and a series of backcrosses is then made using the variety as a recurrent parent. Each of the backcross F_1 (BCF_1) progenies is evaluated for resistance if the gene for resistance is dominant. Then, only the BCF_1 plants are used for the next backcross. If the gene for resistance is recessive, the BCF_1 plants need to be progeny tested and only those carrying the gene are used in the next backcross. Once backcrossing is complete, the gene being transferred is in a heterozygous condition. At this stage, the backcross progenies are selfed and homozygous individuals are selected. Backcrossing can be speeded up substantially if marker-assisted selection (MAS) is used to select for the recurrent parent genotype at the same time as the required resistance from the donor. Ultimately, backcrossing is a conservative approach to breeding and does not allow for much progress in yield or for traits other than the disease-resistance gene being sought.

Partial backcrossing is a variation of backcrossing, where a single backcross is made to the better adapted parent, followed by selection through subsequent generations by the pedigree or other methods. This method is particularly useful where inheritance is multigenic and/or is inherited as a recessive trait and molecular markers are not available, that is, where normal backcrossing is not possible (Wallwork, 2009).

Recurrent selection

This method is used mainly to promote recombination and to increase the frequencies of favorable genes for quantitatively inherited traits. It is cyclic, with each cycle comprising two phases of breeding—selection of a group of genotypes that possess polygenes for resistance, and crossing among the selected genotypes to obtain genetic recombination. Parents with a reasonable level of resistance are allowed to intercross to produce a heterozygous population that is then exposed to pathogen or insect pressure. Susceptible plants are discarded and resistant plants allowed to interbreed or are artificially crossed with each other. Progeny from the intercrosses are grown and exposed to pathogen or pest pressure, and the cycle of selection and intercrossing is repeated several times, leading to the accumulation of polygenes for resistance from various parents. A disadvantage of the method is that useful gene linkage blocks are often disrupted through the multiple crossing processes, requiring intense selection to maintain good agronomic performance and quality.

Using this method, a maize population with excellent resistance to *Spodoptera frugiperda* was developed (Widstrom *et al.*, 1992). Evidence suggests that it can be successful for some diseases of high heritability, such as foliar pathogens, although it is much less likely to result in well-adapted, quality germplasm (Wallwork, 2009).

F_2 progeny method

This technique combines parts of two general breeding methods—the principle of mass selection for one generation with pedigree breeding over subsequent generations. The difference here is that populations rather than single plants are subject to assessment. The main aim is to delay heavy selection for the desired trait until F_2-derived lines can be replicated in trials, thereby allowing primary selection to occur on the greatest diversity of lines using more objective measurements of adaptation. The method avoids for selection for resistance in the F_2, when recessive genes are expressed poorly, but may allow for mass selection based on traits such as yield. It has a great advantage for traits of low heritability.

6.2.3   *Resistance in practice*

The methods described above have been used by plant breeders to introduce novel sources of resistance to many pests and diseases. However, there are many examples of the protection provided by this genetic resistance proving to be short-lived under field conditions. A good example is the development of barley cultivars resistant to powdery mildew, *B. graminis* f.sp. *hordei*, where the introduction of a host gene for resistance is regularly followed by the appearance of a new race of the pathogen, capable of overcoming the effects of the resistance gene (Lucas, 1998) (Figure 6.1). The speed with which a pest or a pathogen responds to such newly created selection pressures is dependent on a number of factors, including the crop area occupied by the new plant cultivar. Thus, the planting of huge areas of a cultivar containing one or a few resistance genes tends to favor genotypes of the pest or pathogen, possessing matching virulence genes.

6.2.4   *Types of resistance*

The topic of plant resistance to pathogens (and pests) is beset with a diversity of terms used to describe the different forms of expression of resistance. Although resistance is usually

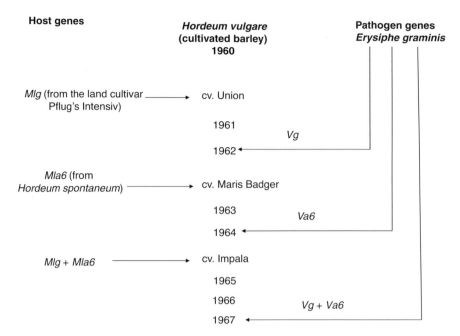

Figure 6.1   Successive introductions into the UK barley crop of cultivars containing genes for resistance to powdery mildew (*Erysiphe graminis*) and the subsequent selection for virulence genes in the *Erysiphe* population. (Reproduced from Lucas (1998), with permission of Blackwell Publishing Ltd.)

divided into two main types, one controlled by one or a few genes and the other controlled by many genes (see Chapter 2), in reality, there is a continuum of variation between these two extremes. Below we consider briefly the various terms used to describe plant resistance.

6.2.4.1   Monogenic resistance

Some plant cultivars are resistant to some races of a pathogen, but susceptible to other races of the same pathogen. This is known as race-specific resistance and differentiates clearly between races of a pathogen, because it is effective against specific races and is ineffective against others (Figure 6.2a). As indicated in Chapter 2, this type of resistance is also sometimes known as strong, major, qualitative, differential, or vertical resistance. Because this type of resistance is controlled by one or a few genes, it is also referred to as monogenic or oligogenic resistance. Monogenic resistance is liable to break down suddenly due to the development of new races of the pathogen (Table 6.1).

Genetic resistance to broomrape (*Orobanche cumana*), based on single, dominant genes, was introduced into sunflower cultivars from wild *Helianthus* more than 40 years ago and has proved very useful (Fernández-Martínez *et al.*, 2000; Rispail *et al.*, 2007). Similarly, in cowpea, useful resistance to *Striga gesnerioides*, again controlled by single, dominant genes, was identified in several cultivars and breeding lines (Berner *et al.*, 1995). However, widespread use of the resistant cultivars of cowpea and sunflower led to new races of broomrape and *S. gesnerioides* appearing (Berner *et al.*, 1995; Fernández-Martínez *et al.*, 2000; Rispail *et al.*, 2007).

Table 6.1 Differences between monogenic and polygenic resistance

	Monogenic	Polygenic
Genetic control	One or few genes involved	Usually many genes involved
Description	Generally, clear-cut; functions throughout the life of the plant, or might be expressed only in mature plants	Variable; seedlings are generally less resistant, but resistance increases as plants mature
Mechanism	Generally, a hypersensitive host reaction	Reduced rate and degree of infection, development, and/or reproduction of pathogens
Efficiency	Highly efficient, but often extremely low resistance to other pathogen races	Variable, effective against all races of the pathogen
Stability	Liable to sudden break down due to the development of new physiological races of the pathogen	Not affected by changes in virulence genes of the pathogen
Commonly used approximate synonyms	Vertical, major gene, seedling, race-specific	Horizontal, minor gene, adult plants, race nonspecific, field, rate-reducing

Source: Reproduced from Hayes & Johnston (1971), with permission.

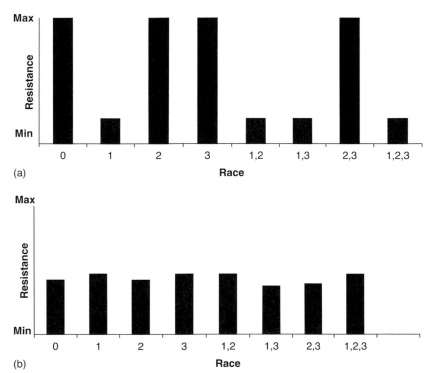

Figure 6.2 Comparison of vertical and horizontal resistance. (a) A cultivar with a single gene for vertical resistance shows a high level of resistance to some pathogen races, but little to others. (b) A cultivar with horizontal resistance shows an intermediate level of resistance to all races of the pathogen. (Reproduced from Lucas (1998), with permission of Blackwell Publishing Ltd.)

Monogenic resistance is also found in plant–insect interactions, and here, instead of being termed race-specific resistance, is known as biotype-specific resistance (Panda & Khush, 1995). As with pathogens, insects can break the resistance by developing new biotypes that possess the genetic capability to overcome the resistance. It can take as few as three insect generations to select resistance-breaking biotypes. Sometimes, the potential of the insect to overcome the plant resistance is so great that the effect of the resistance is nullified before the resistant cultivar is used widely. This was the case with cultivars of Brussels sprouts resistant to the cabbage aphid, *Brevicoryne brassicae* (Dunn & Kempton, 1972).

6.2.4.2   Polygenic resistance

All plants will possess some resistance to each of their pathogens. As a result, this type of resistance is known as partial resistance, although other terms that have been used to describe it include quantitative, adult plant, field, and durable. The lower level of resistance provided by partial resistance tends to be effective against most races of a pathogen, hence use of the term "race" nonspecific to describe it (Figure 6.2b). Because it is probably controlled by several genes, it is also known as polygenic or multigenic resistance. Individually, these genes might play a minor role in the resistance observed (hence the term "minor gene resistance"), but together, they can act additively. For example, in the barley–brown rust interaction, resistance was found to be governed by 6–7 minor genes with additive effects, and was correlated with increased latent period, and reduced infection frequency, pustule size, infectious period, and sporulation (Parlevliet & Van Ommeren, 1975). Polygenic resistance is not affected by changes in virulence genes in the pathogen (Table 6.1). In plant–insect interactions, race nonspecific resistance is known as biotype nonspecific resistance (Panda & Khush, 1995). According to Schoonhoven *et al.* (2005), polygenic resistance is probably not any more durable per se than monogenic resistance, especially if it is based on the concentration of a single compound. If, however, the resistance is based on multiple chemical, physiological, or morphological mechanisms, there is a reduced risk of the resistance being broken by the insect.

Screening wild relatives for resistance to parasitic plants is proving to be a promising approach for detection and transfer of novel resistance to crops. Good examples are the novel resistances found in *Tripsacum dactyloides* and in some *Viciae* species (Gurney *et al.*, 2003; Sillero *et al.*, 2005). The quantitative resistance resulting from selection procedures has led to the release of cultivars with useful levels of resistance, which, because it is not based on a single gene, is likely to be long-lasting.

The situation with this type of resistance can be confusing, however, because there are cases where quantitative and race nonspecific resistance is determined by single genes.

6.2.4.3   Durable resistance

The fact that new pest or pathogen resistance, once deployed in the field, can break down quite quickly, has led to the search for more durable sources of resistance. Durable resistance can be defined as resistance that continues to provide control even after exposure to the pathogen over an extended period (Johnson, 1981). The concept of durability makes no implication about the genetic control, mechanism, degree of expression, or race-specificity of the resistance. Practically, what is important is that it remains effective for a long time.

Specific, well-studied examples include the genes *Sr2* and *Rpg1* for partial resistance to stem rust in wheat and barley, respectively, the genes *Lr34* and *Yr18* that provide partial resistance to leaf and stripe rust in wheat, and the *mlo* locus, which provides resistance to powdery mildew in barley (Park, 2008). Thus, the *mlo* gene was introduced in 1979, and still gives satisfactory control of barley powdery mildew in Europe. Similarly, the *Lr34* gene for resistance in wheat to leaf rust, *Puccinia triticina*, identified originally in South American cultivars, has remained effective for many years in several countries. In Australia, *Lr34* was first released in 1983 and remained effective until 2000, when a virulent pathotype of *P. triticina* was detected, initially in South Australia and subsequently in several other territories (Park, 2008). Nevertheless, *Lr34* remains important in Australia because it is still effective against all known pathotypes of *P. triticina*, when combined with *Lr13* and *Lr37*.

The basis for such durability is not fully understood, although it has been suggested that the genes concerned code for resistance mechanisms, which are not easy to overcome by changes in the pathogen. Thus, the *mlo* gene is responsible for cell wall reinforcement (formation of a papilla, see Chapter 2) at sites of attempted penetration, while the *Lr34* gene appears to reduce infection and delay pathogen development and also appears to interact with other genes to confer resistance (Lucas, 1998). Interestingly, more recent work indicated that resistance mediated by *Lr34* was associated with a high demand for energy, resulting in the stimulation of several metabolic pathways. This stimulated energy demand did not last for more than 7 days after inoculation, which possibly explains why *Lr34* does not inhibit the pathogen fully, but does increase the latent period (Bolton *et al.*, 2008).

The effective working life of resistance genes can be increased if, instead of deploying them singly, they are deployed in combination with a second or even third resistance gene, so the pathogen is required to evolve multiple avirulence genes simultaneously. This type of pyramided resistance is likely to be quite durable (Wallwork, 2009). Apparently, the durable resistance in wheat to rye powdery mildew is due to such an effect (Matsumura & Tosa, 2000). The availability of tightly linked molecular markers and cloned resistance genes could lead to more rapid development of resistant varieties, although this might reduce the diversity of resistance genes being deployed, with greater vulnerability to losses should one or more of the resistances be overcome (Wallwork, 2009).

6.2.4.4 *Gene-for-gene concept*

As pointed out by Lucas (1998), the speed with which a resistance gene loses its effectiveness in the field might be due, in part, to a legacy of breeding for particular types of resistance. Plant breeders have tended to select single genes with a major impact on the pest or pathogen. Such resistance (*R*) genes tend to conform to the gene-for-gene model of host–pathogen/pest interaction.

The gene-for-gene relationship between hosts and pathogens, first proposed by Flor in 1942 and further elaborated in 1956, states that for every gene in the plant conferring resistance, there is a complementary gene in the pathogen determining virulence. Usually, genes for resistance in the plant are dominant (*R*), while genes for susceptibility are recessive (*r*). In contrast, in the pathogen, genes for avirulence are usually dominant (*A*), and genes for virulence are recessive (*a*). The possible interactions between a pair of alleles governing resistance in a plant and the corresponding pair determining virulence in the pathogen are shown in Table 6.2 and Figure 6.3. The gene-for-gene concept was first demonstrated in

Table 6.2 Quadratic check of gene combinations and disease reaction types in a host–pathogen system in which the gene-for-gene concept for one gene operates

Virulence or avirulence genes in the pathogen	Resistance or susceptibility genes in the plant	
	R (resistant) dominant	R (susceptible) recessive
A (avirulent) dominant	AR $(-)$	Ar $(+)$
A (virulent) recessive	aR $(+)$	ar $(+)$

Note that minus signs indicate incompatible (resistant) reactions and plus signs indicate compatible (susceptible) reactions. Reproduced from Agrios (2005). Copyright 2005, with permission of Elsevier.

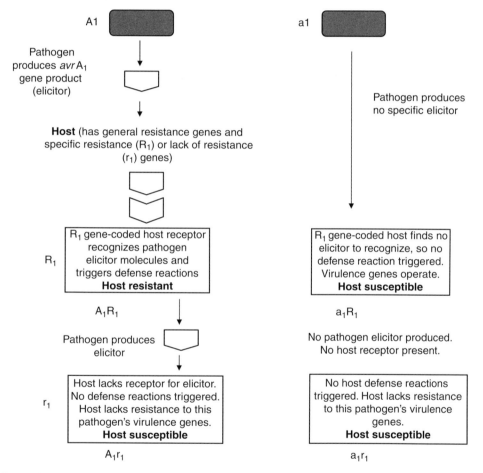

Figure 6.3 Basic interactions of pathogen avirulence (A)/virulence (a) genes with host resistance (R)/susceptibility (r) genes in a gene-for-gene relationship and final outcomes of the interactions. (Reproduced from Agrios (2005). Copyright 2005, with permission of Elsevier.)

the interaction between flax and the rust *Melampsora lini* (Flor, 1942, 1956), but has since been shown to operate in a number of other plant–pathogen interactions (Crute, 1994). The concept as applied to interactions between plants and insects has been demonstrated in the interaction between wheat and the Hessian fly, *Mayetiola destructor* (Gallun, 1977). However, there has been some debate regarding the gene-for-gene concept in plant–insect interactions, with some workers questioning its applicability to all plant–insect systems (Diehl & Bush, 1984), and others suggesting that in all cases of vertical resistance (see Section 6.2.4.1) the relationship exists (Robinson, 1991).

6.2.5   *Making life more difficult for the attacker*

As we have seen above, growing large areas of the same crop cultivar, especially if resistance in the cultivar is monogenically determined, will favor any genetic variants of the pathogen or pest able to attack the cultivar. The longevity of such cultivars can be cut short under these cropping conditions, although if deployed sensibly and with care, monogenic resistance can be effective. Polygenic resistance can be more durable, as can combining (pyramiding) different *R* genes in one cultivar. However, there are other approaches to avoiding directional selection for virulence in pathogens (or parasitic ability in pests). These include deploying *R* genes in an organized manner across regions or farms, or fields, although this requires high levels of coordination and on a regional scale is difficult to organize. Other approaches include the use of multiline cultivars, cultivar mixtures, and intercropping.

A multiline is a series of near-isogenic (genetically identical) plant lines that differ in a single character, such as disease resistance (Wolfe, 1985; Mundt, 2002). Because the lines retain the agronomic characteristics of a normal cultivar, such as uniformity, they can be grown together and harvested like a conventional crop. The presence of different *R* genes in the different lines presents the pathogen with considerable difficulty in terms of overcoming the resistances. A similar approach is the concept of using mixtures of cultivars, where a number of genetically distinct cultivars are grown as a single crop. Disease control in mixtures can be attributed to dilution of the density of susceptible plants, introduction of resistant plants as physical barriers limiting pathogen spread, and induced resistance from incompatible pathotypes trying to infect the different host plants (Chin & Wolfe, 1984). There is a clear relationship between the number of component cultivars in the mixture and disease control. Use of just two component cultivars in the mixture can give disease control, although as Cox *et al.* (2004) demonstrated in wheat, the level of control achieved depends on the disease (Figure 6.4). However, greater levels of disease control and greater yield increases can be achieved with more components in the mixture (Figure 6.5) (Newton, 2009). For disease control, the optimum structure of the mixture will differ depending on the dispersal characteristics of the pathogen and its population structure. Very high levels of disease control can be achieved in some cases. Thus, in one of the largest mixture experiments, where in its second year 3342 ha of rice was grown as row mixtures of susceptible and resistant cultivars, rice blast was reduced by 94% and grain yield increased by 89% (Zhu *et al.*, 2000).

Interestingly, simulation modeling suggested that the smaller the homogeneous genotype area (or genotype unit area, GUA), the greater the efficacy of the mixture. For *Rhynchosporium secalis* on winter barley, the optimum GUA was determined to be 4 m^2, which led to experiments to determine whether a patchy arrangement of component cultivars might

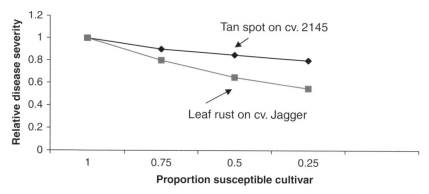

Figure 6.4 Relative effects of mixtures for leaf rust versus tan spot on susceptible wheat cultivars Jagger and 2145. Relative severity of each mixture is expressed as a proportion of the severity in the susceptible monoculture. The two slopes are significantly different. (From Cox *et al.* (2004), with permission of APS.)

be better than homogeneous mixing of the component cultivars (Newton, 2009). Patchy sowing reduced disease and increased yield more than homogeneous sowings of the component cultivars, compared with the mean of monocultures sown alongside (Newton & Guy, 2008). Patchiness was easy to deploy in practice, since no premixing of seed was required. Although there has been reluctance on the part of growers to adopt the use of mixtures, there have been several notable large-scale and successful uses of mixtures in Europe and the USA (see Newton, 2009).

A number of studies have examined the impact of using plant mixtures (e.g., dicultures with two different plant species) on insect population dynamics (Hooks & Johnson, 2003; Altieri & Nicholls, 2004). One such study examined flea beetle (*Phyllotreta cruciferae*) infestation on different broccoli/*Vicia* spp. planting systems (Garcia & Altieri, 1992). Flea beetles were introduced to the plots, and the study found that more beetles left the mixed cultures than the monocultures, leading to a faster reduction in flea beetle populations in

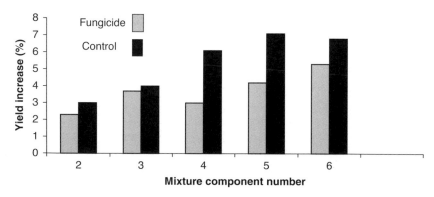

Figure 6.5 Changes in yield in mixtures of winter barley cultivars compared with the mean of their components with different numbers of component cultivars. (Reproduced from Newton (2009), with permission of Blackwell Publishing Ltd.)

the mixed cultures. These dicultures represent a form of intercropping, a practice that is still commonly practiced in subsistence agriculture (Vandermeer, 1989). There are four types of intercropping: (1) mixed cropping, which is the growing of two or more crops simultaneously, with no distinct row arrangement, (2) row intercropping, where one or more crops are grown simultaneously in different rows, (3) strip intercropping, where two or more crops are grown in strips, wide enough to allow independent cultivation of each crop, but narrow enough to allow the crops to interact, and (4) trap-cropping, where one plant species serves as a trap crop to lure the pest away from the major crop (Schoonhoven et al., 2005). Importantly, intercropping need not involve different plant species, since it can be practiced using different cultivars, for example.

6.3   Using plant resistance to protect crops—induced resistance

6.3.1   Introduction

As we saw in Chapter 3, treatment of plants with various agents can promote resistance to subsequent pathogen or pest attack, both locally and systemically. Different types of induced resistance have been defined based on differences in signaling pathways and spectra of effectiveness. Systemic acquired resistance (SAR) occurs in distal plant parts following localized infection by a necrotizing agent, or treatment with certain chemicals, while induced systemic resistance (ISR) is promoted by selected strains of nonpathogenic plant growth-promoting rhizobacteria (PGPR). Induction of resistance can result in the direct activation of defense genes, but can also lead to the phenomenon of priming, wherein plant defenses are not directly activated by the inducing agent, instead, are potentiated for enhanced expression on subsequent attack (Beckers & Conrath, 2007). Numerous biotic and abiotic agents have been reported to activate plant defense responses, thereby rendering plants more resistant to pathogens and pests (Reignault & Walters, 2007; Stout, 2007). The commercialization of some of these resistance inducers has created a new generation of crop protection agents including Bion®/Actigard® (acibenzolar-S-methyl (ASM), Syngenta), Milsana® (extract of *Reynoutria sachalinensis*, KHH BioScience Inc., USA), Elexa® (chitosan, SafeScience, USA), and Messenger® (harpin protein, Eden Bioscience, USA).

6.3.2   Induced resistance for pathogen control

The chemical probenazole was first introduced in 1975 for the control of rice blast (*Magnaporthe grisea*) and bacterial blight (*Xanthomonas oryzae*). Marketed as Oryzemate® (Meiji Seika Kaisha Ltd, Japan), it was found subsequently to activate defense systems in rice (e.g., Watanabe et al., 1977; Iwata et al., 1980). It is widely used in Asia, where, despite continuous use since its introduction, there have been no reports of pathogen insensitivity. Indeed, it still accounted for some 53% of the chemicals used for seedling box treatments of rice in 2005 (Ishii, 2008). Probenazole was not specifically designed as a resistance inducer, but a number of other agents were. One example is ASM, a functional analog of salicylic acid (SA) introduced more than a decade ago. It was shown to activate expression of defense genes in wheat and to protect plants against powdery mildew when applied between 4 and 7 days prior to inoculation with the pathogen (Gorlach *et al.*, 1996). Since then, numerous studies under controlled conditions have confirmed this protective effect. For example, resistance to the fungal pathogen *Colletotrichum destructivum* was induced rapidly in cowpea seedlings following treatment with ASM (Latunde-Dada & Lucas, 2001). The enhanced resistance observed was accompanied by rapid, transient increases in the activities of the defense-related enzyme phenylalanine ammonia lyase (PAL), and accelerated accumulation of the isoflavonoid phytoalexin phaseollidin (Figure 6.6).

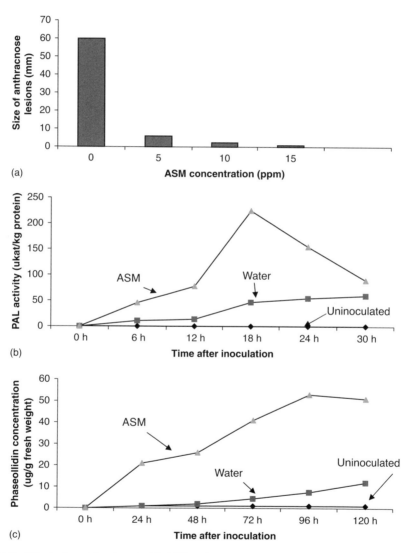

Figure 6.6 Effect of seed treatment with ASM on (a) the size of anthracnose (*C. destructivum*) lesions, (b) PAL activity, and (c) phaseollidin concentration, in cowpea. (Adapted from Latunde-Dada & Lucas (2001), with permission of Elsevier.)

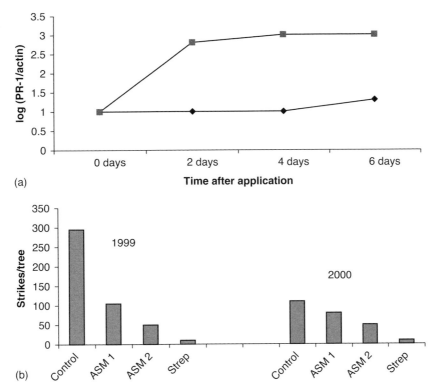

Figure 6.7 (a) Expression of PR-1 in apple leaf samples after seedlings of cultivar Jonathan were sprayed with ASM or water. Concentrations of cDNA for each gene were measured with real-time polymerase chain reaction using actin as a quantity standard. (b) Effect of treatment with ASM (75 mg/L) or streptomycin (100 mg/L) on the incidence of natural fire blight (*E. amylovora*) strikes per tree. Numbers of strikes were evaluated 16 and 12 days after a severe storm in 1999 and 2000, respectively. ASM 1, applied weekly; ASM 2, applied biweekly. (Adapted from Maxson-Stein *et al.* (2002), with permission of APS.)

ASM can also provide effective disease control under field conditions. It was shown to protect the apple cultivar "Jonathan" from infection by the bacterial pathogen *Erwinia amylovora*, the causal agent of fire blight (Figure 6.7). ASM significantly reduced the incidence of natural fire blight strikes on this apple cultivar and was most effective when applied weekly. In this work, application of ASM was also associated with enhanced expression of defense-related genes (Maxson-Stein *et al.*, 2002).

As with SAR, ISR has also been shown to provide protection against pathogens under controlled conditions. Thus, in cucumber, 94 PGPR strains were examined for ISR against the pathogen *Colletotrichum orbiculare* and 6 were found to provide significant disease control (Wei *et al.*, 1991). Similarly, two PGPR strains, *Bacillus pumilus* SE34 and *Pseudomonas fluorescens* 891361, elicited ISR against late blight of tomato (Yan *et al.*, 2002), while PGPR-mediated ISR has also been shown to be effective against viruses. For example, two strains of PGPR induced resistance to cucumber mosaic virus (CMV) in cucumber and tomato (Raupach *et al.*, 1996). Successful field evaluations of PGPR-mediated ISR were

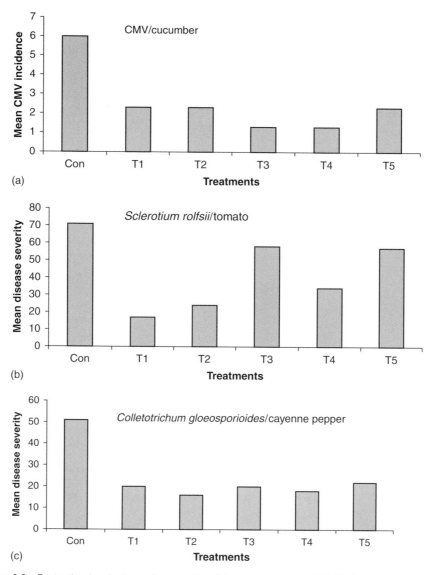

Figure 6.8 Protection by plant growth-promoting rhizobacteria against (a) CMV on cucumber, (b) *S. rolfsii* on tomato, and (c) *C. gloeosporioides* on cayenne pepper. Treatments used: T1, IN937a alone; T2, IN937a + IN937b; T3, IN937b + SE34; T4, IN937b + SE49; T5, T4 + INR7. PGPR strain identifications: strain IN937a, *Bacillus amyloliquefaciens*; strains IN937b, SE34, SE49, T4, and INR7, *B. pumilus*. (Adapted from Jetiyanon *et al.* (2003), with permission of APS.)

carried out in the early 1990s on cucumber, demonstrating reductions in the severity of bacterial wilt (e.g., Wei *et al.*, 1996). In field studies in Thailand in 2001 and 2002, PGPR used as single strains or combinations of strains provided control of various diseases on a range of crops, including CMV in cucumber, *Sclerotium rolfsii* on tomato, and *Colletotrichum gloeosporioides* on cayenne pepper (Figure 6.8) (Jetiyanon *et al.*, 2003).

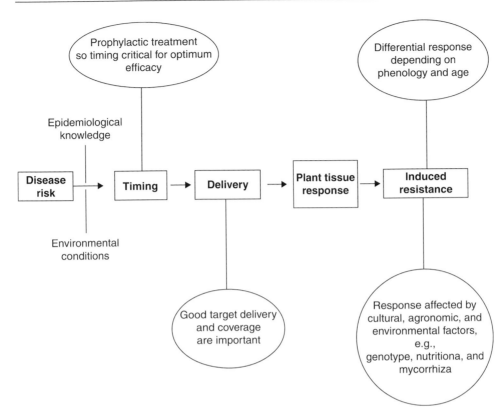

Figure 6.9 Factors affecting the efficacy of induced resistance. (Reproduced from Reglinski *et al.* (2007), with permission of Blackwell Publishing Ltd.)

However, the efficacy of induced resistance, whether SAR or ISR, or induced by other agents such as Elexa or β-aminobutyric acid (BABA), is variable. Disease control is rarely complete, and in some cases, induced resistance provides no disease control (Walters *et al.*, 2005). This lack of consistency and incomplete disease control should not be surprise, since induced resistance is a plant response and as such will be affected by many factors, including the abiotic environment, host genotype, and the extent to which plants in the field are already induced (Figure 6.9) (Walters *et al.*, 2005; Walters, 2009). Because of this, induced resistance is most likely to be implemented in controlled production conditions, such as glasshouses, where environmental and cultural variables can be minimized. In arable cropping systems, inducing agents could be used in crop protection programs, together with fungicides, to maximize disease control. Here, inducing agents applied early in the growing season could be used to reduce pathogen inoculum levels, thereby necessitating use of less fungicide (Havis *et al.*, 2009). Seed treatment is also a possibility, and indeed, work on cotton has shown that ASM applied to seeds prior to sowing in the field, or spraying into the furrow on top of the seeds at sowing, can reduce the severity of black root rot disease (Mondal *et al.*, 2005). In perennial crops, induced resistance is most likely to be implemented by the application of inducing agents to foliage, flowers, and fruits (Figure 6.10).

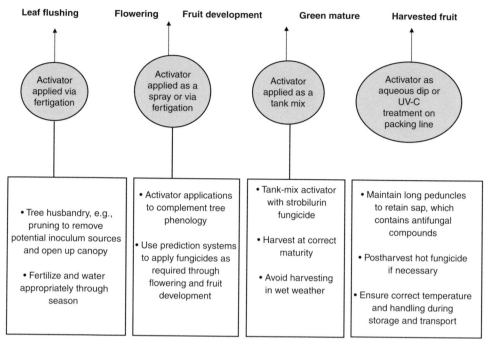

Figure 6.10   A theoretical example of how activators of induced resistance (text ovals) could be combined with existing mango production practices (text rectangles) through the season to provide effective protection against field and postharvest diseases. (Reproduced from Reglinski *et al.* (2007), with permission of Blackwell Publishing Ltd.)

6.3.3   *Induced resistance for control of herbivorous insects*

Karban & Baldwin (1997) considered that introductions of avirulent organisms that induce resistance against subsequent attacks would be the most straightforward approach to using induced resistance in agriculture. Thus, they speculated that the activities of one herbivore species might induce resistance against a much more damaging herbivore, and this effect might occur naturally, or might be facilitated by agronomic manipulations. One example of such an approach concerned wine grapes in California's central valley, which are attacked by two species of mite: the economically damaging Pacific mite (*Tetranychus pacificus*) and the economically less important, but native, Willamette mite (*Eotetranychus willamettei*). The two species of mite are often negatively associated, and vineyards with Willamette mites early in the season are less likely to have problems with Pacific mites. Indeed, introductions of Willamette mites into vineyards suffering with problems caused by Pacific mites led to reduced populations of the Pacific mites and increased yields. Such "vaccinations" only worked if the Willamette mites were present early in the season as the new shoots were expanding (English-Loeb & Karban, 1988; Hougen-Eitzman & Karban, 1995). Although several mechanisms could have accounted for these effects, induced resistance was found to be the strongest and most consistent (Karban & Baldwin, 1997).

A number of natural and synthetic compounds have been shown to induce plant defenses against insects, including *cis*-jasmone (CJ) and BABA. CJ occurs as a component of flower

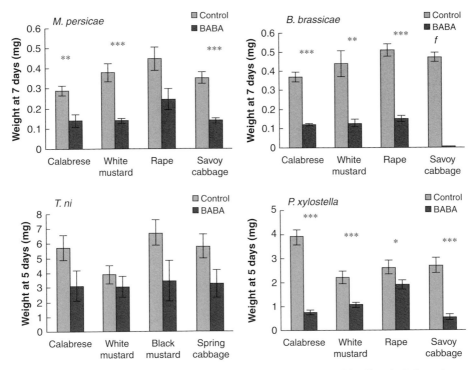

Figure 6.11 Weights of *Myzus persicae* and *B. brassicae* (7 days) and *Plutella xylostella* and *Trichoplusia ni* (5 days) after feeding on various brassicas treated with 10 mM β-aminobutyric acid solution (mean ± SE) (within-plant comparisons significant at *$P < 0.05$, **$P < 0.01$, ***$P < 0.001$; f, no survivors on BABA-treated plants). (Reproduced from Hodge *et al.* (2006), with permission.)

volatiles, but can also be produced by damaged vegetative tissues, and appears to have a role in plant defense (Birkett *et al.*, 2000). CJ was found to be directly repellent to the lettuce aphid, *Nasonovia ribis-nigri*, and the damson-hop aphid, *Phorodon humuli*. Moreover, CJ-treated wheat plots had reduced populations of cereal aphids, and as if that was not enough in terms of its effects, CJ was also directly attractive to the predatory seven-spot ladybird (*Coccinella septempunctata*) and the aphid parasitoid, *Aphidius ervi* (Birkett *et al.*, 2000).

The nonprotein amino acid BABA, applied as a root drench to tic bean (*Vicia faba* var. *minor*), increased mortality of the aphid, *Acyrthosiphon pisum*, caused a reduction in the growth rate of individual insects and reduced the intrinsic rate of population increase (Hodge *et al.*, 2005). Applied to a range of brassicas, BABA reduced the performance of two aphid species and larvae of two species of Lepidoptera (Figure 6.11). However, although it seems likely that the effects observed were the result of induced resistance, the underlying mechanisms remain to be elucidated.

6.3.4   Induced resistance for control of nematodes and parasitic plants

A number of studies have demonstrated that induced resistance can provide control of plant parasitic nematodes. For example, BABA applied to tomato plants as a foliar spray or a root drench reduced galling and numbers of eggs produced by *Meloidogyne javanica*, while fewer second-stage juveniles invaded roots of BABA-treated plants (Oka *et al.*, 1999). On wheat, foliar sprays and root drenches with BABA also induced resistance against the cyst and root-knot nematodes *Heterodera avenae* and *H. latipons*, respectively, and inhibited development of the nematodes within roots (Oka & Cohen, 2001). ASM has also demonstrated potential for control of *Meloidogyne* spp. on tomato, especially in combination with thymol. In fact, ASM/thymol combinations were proposed as an alternative to methyl bromide for managing soilborne diseases on tomato (Ji *et al.*, 2007).

Induced resistance has also been proposed as a useful strategy for the control of certain parasitic plants. For example, soaking of sunflower seeds in a solution of ASM reduced the number of attachments of *O. cumana*, while drenching the soil with ASM reduced the attachments of *Orobanche ramosa* in hemp and sunflower and *O. cumana* in sunflower (Sauerborn *et al.*, 2002; Gonsior *et al.*, 2004; Müller-Stöver *et al.*, 2005). In addition, attachments of *Orobanche crenata* in pea were reduced following foliar application of ASM (Pérez-de-Luque *et al.*, 2004).

6.4   Using plant resistance to protect crops—biotechnological approaches

6.4.1   Introduction

Advances in the techniques of plant molecular biology and genetic engineering have made it possible to transfer genes from unrelated sources into crop plants. These novel genes either reinforce existing functions or add new traits to the transformed or transgenic plants. The foreign DNA must be integrated stably into the genome of the recipient species and must express itself in the new genetic background. Moreover, the transgenic plants so generated must be fertile and otherwise normal in all respects. Foreign genes can be transferred into plants using vector-mediated or direct DNA transfer methods, and the transformation technique chosen will depend on the tissue culture methods used. Methods used for transferring foreign genes into plant tissues include (1) *Agrobacterium*-mediated DNA transformation, which utilizes the natural capacity of *Agrobacterium* to introduce the Ti plasmid into plant cells. The gene to be transferred is inserted into a modified plasmid, which is then transferred when the bacterium infects the plant. (2) Biolistics, where the foreign DNA is coated onto small particles that are then shot into the plant tissue. (3) Microinjection, where the foreign DNA is delivered into protoplasts using a glass micropipette. This is a useful technique for delivery of DNA into target cells. (4) Electroporation, which involves the application of high electrical pulses to solutions containing protoplasts and the foreign DNA in the form of plasmids. Here, the DNA enters the protoplasts through reversible pores created in the protoplast membrane by the electrical pulses.

6.4.2   Engineering resistance to pathogens

Several approaches have been used to develop transgenic plants with enhanced resistance to fungal pathogens using genetic manipulation (GM) techniques: (1) expression of gene products that could enhance the structural defenses of the plant, for example, peroxidase and lignin, (2) expression of gene products that are directly toxic to, or reduce the growth of, the fungus, for example, chitinases, glucanases, and phytoalexins, (3) expression of gene products that neutralize some part of the pathogen's means of entering the plant, for example, cell wall-degrading enzymes and lipases, (4) expression of products that release signals for the regulation of plant defenses, for example, SA or ET, (5) expression of *R* gene products involved in *R/Avr* interactions (Punja, 2004).

There are many examples of plants that have been genetically engineered for increased pathogen resistance (Punja, 2004). For example, hydrolytic enzymes such as chitinases and glucanases are pathogenesis-related (PR) proteins, which act on chitin and glucans in the cell walls of fungi. It was thought that overexpression of these enzymes in plants would result in hyphal lysis, thereby reducing fungal growth. Indeed, transfer of a chitinase from the biocontrol fungus *Trichoderma harzianum* to tobacco and potato provided a high level of broad-spectrum disease control. Selected transgenic lines exhibited greatly increased levels of resistance to *Alternaria alternata*, *Alternaria solani*, *Botrytis cinerea*, and the soilborne pathogen *Rhizoctonia solani* (Figure 6.12) (Lorito *et al.*, 1998). Results are not always so impressive however, since chitinase overexpression can be ineffective against certain pathogens, for example, *Cercospora nicotianae* and *Colletotrichum lagenarium* (see Punja, 2004). However, it has been reported that chitinase and glucanase act synergistically (e.g., Melchers & Stuiver, 2000), and indeed, combined expression of the two hydrolases in various plants has proved to be more effective in disease control than either enzyme on its own (Zhu *et al.*, 1994; Jongedijk *et al.*, 1995).

Engineering the complex biochemical pathways leading to phytoalexin formation is known to be difficult (Dixon *et al.*, 1996). Nevertheless, overexpression of genes coding for certain phytoalexin biosynthetic enzymes led to delayed disease development and reduced symptoms in transgenic plants (Punja, 2004). For example, overexpression of resveratrol synthase in alfalfa, leading to enhanced phytoalexin accumulation, resulted in reduced lesion size and sporulation of *Phoma medicaginis* (Hipskind & Paiva, 2000).

Many resistance genes have now been isolated from a broad range of hosts (Hulbert *et al.*, 2001). One example is the rice bacterial blight gene *Xa21*, which was sequenced and cloned, and when transformed into rice was found to stably express resistance against a range of pathogen isolates (Wang *et al.*, 1996). Currently, transgenic rice lines containing *Xa21* are still undergoing field testing, and other transgenic lines are being developed that carry cloned resistance genes for sheath blight (Brar & Khush, 2006). However, *Xa21* has already been bred into rice using plant breeding approaches, and new strains of bacterial blight with virulence on *Xa21* have already been discovered in Korea and the Philippines (Wallwork, 2009). Nevertheless, it is hoped that in the future, GM technologies will be used to pyramid multiple resistance genes into a single cultivar, thereby providing more durable resistance. Transferring major resistance genes in this way might be limited to closely related species, because to date there have been no reports of resistance genes being effective when transferred to different families, except, for example, where the corresponding avirulence product was artificially provided (Wallwork, 2009). Rather than transferring novel resistance genes into a plant, another option might be to modify the expression of existing genes, for example, those involved in the resistance signaling pathway (see Rommens & Kishore, 2000). This was demonstrated successfully by overexpressing the *NPR1* gene (see Chapter 3) in *Arabidopsis*, leading to increased resistance to a range of pathogens (Cao *et al.*, 1998).

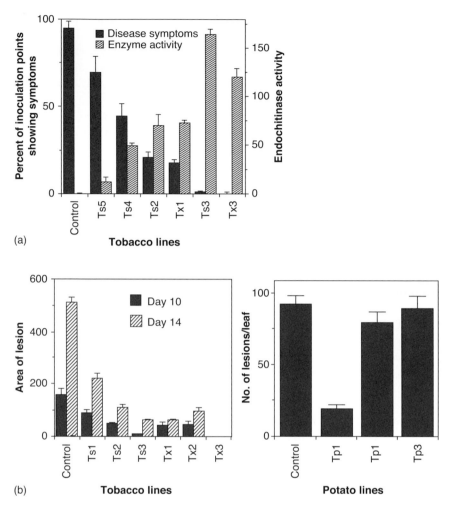

(a)

(b)

Figure 6.12 (a) Resistance of *Trichoderma* endochitinase tobacco plants to *A. alternata*. Disease symptoms 11 days after inoculation and endochitinase activity in leaf protein extracts of representative transgenic lines. C, control plants (nontransformed or empty vector transformed): Cs, tobacco cv. Samsun NN; Cx, tobacco cv. Xanthii; Cp, potato cv. Desiree. T, transgenic lines: Ts (number of plant line), tobacco cv. Samsun NN transformed with p35S-ThEn42; Tx (number of plant line), tobacco cv. Xanthii transformed with p35S-CHIT42 (Tx1) or with p35S-psCHIT42 (Tx2 and Tx3); Tp (number of plant line), potato cv. Desiree transformed with p35S-ThEn42; E, pure endochitinase from *T. harzianum* culture. Control is an average of different lines of Cs and Cx. (b) Resistance of *Trichoderma* endochitinase tobacco plants to *B. cinerea* and of potato plants to *A. solani*. (Left graph) Size of lesions (mm^2) produced on leaves of different tobacco transgenic lines (see above) and controls (average of Cs and Cx lines), 10 and 14 days after inoculation with *B. cinerea* agar plugs. (Right graph) Number of lesions observed on different potato transgenic lines and control, 9 days after spray inoculation with *A. solani*. (Adapted from Lorito *et al.* (1998), with permission of the National Academy of Sciences, USA.)

One option for improving resistance to viruses is to transform plants with the gene that codes for the viral coat protein. This approach was first demonstrated with tobacco mosaic virus (TMV), when tobacco plants expressing TMV coat protein were found to be resistant to TMV (Powell-Abel *et al.*, 1986). This approach soon proved to be very useful in practice, when a papaya line resistant to papaya ringspot virus (PRSV) was developed on Hawaii using transformed PRSV coat protein incorporated into the papaya genome (Fitch *et al.*, 1992). This led to the commercialization of two varieties, one of which, Rainbow, was widely planted, helping to save the papaya industry from devastation by PRSV (Gonsalves, 2004).

6.4.3   *Engineering resistance to insects*

Several sources of novel genes for insect resistance have been identified, including genes from insect pathogens such as *Bacillus thuringiensis* (*Bt*). Such genes can be transferred to plants, perhaps rendering them resistant to the insect. In fact, *Bt* genes have been introduced into a number of crop plants, including tomato, rice, and cotton. Thus, transgenic tomato plants expressing a *Bt* gene were highly resistant to larvae of the tobacco hornworm *Manduca sexta*, while transgenic potato plants expressing the *Bt* gene *CryIA(b)* were highly resistant to feeding and tunneling damage by larvae of the potato tuber moth (Fischhoff *et al.*, 1987; Peferoen *et al.*, 1990). In field trials, *Bt*-transformed tomato plants showed very little damage from *M. sexta*, while control plants were almost completely defoliated within 2 weeks (Delannay *et al.*, 1989).

Proteinase inhibitors from plants are also of interest in terms of improving resistance to herbivorous insects. In fact, the first plant gene to be successfully transferred to another plant species, thereby enhancing its resistance to insects, was the cowpea trypsin inhibitor (*CpTi*) gene isolated from cowpea (Hilder *et al.*, 1987). These workers also introduced a serine protease inhibitor gene of cowpea into tobacco. Transgenic plants expressing high levels of *CpTi* showed a significant decrease in insect damage. Other genes of interest include those encoding lectins, such as chitin-binding lectins (see Chapter 2). The deleterious effect of chitin-binding lectins on insect development is mediated by binding to chitin in the peritrophic membrane in the insect midgut, which interferes with nutrient uptake. Genes coding for pea lectin have been expressed at high levels in transgenic tobacco, and such plants exhibited much reduced leaf damage by *Heliothis virescens*, together with greatly reduced larval biomass (Boulter *et al.*, 1990). Expression of snowdrop lectin in transgenic rice plants yielded plants that were partially resistant to the rice brown planthopper and other hemipteran pests (Foissac *et al.*, 2000). Further progress here was limited by concerns regarding possible consequences to higher animals of ingesting snowdrop lectin, although a 90-day feeding trial found no adverse effects on rats of ingesting transgenic rice expressing snowdrop lectin (Poulsen *et al.*, 2007).

Other approaches worth mentioning include new insecticidal proteins and exploiting plant secondary metabolism (Gatehouse, 2008). New insecticidal proteins include bacterial cholesterol oxidase, which has insecticidal activity comparable to *Bt* toxins, and avidin, which exerts a powerful effect on many insects. There are also possibilities in engineering plant secondary metabolism (transfer of biosynthetic pathways between plants) and engineering the metabolism of plant volatiles. The latter could be targeted, for example, at attraction of insect predators. Thus, *Arabidopsis* transformed with the gene encoding

(E)-β-farnesene from mint exhibited enhanced levels of aphid deterrence and was attractive to aphid parasitoids (Beale *et al.*, 2006).

6.4.4   *Prospects for using transgenic resistance*

Globally, the introduction of transgenic crops into commercial practice is taking place at a remarkable rate (Schoonhoven *et al.*, 2005). Four transgenic crops—soybean, maize, cotton, and oilseed rape—now occupy nearly 30% of the total global area occupied by these crops (James, 2004). Although the dominant trait in transgenic crops is herbicide tolerance (>70%), insect resistance is also a popular trait (~20%). However, the introduction of transgenic crops is not without its problems. For example, if only one resistance gene is transferred into a crop plant, there is the potential for this to be overcome by the pathogen or insect, as we have seen earlier in this chapter. This is of concern, although several options are available to improve the durability of transgenic resistance (Schoonhoven *et al.*, 2005): (1) sequential release of cultivars, which has been used to deploy genes for resistance to the brown planthopper in rice, (2) gene pyramiding, where several genes are introduced (see Section 6.2.4.3 above), and (3) gene rotation, where one gene is alternated with another gene.

Constitutive expression of the transgene is likely to exert considerable selection pressure on the pathogen or insect. However, selection pressure might be greatly reduced by targeting the gene(s) to specific plant parts (tissue-specific) or ensuring its expression at certain plant growth stages (temporal-specific), or only in response to, for example, insect feeding (wound-specific) (Panda & Khush, 1995). A good example is the attachment of the *pin-2* promoter to an insecticidal gene, which directs the synthesis of insecticidal proteins in those plant tissues preferentially consumed by larvae of *M. sexta* (Thornburg *et al.*, 1990).

Other concerns include possible environmental side effects of using transgenic plants, such as effects on nontarget organisms (e.g., pollinators, protected rare species), and the possibility of outcrossing of the transgene and introgression into wild relatives. This area is important and rapidly expanding, and is covered in detail elsewhere (Pilson & Prendeville, 2004; O'Callaghan *et al.*, 2005). Work on improving the efficacy and durability of resistance in transgenic crops, together with long-term studies of environmental consequences, is required if GM technologies are to fulfill their enormous potential.

6.5   Conclusions

Genetic resistance is the ideal form of disease and pest control, providing it is durable and that, in its selection, progress in improving other crop traits of economic importance is not compromised. In the past, two major problems with host resistance have been a lack of variation for resistance to some pathogens, pests, etc., and the loss of resistance following the appearance of new races of pathogens or insect biotypes. Novel forms of resistance might arise from more detailed studies of interactions between plants and their attackers. If these novel resistances are involved in general or nonhost resistance mechanisms, they might be less prone to being overcome by the attacker and so might prove to be durable. Of course, existing sources of resistance could be made more effective and long-lasting providing they are deployed sensibly. Ultimately however, crop protection is best achieved using a combinatorial approach, and not relying solely on one method. Diversification is the key here, using a variety of resistances and agronomic techniques.

In this book, we have seen the remarkable ability of plants to defend themselves against attackers. The diversity of mechanisms and the complexity of signaling and coordination are astounding. As great as our knowledge of plant defense is at present, there is much still to know. Such knowledge and understanding not only will excite those of us fortunate enough to study plants, but will be vital to our future ability to protect our crops.

Recommended reading

Berner DK, Kling JG, Singh BB, 1995. *Striga* research and control: a perspective from Africa. *Plant Disease* **79**, 652–660.
Lucas JA, 1998. *Plant Pathology and Plant Pathogens*. Oxford: Blackwell Publishing Ltd.
Newton A, 2009. Plant disease control through the use of variety mixtures. In: Walters D, ed. *Disease Control in Crops: Biological and Environmentally Friendly Approaches*. Oxford: Blackwell Publishing Ltd., pp. 162–171.
Panda N, Khush GS, 1995. *Host Plant Resistance to Insects*. Wallingford: CAB International.
Punja ZK, 2004. Genetic engineering of plants to enhance resistance to fungal pathogens. In: Punja ZK, ed. *Fungal Disease Resistance in Plants: Biochemistry, Molecular Biology and Genetic Engineering*. Binghamton, NY: The Haworth Press, Inc., pp. 207–258.
Rispail N, Dita M-A, González-Verdejo C, Perez-de-Luque A, Castillejo M-A, Prats E, Román B, Jorrín J, Rubiales D, 2007. Plant resistance to parasitic plants: molecular approaches to an old foe. *New Phytologist* **173**, 703–712.
Schoonhoven LM, Van Loon JJA, Dicke M, 2005. *Insect–Plant Biology*. Oxford: Oxford University Press.
Wallwork H, 2009. The use of host plant resistance in disease control. In: Walters D, ed. *Disease Control in Crops: Biological and Environmentally Friendly Approaches*. Oxford: Blackwell Publishing Ltd., pp. 122–141.
Walters D, Walsh D, Newton A, Lyon G, 2005. Induced resistance for plant disease control: maximizing the efficacy of resistance elicitors. *Phytopathology* **95**, 1368–1373.

References

Agrios GN, 2005. *Plant Pathology*, fifth edition. Burlington, MA: Elsevier Academic Press.
Allard RW, 1960. *Principles of Plant Breeding*. New York: John Wiley & Sons, Ltd.
Altieri MA, Nicholls CI, 2004. *Biodiversity and Pest Management in Agroecosystems*. New York: Food Products Press.
Beale MH, Birkett MA, Bruce TJA, Chamberlain K, Field LM, Huttly AK, Martin JL, Parker R, Phillips AL, Pickett JA, Prosser IM, Shewry PR, Smart LE, Wadhams LJ, Woodcock CM, Zhang YH, 2006. Aphid alarm pheromone produced by transgenic plants affects aphid and parasitoid behavior. *Proceedings of the National Academy of Sciences, USA* **103**, 10509–10513.
Beckers GJ, Conrath U, 2007. Priming for stress resistance: from the lab to the field. *Current Opinion in Plant Biology* **10**, 425–431.
Berner DK, Kling JG, Singh BB, 1995. *Striga* research and control: a perspective from Africa. *Plant Disease* **79**, 652–660.
Biffen RH, 1907. Studies on the inheritance of disease resistance. *Journal of Agricultural Science, Cambridge* **2**, 109–128.

Birkett MA, Campbell CAM, Chamberlain K, Guerrieri E, Hick AJ, Martin JL, Matthes M, Napier JA, Pettersson J, Pickett JA, Poppy GM, Pow EM, Pye BJ, Smart LE, Wadhams GH, Wadhams LJ, Woodcock CM, 2000. New roles for cis-jasmone as an insect semiochemical and in plant defense. *Proceedings of the National Academy of Sciences, USA* **97**, 9329–9334.

Bolton, MD, Kolmer, JA, Xu, WW, Garvin, DF, 2008. Lr34-mediated leaf rust resistance in wheat: transcript profiling reveals a high energetic demand supported by transient recruitment of multiple metabolic pathways. *Molecular Plant-Microbe Interactions* **21**, 1515–1527.

Boulter D, Edwards GA, Gatehouse AMR, Gatehouse JA, Hilder VA, 1990. Additive protective effects of different plant-derived insect resistance genes in transgenic tobacco plants. *Crop Protection* **9**, 351–354.

Brar DS, Khush GS, 2006. Breeding rice for resistance to biotic stresses: conventional and molecular approaches. In: Ladha JK, Aggrawal P, Hardy B, eds. *Proceedings of the Second International Rice Congress*, October 9–13, New Delhi, India, http://www.irri.org/science/abstracts/030.asp.

Büschges R, Hollricher K, Panstruger R, Simons G, Wolter M, Frijters A, Van Daelen R, Van Der Lee T, Diergaarde P, Groenendijk J, Topsch S, Vos P, Salamini F, Schulze-Lefert P, 1997. The barley *Mlo* gene: a novel control element of plant pathogen resistance. *Cell* **88**, 57–63.

Cao H, Li X, Dong X, 1998. Generation of broad-spectrum disease resistance by over-expression of an essential regulatory gene in systemic acquired resistance. *Proceedings of the National Academy of Sciences, USA* **95**, 6531–6536.

Chin KM, Wolfe MS, 1984. The spread of *Erysiphe graminis* f.sp. *hordei* in mixtures of barley varieties. *Plant Pathology* **33**, 89–100.

Cox, CM, Garrett, KA, Bowden, RL, Fritz, AK, Dendy, SP Heer, WF, 2004. Cultivar mixtures for simultaneous management of multiple diseases: tan spot and leaf rust of wheat. *Phytopathology* **94**, 961–969.

Crute IR, 1994. Gene-for-gene recognition in plant-pathogen interactions. *Philosophical Transactions of the Royal Society, Series B—Biological Sciences* **346**, 345–349.

Delannay X, Lavallee BJ, Proksch RK, Fuchs RL, Sims SR, Greenplate JT, Marrone PG, Dodson RB, Augustine JJ, Layton JG, Fischoff DA, 1989. Field performance of transgenic tomato plants expressing the *Bacillus thuringiensis* var. *kurstaki* insect control protein. *Bio/Technology* **7**, 1265–1269.

Diehl SR, Bush GL, 1984. An evolutionary and applied perspective of insect biotypes. *Annual Review of Entomology* **29**, 471–504.

Dixon RA, Lamb CJ, Masoud S, Sewalt VJH, Paiva NL, 1996. Metabolic engineering: prospects for crop improvement through the genetic manipulation of phenylpropanoid biosynthesis and defense responses—a review. *Gene* **179**, 61–71.

Dunn JA, Kempton DPH, 1972. Resistance to attack by *Brevicoryne brassicae* among plants of Brussel sprouts. *Annals of Applied Biology* **72**, 1–11.

English-Loeb GM, Karban R, 1988. Negative interactions between Willamette mites and Pacific mites: possible management strategies for grapes. *Entomologia Experimentalis et Applicata* **48**, 269–274.

Fernández-Martínez J, Melero-Vara J, Muñoz-Ruz J, Ruso J, Domínguez J, 2000. Selection of wild and cultivated sunflower for resistance to a new broomrape race that overcomes resistance of the *Or*(5) gene. *Crop Science* **40**, 550–555.

Fischhoff DA, Bowdish KS, Perlak FJ, Marrone PG, McCormic SM, Niedermeyer JG, Dean RA, Kusano-Kretzmer K, Mayer EJ, Rochester DE, Rogers SG, Fraley RT, 1987. Insect tolerant transgenic tomato plants. *Bio/Technology* **5**, 807–813.

Fitch MM, Manshardt RM, Gonsalves D, Slightom JL, Sanford JC, 1992. Virus resistant papaya derived from tissues bombarded with the coat protein gene of papaya ringspot virus. *Bio/Technology* **10**, 1466–1472.

Flor HH, 1942. Inheritance of pathogenicity in *Melampsora lini*. *Phytopathology* **32**, 653–669.

Flor HH, 1956. The complementary genetic systems in flax and flax rust. *Advances in Genetics* **8**, 29–54.

Foissac X, Loc NT, Christou P, Gatehouse AMR, Gatehouse JA, 2000. Resistance to green leafhopper (*Nephotettix virescens*) and brown planthopper (*Nilaparvata lugens*) in transgenic rice expressing snowdrop lectin (*Galanthus nivalis* agglutinin; GNA). *Journal of Insect Physiology* **46**, 573–583.

Gallun RL, 1977. Genetic basis of Hessian fly epidemics. In: Day PR, ed. *The Genetic Basis of Epidemics in Agriculture*. Annals of the New York Academy of Sciences, No. 287. New York: New York Academy of Sciences, pp. 223–229.

Garcia MA, Altieri MA, 1992. Explaining differences in flea beetle *Phyllotreta cruciferae* Goeze densities in simple and mixed broccoli cropping systems as a function of individual behavior. *Entomologia Experimentalis et Applicata* **62**, 201–209.

Gatehouse JA, 2008. Biotechnological prospects for engineering insect-resistant plants. *Plant Physiology* **146**, 881–887.

Gonsalves D, 2004. Transgenic papaya in Hawaii and beyond. *AgBioForum* **7** (1&2), 36–40. http://www.agbioforum.org.

Gonsior G, Buschmann H, Szinicz G, Spring O, Sauerborn J, 2004. Induced resistance—an innovative approach to manage branched broomrape (*Orobanche ramosa*) in hemp and tobacco. *Weed Science* **52**, 1050–1053.

Gorlach J, Volrath S, Knauf-Beiter G, Hengy G, Beckhove U, Kogel K-H, Oostendorp M, Staub T, Ward E, Kessmann H, Ryals J, 1996. Benzothiadiazole, a novel class of inducers of systemic acquired resistance, activates gene expression and disease resistance in wheat. *The Plant Cell* **8**, 629–643.

Gurney AL, Grimanelli D, Kanampiu F, Hoisington D, Scholes JD, Press MC, 2003. Novel sources of resistance to *Striga hermonthica* in *Tripsacum dactyloides*, a wild relative of maize. *New Phytologist* **160**, 557–568.

Hanson CH, Busbice TH, Hill RR, Hunt OJ, Oakes AJ, 1972. Directed mass selection for developing multiple pest resistance and conserving germplasm in alfalfa. *Journal of Environmental Quality* **1**, 106–111.

Havens, JN, 1792. Observations on the Hessian fly. *Society of Agriculture of New York* **1**, 89–107.

Havis ND, Paterson L, Taylor JGM, Walters DR, 2009. Use of resistance elicitors to control *Ramularia collo-cygni* on spring barley. The 2nd European Ramularia Workshop—a new disease and challenge in barley production. *Aspects of Applied Biology* **92**, 127–132.

Hayes JD, Johnston TD, 1971. Breeding for disease resistance. In: Western JH, ed. *Diseases of Crop Plants*. London: Macmillan, pp. 62–88.

Hilder VA, Gatehouse AMR, Sheerman SE, Baker RF, Boulter D, 1987. A novel mechanism of insect resistance engineered into tobacco. *Nature* **330**, 160–163.

Hipskind JD, Paiva NL, 2000. Constitutive accumulation of a resveratrol-glucoside in transgenic alfalfa increases resistance to *Phoma medicaginis*. *Molecular Plant-Microbe Interactions* **13**, 551–562.

Hodge S, Thompson GA, Powell G, 2005. Application of DL-β-aminobutyric acid (BABA) as a root drench to legumes inhibits the growth and reproduction of the pea aphid *Acyrthosiphon pisum* (Hemiptera: Aphididae). *Bulletin of Entomological Research* **95**, 449–455.

Hodge S, Pope TW, Holaschke M, Powell G, 2006. The effect of beta-aminobutyric acid on the growth of herbivorous insects feeding on Brassicaceae. *Annals of Applied Biology* **148**, 223–229.

Hooks CRR, Johnson MW, 2003. Impact of agricultural diversification on the insect community of cruciferous crops. *Crop Protection* **22**, 223–238.

Hougen-Eitzman D, Karban R, 1995. Mechanisms of interspecific competition that result in successful control of Pacific mites following inoculations of Willamette mites on grapevines. *Oecologia* **103**, 157–161.

Hulbert SC, Webb CA, Smith SM, Sun Q, 2001. Resistance gene complexes: evolution and utilization. *Annual Review of Phytopathology* **39**, 285–312.

Ishii H, 2008. Fungicide research in Japan—an overview. In: Dehne HW, Deising HB, Gisi U, Kuck KH, Russell PE, Lyr H, eds. *Modern Fungicides and Antifungal compounds. V. Proceedings of the 15th International Reinhardsbrunn Symposium on Modern Fungicides and Antifungal Compounds, 2007.* Braunschweig, Germany: The German Phytomedical Society (DPG), pp. 11–17.

Iwata M, Suzuki Y, Watanabe T, Mase S, Sekizawa Y, 1980. Effect of probenazole on the activities of enzymes related to the resistant reaction in rice plant. *Annals of the Phytopathological Society of Japan* **46**, 297–306.

James C, 2004. *Preview: Global Status of Commercialized Biotech/GM Crops: 2003.* Ithaca, NY: ISAAA Briefs No. 32.

Jetiyanon K, Fowler WD, Kloepper JW, 2003. Broad-spectrum protection against several pathogens by PGPR mixtures under field conditions in Thailand. *Plant Disease* **87**, 1390–1394.

Ji PS, Momol MT, Rich JR, Olson SM, Jones JB, 2007. Development of an integrated approach for managing bacterial wilt and root-knot nematode on tomato under field conditions. *Plant Disease* **91**, 1321–1326.

Johnson R, 1981. Durable resistance—definition of, genetic control, and attainment in plant breeding. *Phytopathology* **71**, 567–568.

Jongedijk E, Tigelaar H, Van Roekel JSC, Bres-Vloemans SA, Dekker I, Van Den Elzen PJM, Cornelissen BJC, Melchers LS, 1995. Synergistic activity of chitinases and β-1,3-glucanases enhances fungal resistance in transgenic tomato plants. *Euphytica* **85**, 173–180.

Jorgensen JH, 1992. Discovery, characterization and exploitation of *Mlo* powdery mildew resistance in barley. *Euphytica* **63**, 141–152.

Karban R, Baldwin IT, 1997. *Induced Responses to Herbivory.* Chicago, IL: University of Chicago Press.

Khush GS, 1977. Disease and insect resistance in rice. *Advances in Agronomy* **29**, 265–341.

Khush GS, 1989. Multiple disease and insect resistance for increased yield stability in rice. *Progress in Irrigated Rice Research.* Manila, the Philippines: International Rice Research Institute, pp. 79–92.

Latunde-Dada AO, Lucas JA, 2001. The plant defence activator acibenzolar-S-methyl primes cowpea [*Vigna unguiculata* (L.) Walp.] seedlings for rapid induction of resistance. *Physiological and Molecular Plant Pathology* **58**, 199–208.

Lorito M, Woo SL, Fernandez IG, Colucci G, Harman GE, Pintor-Toro JA, Filippone E, Muccifora S, Lawrence CB, Zoina A, Tuzun S, Scala F, 1998. Genes from mycoparasitic fungi as a source for improving plant resistance to fungal pathogens. *Proceedings of the National Academy of Sciences, USA* **95**, 7860–7865.

Lucas JA, 1998. *Plant Pathology and Plant Pathogens.* Oxford: Blackwell Publishing Ltd.

Matsumura K, Tosa Y, 2000. The rye mildew fungus carries avirulence genes corresponding to wheat genes for resistance to races of the wheat mildew fungus. *Phytopathology* **85**, 753–756.

Maxson-Stein K, He S-Y, Hammerschmidt R, Jones AL, 2002. Effect of treating apple trees with acibenzolar-S-methyl on fire blight and expression of pathogenesis-related protein genes. *Plant Disease* **86**, 785–790.

Melchers LS, Stuiver MH, 2000. Novel genes for disease resistance breeding. *Current Opinion in Plant Biology* **3**, 147–152.

Mondal AH, Nehl DB, Allen SJ, 2005. Acibenzolar-S-methyl induces systemic resistance in cotton against black root rot caused by *Thielaviopsis basicola*. *Australasian Plant Pathology* **34**, 499–507.

Müller-Stöver D, Buschmann H, Sauerborn J, 2005. Increasing control reliability of *Orobanche cumana* through integration of a biocontrol agent with a resistance-inducing chemical. *European Journal of Plant Pathology* **111**, 193–202.

Mundt CC, 2002. Use of multiline cultivars and cultivar mixtures for disease management. *Annual Review of Phytopathology* **40**, 381–410.

Newton A, 2009. Plant disease control through the use of variety mixtures. In: Walters D, ed. *Disease Control in Crops: Biological and Environmentally Friendly Approaches*. Oxford: Blackwell Publishing Ltd., pp. 162–171.

Newton AC, Guy DC, 2008. The effect of uneven, patchy cultivar mixtures on disease control and yield in winter barley. *Field Crops Research* **110**, 225–228.

O'Callaghan M, Glare TR, Burgess EPJ, Malone LA, 2005. Effects of plants genetically modified for insect resistance on nontarget organisms. *Annual Review of Entomology* **50**, 271–292.

Oka Y, Cohen Y, 2001. Induced resistance to cyst and root-knot nematode in cereals by DL-β-amino-*n*-butyric acid. *European Journal of Plant Pathology* **107**, 219–227.

Oka Y, Cohen Y, Spiegel Y, 1999. Local and systemic induced resistance to root-knot nematode in tomato by DL-β-amino-*n*-butyric acid. *Phytopathology* **89**, 1138–1143.

Panda N, Khush GS, 1995. *Host Plant Resistance to Insects*. Wallingford, UK: CAB International.

Park RF, 2008. Breeding cereals for rust resistance in Australia. *Plant Pathology* **57**, 591–602.

Parlevliet JE, Van Ommeren A, 1975. Partial resistance of barley to leaf rust, *Puccinia hordei*. II. Relationship between field trials, micro plot test and latent period. *Euphytica* **24**, 293–303.

Peferoen M, Jansens S, Reynaerts A, Leemans J, 1990. Potato plants with engineered resistance against insect attack. In: Vayda ME, Park WC, eds. *Molecular and Cellular Biology of the Potato*. Wallingford, UK: CAB International, pp. 193–204.

Pérez-de-Luque A, Jorrín JV, Rubiales D, 2004. Crenate broomrape control in pea by foliar application of benzothiadiazole (BTH). *Phytoparasitica* **32**, 21–29.

Pilson D, Prendeville HR, 2004. Ecological effects of transgenic crops and the escape of transgenes into wild populations. *Annual Review of Ecology, Evolution and Systematics* **35**, 149–174.

Poulsen M, Kroghsbo S, Schroder M, Wilcks A, Jacobsen H, Miller A, Frenzel T, Danier J, Rychlik M, Shu QY, Emami K, Sudhakar D, Gatehouse A, Engel KH, Knudsen I, 2007. A 90-day safety study in Wistar rats fed genetically modified rice expressing snowdrop lectin *Galanthus nivalis* (GNA). *Food and Chemical Toxicology* **45**, 350–363.

Powell-Abel P, Nelson RS, De B, Hoffmann N, Rogers SG, Fraley RT, Beachy RN, 1986. Delay of disease development in transgenic plants that express the tobacco mosaic virus coat protein gene. *Science* **232**, 738–742.

Punja ZK, 2004. Genetic engineering of plants to enhance resistance to fungal pathogens. In: Punja ZK, ed. *Fungal Disease Resistance in Plants: Biochemistry, Molecular Biology and Genetic Engineering*. Binghamton, NY: The Haworth Press, pp. 207–258.

Raupach GS, Liu L, Murphy JF, Tuzun S, Kloepper JW, 1996. Induced systemic resistance in cucumber and tomato against cucumber mosaic cucumovirus using plant growth promoting rhizobacteria (PGPR). *Plant Disease* **80**, 891–894.

Reignault P, Walters D, 2007. Topical application of inducers for disease control. In: Walters D, Newton A, Lyon G, eds. *Induced Resistance for Plant Defence: A Sustainable Approach to Crop Protection*. Oxford: Blackwell Publishing Ltd., pp. 179–200.

Reglinski T, Dann E, Deverall B, 2007. Integration of induced resistance into crop production. In: Walters D, Newton A, Lyon G, eds. *Induced Resistance for Plant Defence: A Sustainable Approach to Crop Protection*. Oxford: Blackwell Publishing Ltd., pp. 201–228.

Rispail N, Dita M-A, González-Verdejo C, Perez-de-Luque A, Castillejo M-A, Prats E, Román B, Jorrín J, Rubiales D, 2007. Plant resistance to parasitic plants: molecular approaches to an old foe. *New Phytologist* **173**, 703–712.

Robinson RA, 1991. The genetic controversy concerning vertical and horizontal resistance. *Revista Mexicana de Fitopatologia* **9**, 57–63.

Rommens CM, Kishore GM, 2000. Exploiting the full potential of disease resistance genes for agricultural use. *Current Opinion in Biotechnology* **11**, 120–125.

Sauerborn J, Buschmann H, Ghiasvand Ghiasi K, Kogel K-H, 2002. Benzothiadiazole activates resistance in sunflower (*Helianthus annuus*) to the root-parasitic weed *Orobanche cumana*. *Phytopathology* **92**, 59–64.

Schoonhoven LM, Van Loon JJA, Dicke M, 2005. *Insect-Plant Biology*. Oxford: Oxford University Press.

Sillero JC, Moreno MT, Rubiales D, 2005. Sources of resistance to crenate broomrape in *Vicia* species. *Plant Disease* **89**, 22–27.

Stout MJ, 2007. Types and mechanisms of rapidly induced plant resistance to herbivorous arthropods. In: Walters D, Newton A, Lyon G, eds. *Induced Resistance for Plant Defence: A Sustainable Approach to Crop Protection*. Oxford: Blackwell Publishing Ltd., pp. 89–107.

Thormburg RW, Kernan A, Molin L, 1990. Chloramphenicol acetyl transferase (CAT) protein is expressed in transgenic tobacco in field tests following attack by insects. *Plant Physiology* **92**, 500–505.

Vandermeer J, 1989. *The Ecology of Intercropping*. Cambridge: Cambridge University Press.

Wallwork H, 2009. The use of host plant resistance in disease control. In: Walters D, ed. *Disease Control in Crops: Biological and Environmentally Friendly Approaches*. Oxford: Blackwell Publishing Ltd., pp. 122–141.

Walters DR, 2009. Are plants in the field already induced? Implications for practical disease control. *Crop Protection* **28**, 459–465.

Walters D, Walsh D, Newton A, Lyon G, 2005. Induced resistance for plant disease control: maximizing the efficacy of resistance elicitors. *Phytopathology* **95**, 1368–1373.

Wang GL, Song WY, Ruan DL, Sideris S, Ronald PC, 1996. The cloned gene, *Xa21*, confers resistance to multiple *Xanthomonas oryzae* pv. *oryzae* isolates in transgenic plants. *Molecular Plant-Microbe Interactions* **9**, 850–855.

Watanabe T, Igarashi H, Matsumoto K, Seki S, Mase S, Sekizawa Y, 1977. The characteristics of probenazole (oryzemate) for the control of rice blast. *Journal of Pesticide Science* **2**, 291–296.

Wei G, Kloepper JW, Tuzun S, 1991. Induction of systemic resistance of cucumber to *Colletotrichum orbiculare* by select strains of plant growth promoting rhizobacteria. *Phytopathology* **81**, 1508–1512.

Wei G, Kloepper JW, Tuzun S, 1996. Induced systemic resistance to cucumber diseases and increased plant growth by plant growth promoting rhizobacteria under field conditions. *Phytopathology* **86**, 221–224.

Widstrom NW, Williams WP, Wiseman BR, Davis FM, 1992. Recurrent selection for resistance to leaf feeding fall armyworm on maize. *Crop Science* **32**, 1171–1174.

Wolfe MS, 1985. The current status and prospects of multiline cultivars and variety mixtures for disease resistance. *Annual Review of Phytopathology* **23**, 251–273.

Yan Z, Reddy MS, Ryu C-M, McInroy JA, Wilson M, Kloepper JW, 2002. Induced systemic protection against tomato late blight elicited by plant growth promoting rhizobacteria. *Phytopathology* **92**, 1329–1333.

Zhu Q, Maher EA, Masoud S, Dixon RA, Lamb CJ, 1994. Enhanced protection against fungal attack by constitutive co-expression of chitinase and glucanase genes in transgenic tobacco. *Bio/Technology* **12**, 807–812.

Zhu Y, Chen H, Fan J, Wang Y, Li Y, Chen J, Fan JX, Yang S, Hu L, Leung H, Mew TW, Teng PS, Zonghua W, Mundt CM, 2000. Genetic diversity and disease control in rice. *Nature* **406**, 718–722.

Index

Abscisic acid (ABA), 83, 142
Acibenzolar-S-methyl (ASM), 92, 126, 144, 178, 213–14, 220
Adaptive radiation, 170, 183
Albugo candida, 7
Alkaloids, 20–21, 60, 64, 154, 157, 166
  Different types, 21, 23
Allelochemicals, 49, 53
Allelopathy, 64–6
Allocation costs, 154, 174
  Associated with induced responses to herbivory, 175–8
  Associated with induced responses to pathogens, 178–81
Alnus crispa, 60
Alternaria brassicae, 4
Alternaria brassicicola, 32, 127, 129, 138, 140
Antelope, 59
Anthocyanins, 60
Antifungal proteins, 32–5
Aphids, 5
Aposematic, 59
Arabidopsis, 18
  pad3, 32
  pad4, 78
  sid1, 78
  sid2, 78
Arbuscular mycorrhizal fungi (AMF), 143–4
Arginase, 55
Arthropod-inducible proteins (AIPs), 55–6
Asclepias spp., 51
Autotoxicity, 64
  Costs, 174
Auxins, 83
Avenacin, 24
Avirulence (*Avr*) genes, 187, 209
Azadirachtin, 50

β-aminobutyric acid (BABA), 218–20
Bacillus thuringiensis (*Bt*), 223
Backcrossing, 204
Bacterial communities, effects of induced resistance on, 132–3

Basal resistance, 25, 77–84, 88
Biotroph, 2, 6, 36, 91, 143
Biotrophic, 41
  Pathogens, 78, 83, 129
Black walnut, 65
Blumeria graminis f.sp. *hordei*, 26, 81, 203, 205
Bradysia impatiens, 96
Brassica hirta, 45
Brassica nigra, 45
Breeding for resistance, 201–13
Brevicoryne brassicae, 66, 208
Bruchins, 94
Bursaphelenchus xylophilus, 8
Bursera, 51

Caeliferins, 93–4
Caffeine, 63
Callose, 26, 38, 40, 88
Camalexin, 32, 66
Capsella bursa-pastoris, 7
Carbon-nutrient balance (CNB) hypothesis, 160–61
Cardenolides, 51–2, 62–3, 170, 183–5, 188
Carnivores, 58
Catechin, 65
Cell wall, 18
Cell wall appositions (CWAs), 25–6
Cell wall reinforcement, 40
Chalcone synthase (CHS), 35–6
Chemical defenses
  Against insects, 49–58
  Against pathogens, 18–25, 29–37
  Against vertebrate herbivores, 59–64
Chitinase, 34, 143, 221
Cis-jasmone (CJ), 218–19
Clerodanes, 50
Coevolution
  Between plants and pathogens, 187–8
  Classical, 183
  Diffuse, 184
  Geographical mosaic theory of, 184
Coevolutionary arms race, 183–91
COI1 gene, 98
Colletotrichum lagenarium, 88